W9-AJS-730

# THE NATURAL HISTORY OF THE UNIVERSE

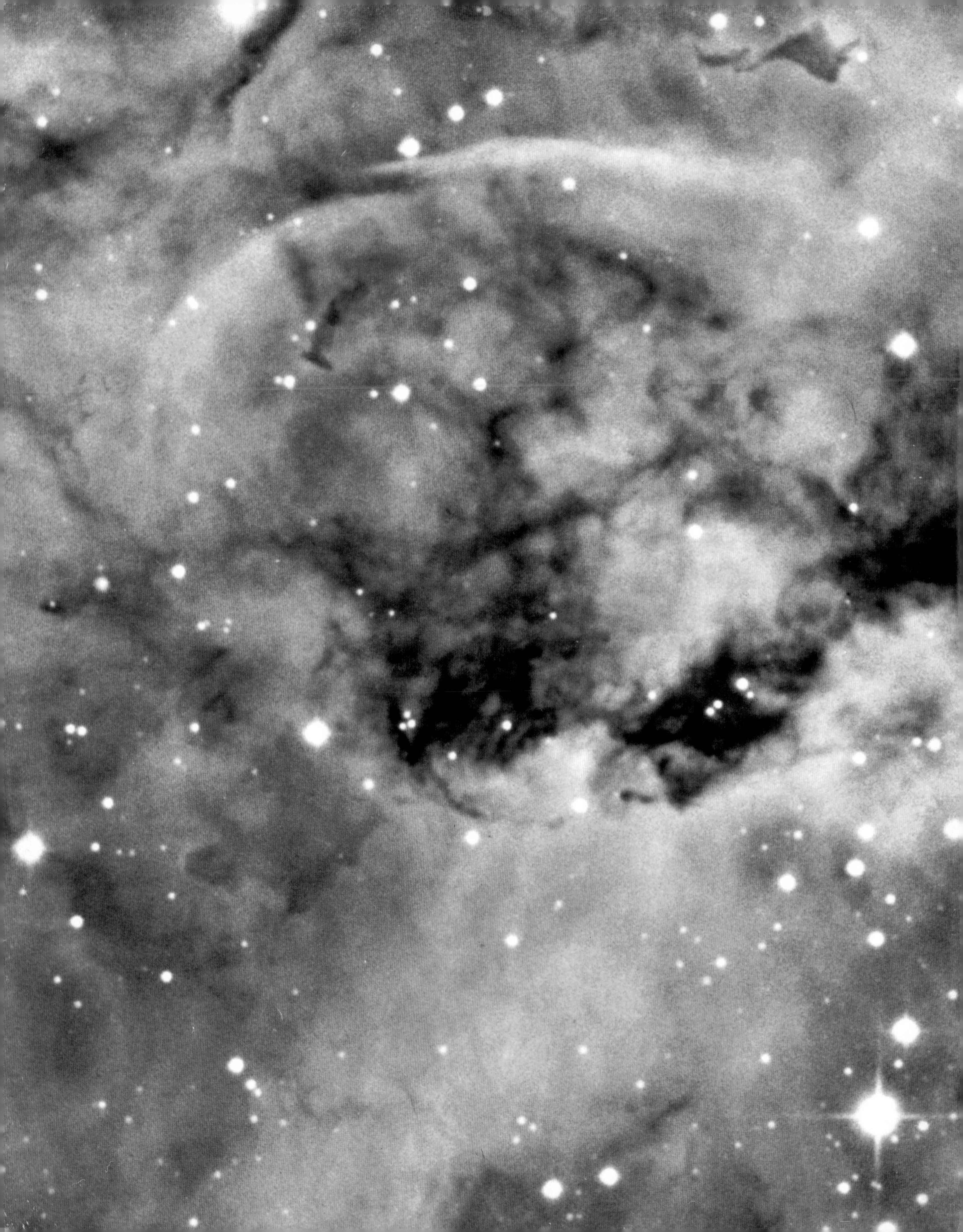

# THE NATURAL HISTORY OF THE UNIVERSE

## FROM THE BIG BANG TO THE END OF TIME

COLIN A. RONAN



MACMILLAN PUBLISHING COMPANY
NEW YORK
Maxwell Macmillan International
NEW YORK OXFORD SINGAPORE SYDNEY

**A Marshall Edition**
This book was conceived, edited and designed by Marshall Editions, 170 Piccadilly, London W1V 9DD



Macmillan Publishing Company
866 Third Avenue
New York, NY 10022

Macmillan Publishing Company is part of the Maxwell Communication Group of Companies.

Library of Congress Cataloging-in-Publication Data

Ronan, Colin A.
The natural history of the universe/Colin A. Ronan.—
1st American ed.
p. cm.
Includes index.
ISBN 0-02-604511-7
1. Astronomy. I. Title.
QB43.2.R64 1991 91-17063 CIP
520—dc20

Macmillan books are available at special discounts for bulk purchases for sales promotions, premiums, fund-raising, or educational use. For details, contact:

Special Sales Director
Macmillan Publishing Company
866 Third Avenue
New York, NY 10022

First American edition

Typeset by MS Filmsetting Limited, Frome, Somerset, UK
Originated by Reprocolor Llovet SA, Barcelona, Spain
Printed and bound in Germany by Mohndruck Graphische Betriebe GmbH

10 9 8 7 6 5 4 3 2 1

Consultants

**Iain Nicolson**
Senior Lecturer, Division of Physical Sciences, Hatfield Polytechnic, UK

**Dr Andy Lawrence**
Lecturer, Department of Physics, Queen Mary and Westfield College, University of London

EDITOR **Christopher Cooper**
EDITORIAL DIRECTOR **Ruth Binney**
EDITORIAL COORDINATOR **Tim Probart**
COPY EDITOR **Lindsay McTeague**
EDITORIAL ASSISTANT **Jon Kirkwood**
ART EDITOR **Lynn Bowers**
ART DIRECTOR **John Bigg**
DESIGN ASSISTANTS **Emma Hutton**
**Paul Tilby**
RESEARCH **Jazz Wilson**
PICTURE RESEARCH **Richard Philpott**
PRODUCTION **Barry Baker**
**Janice Storr**
**Nikki Ingram**

# CONTENTS

Frontispiece: *The spiral galaxy M83, lying 12 million light-years from the Earth.*

Title page: *Part of the huge nebula in the constellation Carina, in the southern skies.*

Left: *Io, the volcanic, sulphur-covered satellite of Jupiter.*

Above: *The fine structure of Saturn's rings, revealed in a false-colour image.*

# INTRODUCTION

From time immemorial the skies have intrigued the human inhabitants of Earth. The pageant of sunrise and sunset, the changing phases of the Moon and the silent procession of the stars across the black dome of heaven have long provided both a spectacle and a puzzle. The spectacle has inspired the artist, the musician and the poet; the puzzle has intrigued philosophers and scientists. What is the meaning of it all? Why is it as it is, and in what manner was it formed?

To the earliest civilizations the sky was a kind of dome which no human could reach. On it were stars, fixed in patterns which they recognized as familiar objects and characters of myth and legend. Yet as time passed, what intrigued many astronomers in the West was why a few of these stars wove paths among the rest. The motions of these "planets", or wanderers, were first investigated in detail in Mesopotamia.

Later, in ancient Greece, the challenge was resumed, and a new view taken of the universe. For aesthetic and mathematical reasons, it was expanded from a dome to a sphere. In doing so, the Greeks took the first step toward realizing that the universe is larger than it appears. They also developed an elaborate mathematical way of describing the cyclic motion of the Moon and planets around the Earth, which, on what seemed good evidence, appeared fixed at the centre of the universe. Indeed, their plan was so satisfactory that it did duty for philosophers and astronomers for more than two thousand years. Not until the 1540s did mathematical reasoning and aesthetic considerations again bring about a change. This time it turned into an intellectual watershed.

The Sun, not the Earth, was put at the centre of the universe, the planets were set in orbit about it, and human beings were dethroned from their privileged position at the centre of all creation. With subsequent mathematical work on planetary motion by Isaac Newton, the foundations of a new scientific universe were laid: a universe whose boundaries seemed infinite.

Since Newton's time the march of science has accelerated. Now, more than three centuries later, its unceasing progress has finally brought us ways of observing the universe which would have astounded earlier astronomers. Yet this is not all. Our understanding of the nature of things has grown immensely, and new theoretical tools have been forged. Mathematics has been developed to an astonishing degree, and our knowledge of the world of physics has leaped ahead in a way that would leave previous generations gasping. As a result, our up-to-date picture is truly amazing compared with anything conceived in previous ages.

A monument to the profound imagination of 20th-century science, it calls on the most advanced and aesthetically elegant concepts of quantum theory to describe the very particles of the material world of which everything in the universe is made. Together with the well-nigh incredible doctrines of relativity, and the stupendous observations brought back from space itself, we have evidence never previously available.

The astonishing picture that emerges is traced in the account that follows. It looks at the universe from its very beginnings to its ultimate end; and assesses the place of humankind in today's totally new scheme of space and time.

***A full Moon shines down*** *on bleak pinnacles of rock in the Australian desert. Earth's companion world is only a step away compared with the immensity of the cosmos known to modern astronomy.*

# CREATION OF THE UNIVERSE

The universe began in a colossal explosion, in which energy, space, time and matter were created. There is little doubt of this among scientists. The evidence that makes them so certain that the story is correct, in broad outline at least, comes from the discoveries of astronomy and subatomic physics, and from relativity and quantum theory, the two revolutionary theories that lie at the heart of modern physics. With the aid of these, theorists can trace the history of the universe to within a minute fraction of a second after the beginning.

There is doubt about precisely when the Big Bang occurred. It was at least 15 billion years ago, and perhaps as much as 20 billion. (Here, as throughout this book, "billion" means "thousand million".) The extreme conditions of the Big Bang do not exist today, but on both the astronomical and the subatomic scales the universe displays some extremely strange aspects. For instance, it almost certainly contains black holes, whose interiors are cut off from the surrounding cosmos. There are stars that spin hundreds of times a second and are made of material 35 billion times as dense as lead. There are stars, and even whole galaxies, that explode.

As for the particles at the heart of the atom, they are composed of still smaller particles called quarks, with such peculiar properties that language has to be stretched to describe them. Words like "colour" and "charm" are pressed into service with new senses – though colour is a meaningless concept at this level and the particles charm the physicist only because they are so puzzling.

Among the astronomical observations supporting the Big Bang theory, three pieces of evidence are specially important. The first is that the galaxies – the vast systems into which stars, gas and dust are grouped – are all moving away from each other. We find that we are living in an expanding universe. A primordial Big Bang can explain this.

The second piece of observational evidence is the discovery of radiation reaching us from every direction of the universe. Moreover, this radiation is of equal intensity from every part of the sky. This fits in well with the idea of a hot Big Bang; what we are seeing is the glow of the primordial universe as it was at a very early date, a few hundred millennia after the beginning. Now, some 15 billion years later, this radiation has cooled to a few degrees above the absolute zero of temperature. This is just the temperature to be expected today if the radiation had originated in a hot Big Bang.

The third item of evidence for the Big Bang comes from nuclear physics. Studies of how the chemical elements would evolve after a Big Bang suggest that in the present-day universe we should find a particular ratio between the amounts of deuterium (a form of hydrogen) and of helium. Astrophysicists have verified that the existing ratio is what the theories predict.

So the hypothesis of a primordial explosion is extremely well founded. Yet there are variations of the theory, and outright alternatives to it, that will merit consideration at the end of this book. Nevertheless, however these unconventional theories fare in the future, the Big Bang theory will remain one of the grandest constructs of 20th-century scientific thinking. It tells a story that spans the microcosmos and the macrocosmos and leads from the beginning of the universe to an ending that can still only be guessed at.

***The tracks of subatomic particles*** *form an intricate pattern in a bubble chamber filled with liquid hydrogen. The reactions brought about by modern particle accelerators mimic processes that occurred in the first split second of the universe.*

# THE SCALE OF THE UNIVERSE

## • *From quarks to superclusters*

The universe that has been opened up to enquiry by development of modern scientific instruments ranges from the wastes of intergalactic space to the interior of the atom. The largest entities of which we now have knowledge are the so-called superclusters – clusters of clusters – of galaxies. The smallest entities are the fundamental subatomic particles, such as quarks. Our knowledge of the latter is indirect, since they have never been observed singly, but is none the less firmly based.

To span this vast range, a progressively increasing scale of length is shown here, each interval representing a distance 10 times larger than the one before. Scientists speak of a tenfold increase or decrease as a change of one "order of magnitude". The scale shown on these pages ranges over more than 40 orders of magnitude. Shorter than the wavelength of light, and therefore invisible, are atoms, molecules, and the smallest living entities, the viruses. Human beings are more than 10 million times, or seven orders of magnitude, larger.

In the same relation to the human body as the human body is to a virus are the gas giants, the largest planets of the solar system. In roughly the same proportion again are the planetary nebulae, the clouds of gas thrown off by dying stars. And in roughly the same ratio to these are the clusters of galaxies.

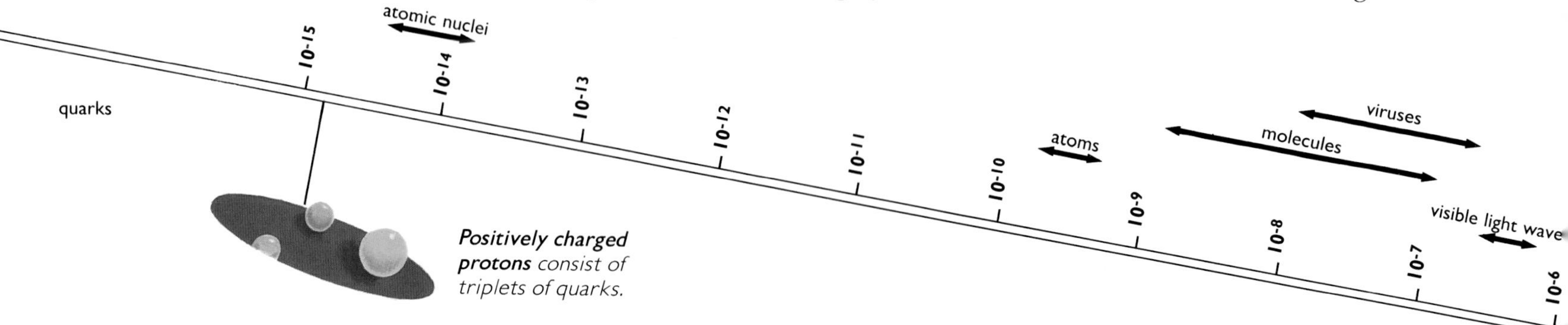


***Positively charged protons** consist of triplets of quarks.*

### Number shorthand

The scale running across these and the following two pages is logarithmic – that is, each division represents a tenfold increase in size over the one before. This convention enables sizes from the subatomic to the cosmic to be encompassed in a small space.

In a similar way, a simple notation enables very large and very small numbers to be written in a compact manner. To begin with everyday numbers: since $1,000 = 10 \times 10 \times 10$, it is written as $10^3$. A number such as a billion billion, which is the result of multiplying 18 tens together and is normally written as 1 followed by 18 zeros, is written in this notation as $10^{18}$.

Very small numbers are written in an analogous way. One millionth is $10^{-6}$ (that is, the result of dividing 1 by $10^6$). The very brief period of time called the Planck time is written as $10^{-43}$ seconds – that is, as one second divided by $10^{43}$.

The price of such compactness of expression is a distortion of scale. It is easy to forget the enormous difference represented by the easy step from, say, $10^2$ to $10^4$ light-years. But scientists and mathematicians find this so-called exponential or powers-of-10 notation indispensable.

***Making numbers manageable.** In the scale, ever greater distances are telescoped into a constant length; for example, lengths of 10 and 100 units (right) are folded into the same distance as a unit length (above). This gives less sense of the difference between objects of these sizes (below).*

***At the limit of visibility*** *to the unaided eye the crab louse nevertheless represents a high level of organization of living matter. Thus its nervous system, primitive though it is, has not yet been fully unravelled.*

***The largest land creature*** *alive today, the African elephant, can stand over 4 m tall. The largest land animals ever, the dinosaurs, could be over 6 m high at the shoulder.*

$10^{-5}$ $10^{-4}$ $10^{-3}$ $10^{-2}$ $10^{-1}$ 1 $10$ $10^{2}$ $10^{3}$ $10^{4}$ $10^{5}$

human beings

***Single-celled protozoa*** *are about a tenth of a millimetre across. They are the simplest of living creatures (discounting viruses, which live parasitically). A human being is 10,000 times as big as a protozoan, which in turn is 10,000 times as big as a large molecule.*

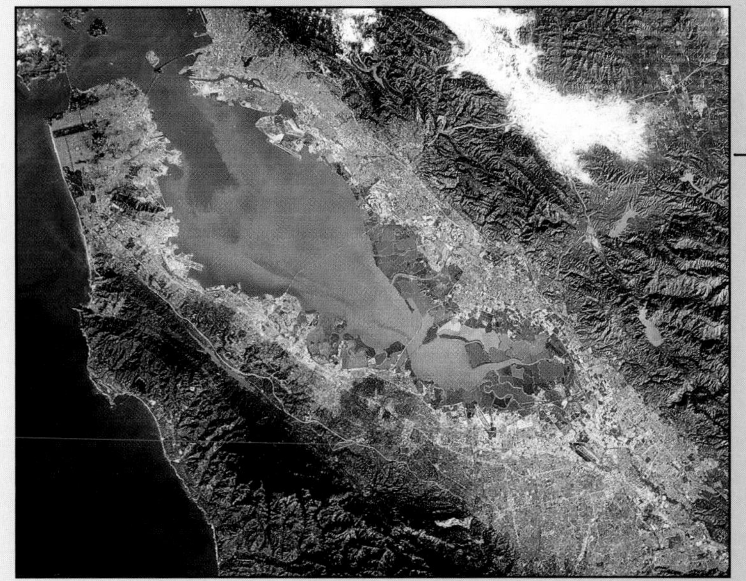

***The largest cities*** *are about 50 km across. One of these great gatherings of human beings approaches half of one percent of our planet's diameter.*

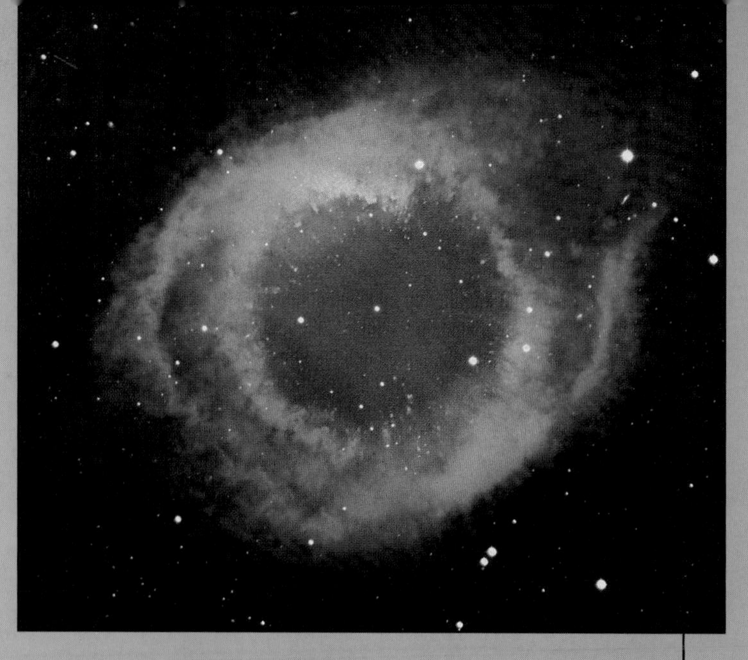

***A planetary nebula**, because of its vast size, appears in a telescope as a disc, whereas a star remains merely a point of light. This type of nebula is a globe of gas puffed off by a dying star.*

***Dominating the solar system**, the Sun has a diameter over 100 times that of the Earth. It is a very ordinary star – many others have diameters hundreds of times as great again.*

largest asteroid

$10^{6}$ $10^{7}$ $10^{8}$ $10^{9}$ $10^{10}$ $10^{11}$ $10^{12}$ $10^{13}$ $10^{14}$ $10^{15}$ $10^{16}$

Earth

gas giants

Earth's orbit

red supergiants

solar system

planetary nebulae

distance to nearest star

***The Earth's satellite**, the Moon, is 3,476 km across. It is one of the largest satellites in the solar system, larger than the planet Pluto.*

***Galaxies are huge assemblages*** *of stars, gas and dust, bound together by gravitation. Many are 100,000 light-years across. A galaxy bears roughly the same proportion to the Earth's orbit that the human body does to an atom.*

***The birthplaces of the stars*** *are the diffuse nebulae, scores of light-years across. These masses of hydrogen gas and grains of dust are more rarefied than many a laboratory vacuum.*

***Clusters of galaxies*** *may have millions of members. Even these are not the largest unit into which matter organizes itself: clusters form superclusters, which can be hundreds of millions of light-years across.*

$10^{17}$ $10^{18}$ $10^{19}$ $10^{20}$ $10^{21}$ $10^{22}$ $10^{23}$ $10^{24}$ $10^{25}$ $10^{26}$ $10^{27}$

galaxies

radio galaxies (across lobes)

Local Supercluster

observable universe

# MATHEMATICS AND REALITY

## *Insight through symbols*

The theory of the Big Bang has much observational evidence to support it. Yet there is another side to the study of the universe. Mathematics often enables cosmologists, like other scientists, to work out in theory what nature should be like *before* experimentalists and observers have confirmed that such is the case.

Mathematical reasoning is the only way to grasp the fundamentals that lie behind what we observe. This is so because mathematics is a language in which ideas can be formulated in a precise way and which allows the mind to work out logical consequences at profound depths, where mere words would present quite insurmountable obstacles. Time and again mathematical reasoning has provided insights available in no other way.

Unfortunately many people suffer a mental block immediately they see a simple equation; they find even one as simple, and as fundamental, as $E = mc^2$ to be elusive. So what does this particular equation mean?

The relationship was first derived by Einstein in the early years of this century in connection with his theory of relativity (pp. 16–17). It concerns the energy obtainable by the complete annihilation of a quantity of matter, which he derived by mathematical analysis. In the equation, $m$ refers to the mass of the material, $E$ refers to the energy, and $c$ is the velocity of light. The latter is an enormous quantity – 300,000 kilometres per second. The equation asserts that, in suitable units, the energy is equal to that mass multiplied by $c^2$ – that is, $c$ multiplied by itself.

The equation therefore specifies precisely the quantity of energy equivalent to a given mass. Since $c$ is such a huge number, it implies that the amount of energy released is immense. This explains why an atomic bomb creates such a tremendous explosion and why the Sun shines so intensely – for the Sun, and all the stars, shine by converting mass into energy.

Expressing the meaning of $E = mc^2$ in words is far more long-winded than writing it in the form of an equation. Furthermore, once you understand the symbols, the equation's meaning is far more readily understood than the words.

Yet there is still more to the use of symbols. From everyday algebra we can write a new equation thus: $m = E/c^2$. This tells us how much mass is equivalent to a given amount of energy, and further suggests that mass can be created from energy. This is highly significant for cosmologists, for it tells them that all the matter in the universe could have come from the energy of the hot Big Bang.

With good reason, it has been said that mathematics is an art, and mathematicians certainly like to see symmetry, elegance and simplicity in their equations. Furthermore, these aesthetic features seem to guide scientists in discovering facts about the real world.

A classic example of the power of symbols is afforded by the equations derived by the 19th-century Scots physicist James Clerk Maxwell. Maxwell was intrigued by the electrical researches of the brilliant self-taught experimentalist Michael Faraday. Faraday was no mathematician, and Maxwell set himself the important task of expressing in mathematical form the relationships Faraday had discovered experimentally.

Faraday had introduced the idea of fields to explain how electric and magnetic effects could be detected over a distance. A field was the pattern of electrical or magnetic influence filling space around a magnetized or electrically charged object. The field was depicted by drawing lines representing the direction of the electrical or magnetic force at each point of space. Maxwell was able to derive equations that expressed the observed effects, and he related them to the characteristics of the magnets and electric currents.

On examining these equations, Maxwell recognized that they gave results of exactly the same form as those expressing the motion of waves in a fluid. He therefore concluded that there are waves in electric and magnetic fields. From the known values of certain quantities that had been measured in laboratory experiments, Maxwell was able to calculate the speed of the waves. It turned out to be 300,000 kilometres per second, the speed of light.

Maxwell concluded that light consisted of electrical and magnetic waves. Furthermore, there should be a whole range of such waves, some longer and some shorter than the familiar waves of visible, infrared and ultraviolet light.

Some 30 years later, Heinrich Hertz in Germany succeeded in producing electromagnetic waves some 30 centimetres long. Hertz showed that these radio waves, as we now call them, behaved as Maxwell's equations had predicted; they travelled at the velocity of light and were reflected or refracted (bent) in the same way as ordinary light waves.

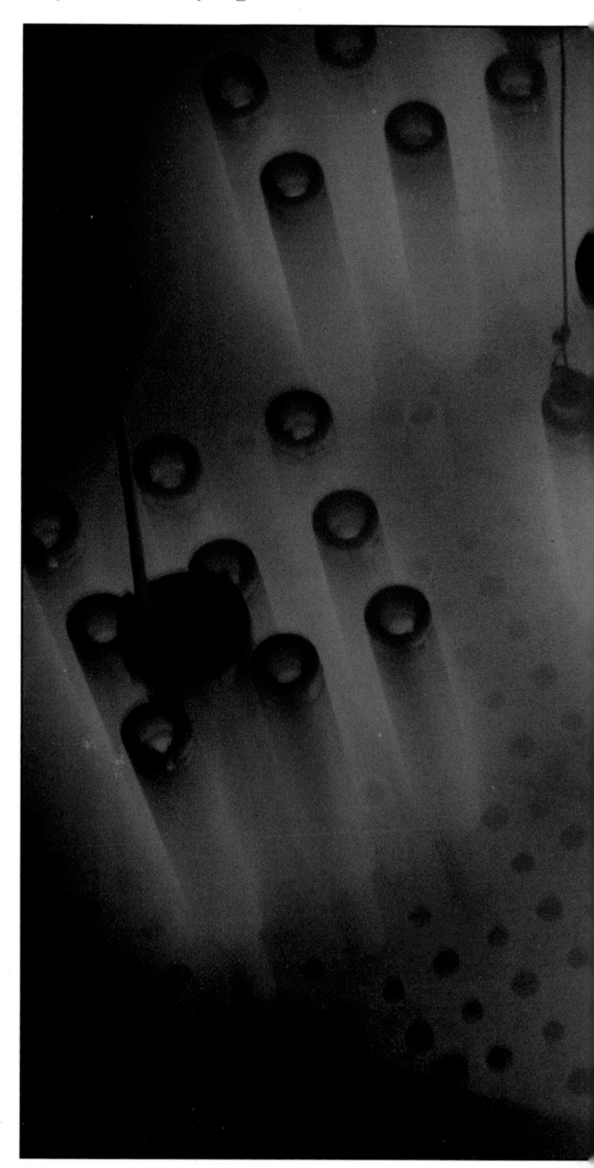

Thus mathematics, devised to express already existing results of experiments, had pointed to totally new experiments and discoveries – a feat that words alone would almost certainly have been quite unable to achieve.

Furthermore, once he had formulated his equations, Maxwell went on to make them more self-consistent, elegant and symmetrical. He introduced an additional term, now known as the displacement current, which proved to play a part in the growth and contraction of a magnetic field associated with a changing electric field. The urge to achieve mathematical symmetry had important consequences. Today it plays an equally crucial part in helping cosmologists fathom what happened during the earliest moments of the universe.

*A blue glow reveals that energy is being released from nuclear fuel rods stored under water (below). The energy is emitted when heavy atoms of uranium or plutonium fission (split) into smaller atoms. The combined masses of the fragments are less than the mass of the original atom. The missing mass is converted into energy in acccordance with Einstein's equation $E=mc^2$.*

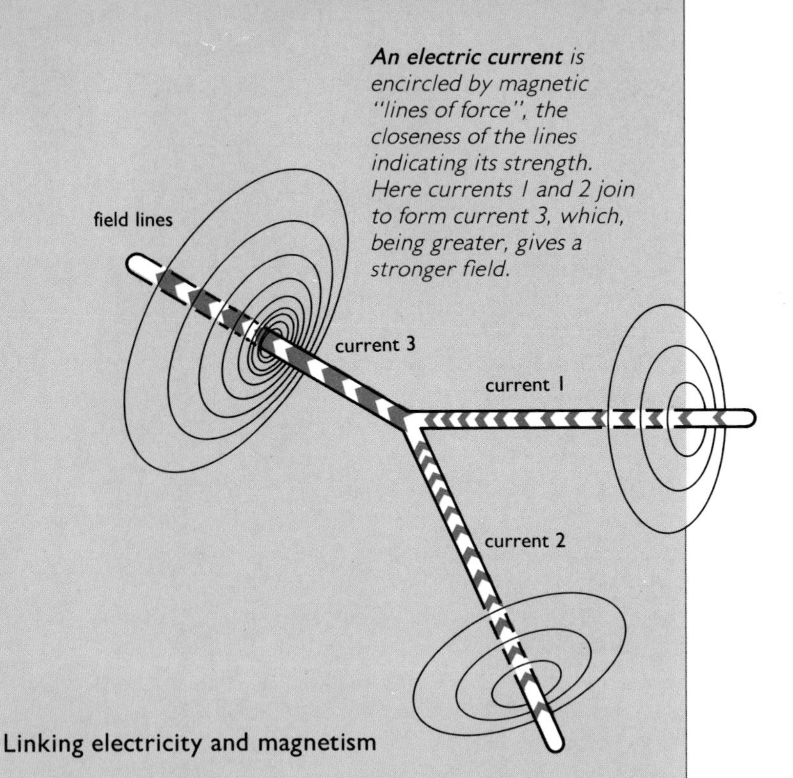


*An electric current is encircled by magnetic "lines of force", the closeness of the lines indicating its strength. Here currents 1 and 2 join to form current 3, which, being greater, gives a stronger field.*

### Linking electricity and magnetism

An example of what Maxwell achieved with mathematics may be illustrated by the first two of his four equations. The first equation reads:

$$\textbf{curl } \mathbf{E} = -\frac{\partial \mathbf{B}}{\partial t}$$

This is written in modern vector notation. A vector is a quantity, such as velocity or force, which has both magnitude and direction. The vector **B** represents the magnetic field, **E** the electric field. The symbol $t$ refers to time, and $\frac{\partial \mathbf{B}}{\partial t}$ means "rate of change of **B**". The equation says that **curl E** – a certain property of **E**, which may be called its "rotation" – is proportional to the rate at which **B** changes.

Maxwell's second equation reads:

$$\textbf{curl } \mathbf{H} = \mathbf{J} + \frac{\partial \mathbf{D}}{\partial t}$$

This says that the "rotation" of a magnetic field **H** gives rise to an electric current **J** and vice versa. The new term $\frac{\partial \mathbf{D}}{\partial t}$ added a certain symmetry previously lacking in the first equation. The equation shows that the magnetic field depends not only on the current **J** – which was well known before Maxwell – but also on the rate of change of a hitherto unrecognized "displacement" current, **D**, spreading throughout space. This asserted the distribution of electromagnetism throughout space. Later experiment showed Maxwell to be correct in this matter.

# SPACE AND TIME

## *The meaning of relativity*

The laws expressed by James Clerk Maxwell were to play a key role in the scientific revolution that culminated in the work of Albert Einstein in the early years of the 20th century. It was a revolution that shook our very ideas of the nature of space and time. The classical physics that had developed from Newton's work had held that events everywhere took place within a shared universal time and an absolute and universal space. But by the beginning of the century it had become evident that mass, distance, energy and the passage of time itself should vary according to the motion of the observer.

When Einstein first examined these ideas, he considered only observers moving relative to each other at unchanging velocities – at constant speeds in straight lines. This first version of his theory, called the special theory of relativity, was put forward in 1905. It was founded on the postulate that the laws of physics must be identical for every observer moving in this way – that is, for every frame of reference or viewpoint from which the measurements of physics are made. No frame of reference could be picked out as being at absolute rest.

In particular, Maxwell's laws of electromagnetism (pp. 14–15) must be the same in all frames of reference. Since the speed of light (as measured in a vacuum) is determined by these equations, the speed of light must also be the same for all observers.

Paradoxical as this assertion seems, it already had experimental backing, for no variation in the speed of light had ever been detected. Light from double stars, for example, which orbit each other at significant speeds, is neither delayed nor advanced by the approach or recession of the source stars. If it were, we should sometimes see such stars in more than one place at once. And a classic experiment by A. A. Michelson, later with the help of E. W. Morley, was designed to measure the speed of the Earth through space by its effect on light waves. But it completely failed to detect any of the expected effects.

The consequences that Einstein was able to deduce from the principles of relative motion and the constancy of the speed of light were astounding. First of all, he showed that the length of a body was relative to the frame of reference in which it was measured. If the body were moving at high speed in relation to the observer, it would appear to contract in the direction of its motion, while remaining unaffected in other directions.

The mass of a body also proved to be relative to the frame of reference adopted. (Mass is the "quantity of matter" in a body: the greater its mass, the greater its resistance to changes in its motion.) As a body moves faster, its mass increases relative to an observer at rest. As it approaches the speed of light, more and more energy is needed to give it an extra unit of speed, with the result that no material body can ever reach the speed of light.

Even more astonishing was the discovery that time also is relative. On an object that is moving fast relative to the observer it seems to pass more slowly. So there is no universal standard of time. Our time literally does not pass at the same rate as that of another observer in a different frame of reference. There is no absolute simultaneity in the universe.

By 1915 Einstein had made another enormous advance: he had achieved his general theory of relativity, which was able to deal with changing velocities – that is, with frames of reference in accelerated motion. And because gravity makes bodies accelerate, this new theory became a theory of gravity as well.

The new theory showed that the motions of bodies have to be considered as motions through space-time: a four-dimensional amalgam of the three dimensions of space and the single dimension of time, all intimately connected. From the mathematics of the theory came the result that space-time is distorted – curved – around any mass, a distortion that compels other bodies to follow curved paths in response. Gravitational force is replaced by curvature of space-time.

Einstein followed a clue offered by what had seemed to be an isolated fact in Newton's theory of gravity. There are

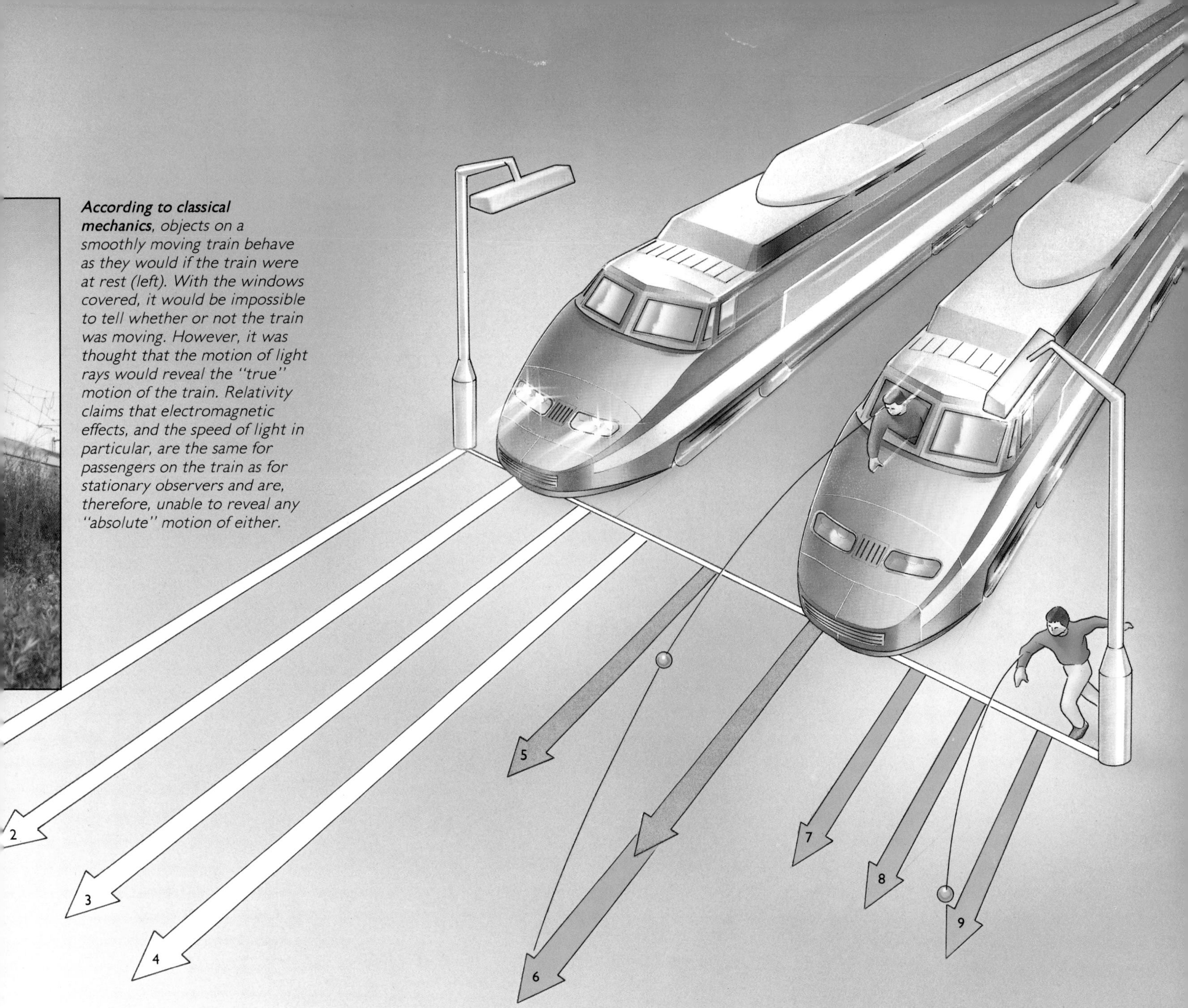

***According to classical mechanics****, objects on a smoothly moving train behave as they would if the train were at rest (left). With the windows covered, it would be impossible to tell whether or not the train was moving. However, it was thought that the motion of light rays would reveal the "true" motion of the train. Relativity claims that electromagnetic effects, and the speed of light in particular, are the same for passengers on the train as for stationary observers and are, therefore, unable to reveal any "absolute" motion of either.*

two ways of measuring mass: by its inertia – its resistance to being accelerated – and by its weight. Inertial mass is always found to be proportional to gravitational mass, and this is why heavy bodies fall at the same rate as light ones: if one body weighs twice as much as another, it will also have twice the inertia, and the one exactly compensates the other. The theory at which Einstein arrived made it seem natural that a heavy and a light object should move together in a gravitational field; whatever the mass of a body, its movement depends only on the local curvature of space.

Einstein's most famous discovery emerged from the special theory. This was the equivalence, and interconvertibility, of mass and energy, expressed in the equation $E=mc^2$ (pp. 14–15). The equation was established in the popular mind, firmly linked with Einstein, by the advent of nuclear bombs and the development of atomic energy. It remains a cornerstone of physics and cosmology today.

***The speed of light*** *has an enormous value and, for a stationary observer, is always the same from any fixed source (1), whereas the speed of a material object, such as a moving ball, is variable (9). Remarkably, the speed of light from a moving source, such as a train, is the same for a stationary observer (2), for an observer on the train (3), and for an observer moving in relation to the train (4). This contrasts markedly with the observed speed of the ball. If thrown from the train, its speed relative to a stationary observer (6) is the sum of the speed of the train (5) and the speed of the object relative to the train (7). The latter is, in turn, equal to the speed of the object measured in relation to the person throwing it (8).*

# CURVED SPACE-TIME

## • *The geometry of relativity*

The general theory of relativity is a theory of gravitation, which is viewed as a distortion of the space and time surrounding a body. In this sense, relativity is a geometrical theory because the mathematical study of space, whether curved or flat, is geometry.

The type of geometry with which most people are familiar is "Euclidean" geometry, named after the Greek philosopher Euclid, who lived about 290 BC. Euclid systematized all the geometrical knowledge of his time in a book now known as the *Elements*, which remained the undisputed touchstone of all geometrical knowledge until the late 19th century. Euclidean geometry was believed to be the only way in which to describe the space that everyone experiences.

Not until 1823 did Euclidean geometry begin to lose its unique position. In that year, the Hungarian mathematician Janos Bolyai discovered that there could be totally consistent geometries different from Euclid's. One of these later proved to be closely linked with general relativity.

In non-Euclidean geometry, the concept of a straight line is replaced by the more general idea of a geodesic, the line that follows the shortest distance between two points. On the curved surface of the Earth, any section of the equator or a line of longitude (a circle passing through both poles) is a geodesic. In Euclidean geometry – the geometry of flat surfaces – the geodesics are straight lines.

Consider a three-sided figure whose sides are geodesics. In Euclidean geometry this is a triangle whose internal angles always add up to 180°. But in the geometry discovered by Bolyai (and independently by the Russian Nikolai Lobachevsky), called hyperbolic geometry, the internal angles add up to less than 180°. The smaller the triangle is, the closer the sum approaches to 180°, but a very large triangle can have very small internal angles.

The second alternative to Euclidean geometry was worked out in the 1850s by a Swiss mathematician, Ludwig Schläfli, and a German, Bernhard Riemann. In such a "Riemannian" geometry, as it has become known, the internal angles of a triangle always add up to more than 180°. Again, the difference from 180° increases with the size of the triangle.

To visualize this new geometry, Schläfli described it as being the geometry of the surface of a hypersphere – the analogue of a sphere drawn in four dimensions. The geodesics then actually do appear as straight lines. (The subject of the fourth dimension – and higher ones – is pursued on pp. 172–75.)

A simple analogy with hyperbolic and Riemannian geometries can be obtained by considering them in two dimensions only – that is, by concentrating on just the surfaces of solid bodies. Hyperbolic geometry is, then, the geometry of a saddle-shaped surface; while Riemannian geometry is that of a spherical surface. In the latter case, for example, the angles of a triangle clearly grow larger as the triangle increases in size.

Riemannian geometry extends our ideas of space. Indeed, there can be some striking differences from the familiar space of Euclid; a geodesic can, so to speak, curve over on itself. Riemannian space can be finite in volume, though lacking any boundary. (This can be grasped by analogy with the surface of a sphere, which is finite but boundless.) Space extends outward without limit – we can travel up and down, or side to side, as far as we like, and we shall never reach a boundary. Yet this space is finite in volume.

Furthermore, travellers continuing in a straight line, without deviation, will eventually return to their starting point from the direction opposite to that in which they set out – just as an imaginary creature confined to the surface of a sphere would do.

Einstein maintained that not only matter but light would follow a curved path through space-time distorted by the presence of matter. This prediction of the bending of light by gravity was triumphantly vindicated by observations of an eclipse of the Sun in 1919. The new theory also accounted for anomalies in the motion of Mercury.

People have great difficulty in grasping the idea of a beginning of the universe. They want to know what was there before it started. And when they consider the expanding universe, they find themselves asking the question "What lies outside it?" In the Riemannian geometry of relativity, it is possible to answer that space and time did not exist before the Big Bang, and that nothing exists outside the present expanding universe because there is no

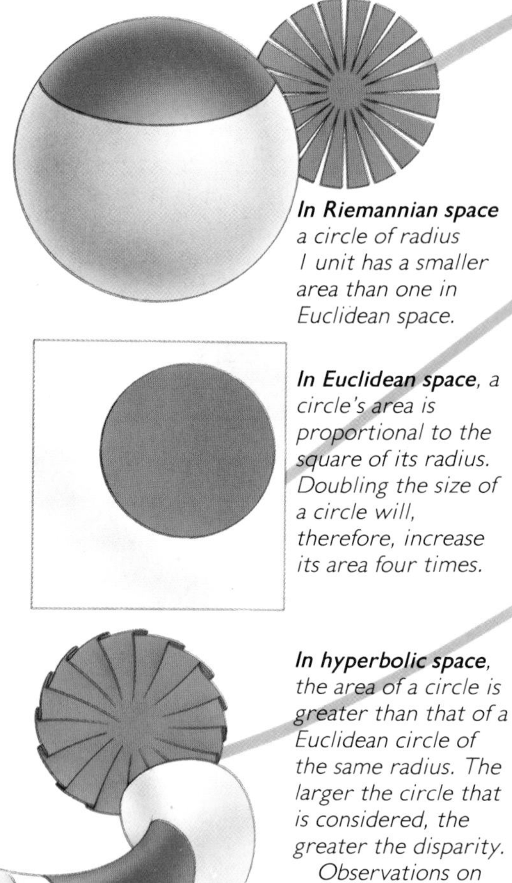

***In Riemannian space*** *a circle of radius 1 unit has a smaller area than one in Euclidean space.*

***In Euclidean space****, a circle's area is proportional to the square of its radius. Doubling the size of a circle will, therefore, increase its area four times.*

***In hyperbolic space****, the area of a circle is greater than that of a Euclidean circle of the same radius. The larger the circle that is considered, the greater the disparity.*

*Observations on an astronomical scale can, in principle, reveal whether the overall geometry of our universe is hyperbolic, Riemannian or Euclidean.*

***The elliptical orbit*** *of Mercury slowly revolves, or precesses, around the Sun. The gravitational influence of other planets would, according to Newtonian theory, produce a small effect (1), but the observed precession is greater (2). Einstein's general theory of relativity was able to explain the discrepancy.*

***On a small scale*** *it is not possible to detect the curvature of geodesics, and space necessarily appears Euclidean.*

outside. We live in a four-dimensional hypersphere. (For two-dimensional creatures living on the surface of the sphere there is no outside – the third dimension does not exist for them.)

The trouble is that it is natural to generalize from everyday experience, and this is not a good guide. For a long time it was generally accepted that the Earth was flat; now that we navigate the oceans and fly in the air it is a matter of practical importance to take into account the curvature of the Earth.

For example, the railway tracks for the dishes of the 5-kilometre radio telescope at Cambridge in the UK are not laid flat on the ground – they are a couple of metres above the ground at one end. This is because they are truly straight, whereas the surface of the ground is curved. In a similar way, although Euclidean geometry is adequate over terrestrial distances, it becomes inadequate over the vast spaces of the universe.

***The distortion of space*** *by the huge mass of the Sun bends light rays from a star as they pass nearby (3). This alters the star's apparent position because we assume that light travels in straight lines.*

on. It also makes it possible to determine the chemical elements that they contain.

Bohr gave rules that specified which orbits are allowed and which are not, and calculated their energy levels. He could then explain the wavelengths of light emitted by hydrogen and other simple atoms. His success in doing this triumphantly vindicated his model, yet it was soon to be superseded by a still more radical revision of physics.

Bohr's model of the atom implied that electrons are never between one orbit and the next. We may feel that they "must" move from one orbit across space to the next, but that is because we are used to seeing this in our everyday experience. This idea is so ingrained in us that when we watch a movie we are sure we see things and people moving continuously across space; in reality we are watching a series of separate images that give only the illusion of continuous movement. In quantum theory we have to suppose that in certain circumstances electrons move in jumps from one part of space direct to another – an example of how the microcosm of the atom differs markedly from our macroscopic world.

Stranger things were to come. They stemmed in part from Einstein, who in 1905 had shown that the electrical voltage produced when light falls on certain metals – the effect used in a camera's exposure meter – could be satisfactorily explained only if light is considered to behave in these circumstances as a stream of particles, not waves. Such light particles are called photons. They are, in fact, examples of Planck's energy quanta. The work of Einstein and Planck made it clear that at the atomic level Maxwell's unified electromagnetic theory was not sufficient – any more than Newton's theory of gravitation – to explain all the behaviour of matter in the universe.

Indeed the existence of photons raised a serious problem, because light energy also behaves like waves travelling through space. In the years around 1800 Thomas Young and others had produced a host of laboratory results that gave firm evidence for the wave theory of light. The most telling are those that demonstrate that beams of light can interfere with each other (see box). The results of Einstein and Planck suggested that light is both particles and waves, depending on the observations being made. Photons are in fact "wavicles".

What is true of light, which we normally think of as consisting of waves, is true also of what we normally think of as particles. If a beam of electrons is passed through a tiny hole in a piece of metal foil, they spread out as waves do, forming a pattern of light and dark rings. Indeed, beams of electrons can be focused and used to form images in an electron microscope. Interference patterns can be displayed by superimposed beams of electrons. And the same is true of all subatomic particles.

Even before these demonstrations of the wave nature of the electron, Erwin Schrödinger, an Austrian physicist, was

### Interference

When two sets of ocean waves overlap, they can interfere with each other. If crests of one set of waves coincide with crests of the other, and troughs coincide with troughs, a more intense combined wave will be produced, with extra-high crests and extra-low troughs. If crests coincide with troughs, the waves will cancel each other out. Analogous effects can be observed in any form of wave motion, including sound and light. When light waves reinforce each other they produce brighter light; when they cancel each other out they give darkness.

Beams of particles can be made to interfere in this way, revealing the wave nature of matter. In this case, reinforcement of the beam corresponds to finding more particles passing every second. Where the beam is cancelled out by the interference effects, no particles are to be found.

Another effect of the wave nature of light is that a parallel beam of light striking the edge of an object spreads out, giving a fuzzy edge to the shadow. When examined closely the edge of the shadow shows a diffraction pattern of alternate light and dark bands. Again, this effect is shown by beams of particles.

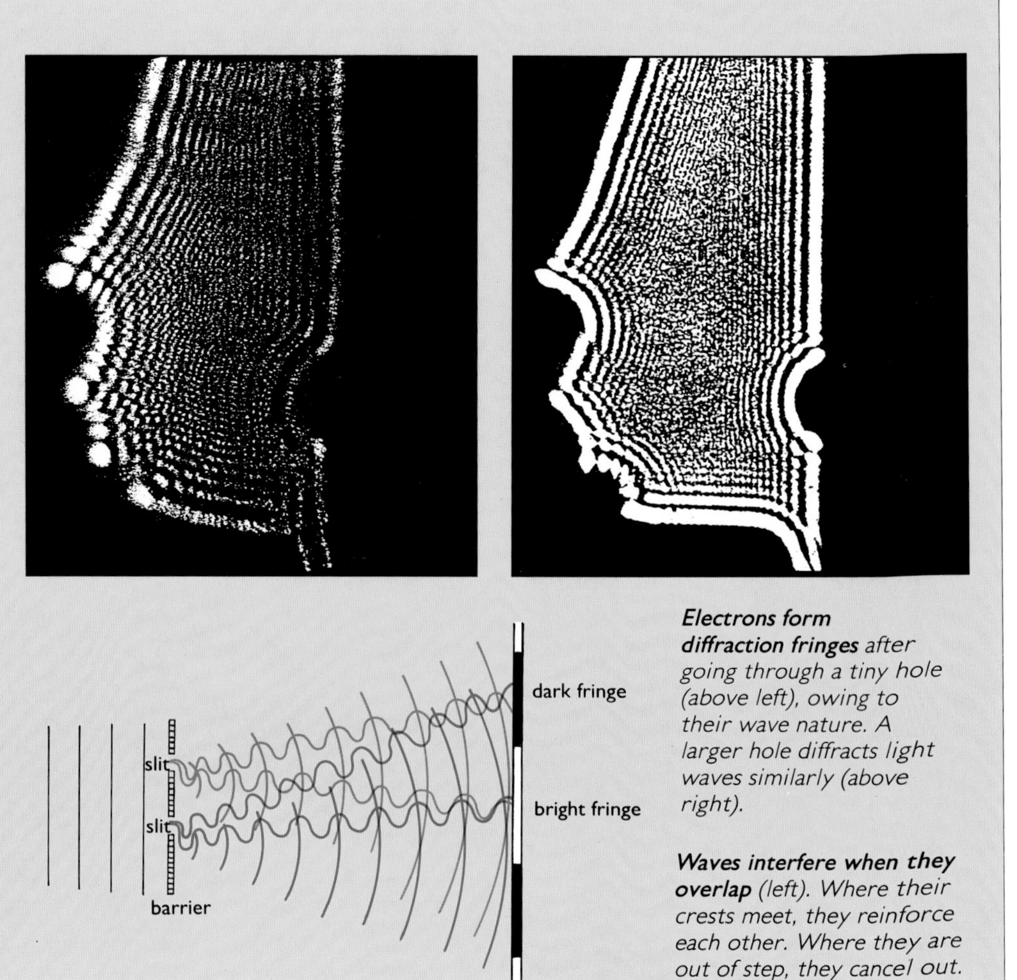


***Electrons form diffraction fringes*** *after going through a tiny hole (above left), owing to their wave nature. A larger hole diffracts light waves similarly (above right).*

***Waves interfere when they overlap*** *(left). Where their crests meet, they reinforce each other. Where they are out of step, they cancel out.*

able to use probability waves to explain the energy levels in the Bohr atom. The waves associated with the electrons were regarded as being standing waves, or resonances, formed much as vibrations of certain frequencies are set up in violin strings. The well-defined paths of electrons in the Bohr atom were replaced by "clouds of probability".

An electron is more probably to be found where the cloud is dense than where it is thin, but we cannot say precisely where and with precisely what speed or energy. It is like being told that a wave of burglaries is affecting a certain town: it is impossible to say whether a break-in will happen at a particular house on a specified day; we can only assess the probability that one will occur in a given period over a given area. Such statistical probabilities are real enough, as insurance companies know. Although the electrons cannot now be viewed as having definite orbits, they do have definite energies.

The fact that particles are wavelike has other enormously important consequences. It means, for example, that in general there cannot be complete certainty about both the position *and* the speed of a particle at a particular instant. We can specify only one factor at a time – the other one remains uncertain. If a physicist performs an experiment to determine the particle's precise position, he will disturb it so that its motion will become completely uncertain. And if he tries to determine the speed precisely, he will disturb the particle so that its position becomes unknown.

This crucial feature of the new physics, first stated by the German physicist Werner Heisenberg, has become known as the uncertainty principle. Another aspect of it is that we cannot specify both the energy involved in a certain event *and* the time of occurrence of that event to arbitrary accuracy. Such uncertainties are a consequence of the nature of the atomic world itself, not of any inability on our part to measure precisely enough.

***According to quantum mechanics*** *position and velocity cannot be precisely specified at the same time. An observation that reveals the speed and direction of movement of a particle "smears" its position – rather as the time exposure (above) reveals the dancer's motion but makes his position less definite. An observation that fixes a particle's position leaves its motion completely uncertain – just as a high-speed exposure (right) makes the dancer's position clear and precise, while giving no information at all about movement.*

In 1925 the Austrian-American physicist Wolfgang Pauli proposed his exclusion principle, which states that in any system only one electron can be in a given quantum "state". The state is defined by such quantities as energy and position, and also by spin, which is in some ways analogous to the spin of a macroscopic object, but with some differences which show that subatomic particles "see" a different world from our own. Spin comes in discrete values, intrinsic to a given particle. Some particles have whole-number spin values, labelled 0, $\pm 1$, $\pm 2$ and so on. Others, including the proton, neutron and electron, have fractional, or non-integral, values: $\pm \frac{1}{2}$, $\pm \frac{3}{2}$ and so on.

An electron in the innermost orbit of an atom can be in either of two states, according to the direction of its spin. Therefore two electrons can occupy this orbit. Limited numbers of states are permitted in other orbits. This is why electrons do not all congregate at the lowest energy state.

The creation of quantum mechanics led to an understanding of the architecture of the atom's outermost layers. But as physicists probed deeper, into the nucleus, they were to encounter new forces and new particles.

# THE SUBATOMIC WORLD

## • *A profusion of particles*

A bewildering variety of subatomic particles has been discovered by modern physicists. Every advance in the power of "atom-smashers", or particle accelerators (see box), leads to the production of yet more, so that today thousands have been catalogued. But in recent years theorists have made great strides in bringing order into this seeming chaos.

Every type of particle is now known to have its own antiparticle (except for a few cases where particle and antiparticle are identical). The properties of an antiparticle are opposite to those of its corresponding particle. Thus the antiparticle of the ordinary negatively charged electron is the positron, which has the same mass but a positive charge equal to the electron's negative charge.

This practically halves the number of distinct types of particle. One way of simplifying things further is to consider how particles interact with each other via "messenger particles".

Consider the electromagnetic force. When an electron in an atom jumps from a higher-energy outer orbit to a lower-energy one closer to the nucleus, the energy difference escapes in the form of a photon – a quantum of electromagnetic radiation. The photon can be considered to be a messenger, which carries the energy away.

Now imagine what happens when moving electrons pass near each other. Both have negative charges and will therefore repel each other. In quantum physics this force is regarded as being due to the emission of a virtual photon by one and its absorption by the other. (A virtual particle is one so short-lived that its existence can be inferred only indirectly.)

An analogy in the macroscopic world occurs when one person throws a ball to another. As the ball is released the thrower experiences a reaction and is impelled backward. The catcher experiences a force, too, also pushing backward. In the quantum world the virtual photons act like balls – messengers that carry electromagnetic forces.

This kind of explanation can account for other subatomic forces. One of these is the weak nuclear force. An example of a process involving the weak force is the decay of a neutron into a proton, an electron and an antineutrino.

The messenger particles of the weak nuclear force are known as W and Z particles. These particles can be detected if they are given enough energy in a particle accelerator to enable them to shoot away and cease being virtual.

However, physicists need another force to explain what holds the particles of an atomic nucleus together, overcoming the repulsion between the positive electric charges of the protons. The weak force is billions of times too weak, and gravity still more so. So the additional force is called the strong nuclear force but, although it is powerful, it extends only a very short distance – no more than $10^{-12}$ millimetres – and so is limited to the atomic nucleus.

To understand more about the messenger particles of the strong force, it is necessary to know more about the particles on which they act. In 1963 the American physicists Murray Gell-Mann and George Zweig proposed that protons and neutrons are not basic particles but are made of smaller entities. They called these quarks, a word coined by the Irish writer James Joyce in his novel *Finnegan's Wake*. According to Gell-Mann and Zweig, particles such as the proton and neutron are composed of three quarks. These are called baryons ("heavy particles"). Mesons are each made up of a quark and an antiquark.

When particles take part in reactions via the strong force, it is actually the quarks of which they are composed that are interacting. The messenger particles of the strong force are called gluons. They are responsible for the remarkable fact that quarks are never observed singly. The strong force grows more and more powerful as two quarks are pulled apart, until it overwhelms the separating force.

Despite the great differences between the basic forces, physicists are seeking to unify them – to show how they could be derived from a single fundamental force. Nor is this merely a theoretical exercise. They believe that in the first moments of the universe the basic forces were literally merged into one.

### Into the heart of the atom

***Part of the Large Electron-Positron collider*** *at CERN, Switzerland. The 27-km-long tube of the machine is 100 m underground in a tunnel straddling the Swiss-French border. Rapidly oscillating electric and magnetic fields, created by electromagnets along the tube, give the circulating particles energy equivalent to a single boost of 50 billion volts.*

Fundamental particles are probed by smashing them into each other at high speed and studying the fragments that result from the collisions. Nearly all the most powerful accelerators are synchrotrons – ring-shaped tubes, up to 27 kilometres long, with a high vacuum inside. An initial burst of electrons or protons (or their antiparticles) is injected into the synchrotron. Powerful electromagnets create rapidly varying fields that whirl the particles around the ring thousands of times at close to the speed of light.

The high-speed particles may then be flung at a stationary target, such as a block of aluminium. Higher energies of impact

positron

electron

neutrino

muon

proton

***When subatomic particles collide**, new ones are created. The higher the energy of collision, the greater the number and variety of new particles. Some are hadrons, consisting of pairs or triplets of quarks, so tightly bound together by the strong nuclear force (red discs) that they are never seen singly. Other particles, which do not feel the strong force, are called leptons, and include electrons and neutrinos.*

are achieved by accumulating two streams of particles that circulate opposite ways in storage rings, and then letting the two streams collide head-on.

Much of the energy of the collision is converted into a cascade of particles, many of which come into existence at the moment of the interaction. Nearly all the particles are ephemeral. The charged particles among them leave tracks in special detectors (the presence of uncharged particles has to be inferred indirectly).

Nearly all known particles fall into two main groups: hadrons and leptons. Leptons seem to have no structure and do not experience the strong force. They include electrons, neutrinos and muons.

neutron

Hadrons, on the other hand, consist of pairs or triplets of quarks and are subject to the strong nuclear force. Quarks come in six "flavours", and each of these in three "colours". The two building blocks of atomic nuclei, the proton and neutron, are triplets of quarks. The neutron consists of one up and two down quarks. In isolation it rapidly breaks down into a proton, electron and antineutrino, but in the nucleus it is stable. The proton consists of two up quarks and one down quark. Its lifetime is enormously long, possibly infinite.

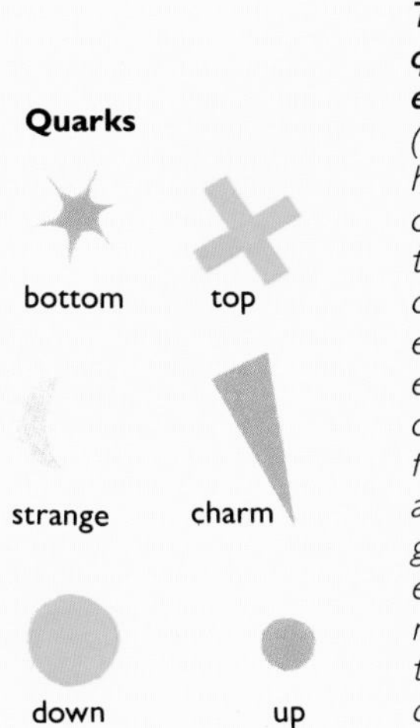


***The six types of quark believed to exist**, though the top (or "truth") quark has yet to be detected. Strikingly, they have electric charges that are either $\frac{1}{3}$ or $\frac{2}{3}$ of the electron charge – once thought to be fundamental. Quarks always combine to give charges that are either zero or whole-number multiples of the electron's electric charge.*

# THE FIRST MOMENT

## *The birth of space, time and matter*

At the birth of the universe all matter and energy were compressed into a fiery mass of unimaginable density. The earliest moment that can be spoken of with certainty came after a period called the Planck time – the incredibly brief interval of $10^{-43}$ seconds.

The entire universe that is observable today – which may be only part of some unknown whole – occupied a space $10^{-20}$ times smaller than an atomic nucleus. And it is believed that at an even earlier moment the four basic forces – gravitation, the strong and weak nuclear forces, and electromagnetism – were unified into a single force (p. 28). However, present theories break down at this point because it is not yet known how gravitation and the other forces were united.

Present theories can be tentatively applied after the Planck time. But detailed information about the state of the universe at this time was irrevocably altered by the next major event in the history of the universe. When it was $10^{-35}$ seconds old the era of inflation began: a period of fantastically rapid increase in size, during which the universe swelled up to at least $10^{50}$ times its previous size. Although the characteristics of inflation are incredible, theorists believe this idea – devised originally by Alan Guth in the US – must be correct, since it explains a number of puzzling features of today's universe. First, it accounts for the fact that the closely packed infant universe expanded neither so slowly that gravitation could crush it back into nothingness, nor so fast that it could thin out before galaxies and stars could form. Relativity tells us that space is in general curved, and that the amount of curvature depends on the amount of mass present in unit volume – the density. Too great a density and the universe would close round on itself and collapse; too low a density and space would open out uncontrollably.

Mathematical analysis shows that to account for this fine balance the density of the universe must have had a particular, or critical, value, and no other, at a very early time. Indeed, at $10^{-33}$ seconds

*__Matter, energy, space and time were created__ in an inferno of particles and forces unknown in the modern universe. Within a billionth of a second the initial force affecting all matter had frozen out into the forces known today. It took three minutes for the kaleidoscope of ephemeral particles to give way to stable atomic nuclei. Millennia were to pass before complete atoms were formed.*

after the Big Bang, it could not have differed from the critical value by more than one part in $10^{-49}$. Such a precise agreement is predicted by the theory of inflation, and no other way of explaining it is known.

Inflation also solves the "horizon" problem. In our present universe we see galaxies moving away from each other into the depths of space. There is a horizon beyond which we cannot see, since galaxies at that distance are receding at the speed of light. The horizon is different for each galaxy. Consider two galaxies, close to our horizon but lying in opposite directions from us: they will both be able to see us, lying at *their* respective horizons, but not each other.

The problem that arises is this: such galaxies, which cannot receive signals from each other, are very similar in what they contain, and in the density and distribution of their matter. The cosmic background radiation, too, is at exactly the same temperature, from whichever part of the sky it comes (p. 36). Why should the universe be so homogeneous when each part of it sees only a small part of the whole?

Certainly, as we go back ever closer to the Big Bang itself, these widely separated parts of the universe were closer together than they are now. But the time that had elapsed from the beginning was shorter too. There still had not been time for radiation to travel between them, evening out any possible differences. On the basis of the galaxies' present motions, there would never have been any contact between these regions.

But the theory of inflation solves the problem. Before inflation occurred the presently observable universe filled a tiny volume, far smaller than the horizon distance of that time. All parts of it reached temperature equilibrium, with any differences evened out. We see this equilibrium today, long after the most widely separated parts of the universe have ceased to influence each other.

The universe we see now is indeed very "smooth" on the large scale. Inflation shows how any initial irregularity in the density of matter and energy would be ironed out to a vanishingly low value. The problem now is to explain how sufficient irregularity survived inflation to make it possible for matter to clump into galaxies and stars.

Lastly, physicists have discovered that presently the universe contains between 100 million and a billion photons for each atom in the universe. Why it should be this number rather than any other is explained by the idea of inflation.

Inflation started when the strong force was beginning to separate from the electroweak (combined electromagnetic and weak) force, but before their separation was complete. By the time this had happened the temperature of the universe had dropped to a ten-thousandth of what it had been at the end of the Planck time, although it was still at the enormous value of $10^{28}$ K. Inflation was triggered by extreme supercooling of the infant universe.

Supercooling occurs when a system falls below a temperature at which it normally changes state, yet fails to undergo that change. For example, steam normally changes to water when it is cooled below 100°C. This is because its particles move increasingly slowly as the steam cools, until they begin to coalesce. In practice, this happens only when grains of dust or ions are around to act as centres on which liquid drops can grow. If there are no such centres, the steam can be supercooled below 100°C while remaining in its gaseous state.

At $10^{-35}$ seconds the universe was on the brink of a change of state marked by the separation of the strong and electroweak forces. But it continued to cool until it was about $10^{-32}$ seconds old before this transition occurred. While it was in its supercooled condition, a false vacuum was formed, the properties of which were very different from those of the true vacuum.

Usually the density of energy in any system – whether in the form of radiation or of matter – decreases when the volume of the system increases and its particles become less densely packed. But the false vacuum state is one in which the energy density stays constant during expansion. Relativity theory shows that the existence of the false vacuum, with its constant energy density, would have caused a vast force of repulsion. Inflation occurred.

While the inflation period lasted the universe doubled in size every $10^{-35}$ seconds. It doubled hundreds of times, with the result that its volume grew at least $10^{50}$ times. The temperature plummeted from $10^{28}$ to $10^{23}$ K.

Inflation ceased when the change of state had occurred and the strong and electroweak forces had separated. The energy density of the false vacuum was released, rather as the latent heat stored in steam is released when it changes state to become water. This burst of energy created a vast number of atomic particles, which reheated the universe to the temperature it had been at when inflation began. In this chaos of radiation and exotic particles began the building of matter as we know it today.

**The meaning of temperature**

The temperature of matter is judged by the movement of the particles of which it is composed. At high temperatures they move swiftly, with great energy; at low temperatures they slow down and their energy is less.

There is a relationship between the temperature, volume and pressure of a gas. At a given temperature, the pressure is greater the smaller the volume; and at a given pressure, cool gas occupies a smaller volume than it does when at a higher temperature.

In fact, for every drop in temperature of 1°C, a gas shrinks by $\frac{1}{273}$ of the volume it occupies at 0°C, the freezing point of water. In the 19th century the British physicist Lord Kelvin argued that at −273°C the energy of movement of molecules would be zero. Actually the molecules would still possess *some* energy of motion – they would vibrate a little – but they would have no spare energy to impart to any other substance.

The absolute scale of temperature is based on this fact. It begins from "absolute zero" (which is actually −273.18°C) and each unit, called a kelvin, is by definition equal to one degree of the Celsius scale. Thus 0°C is equivalent to 273.18 K.

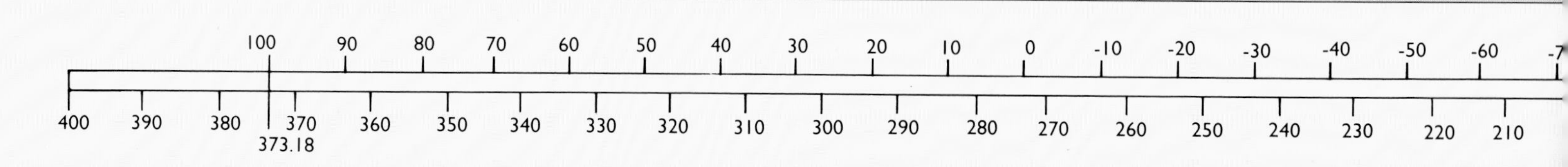

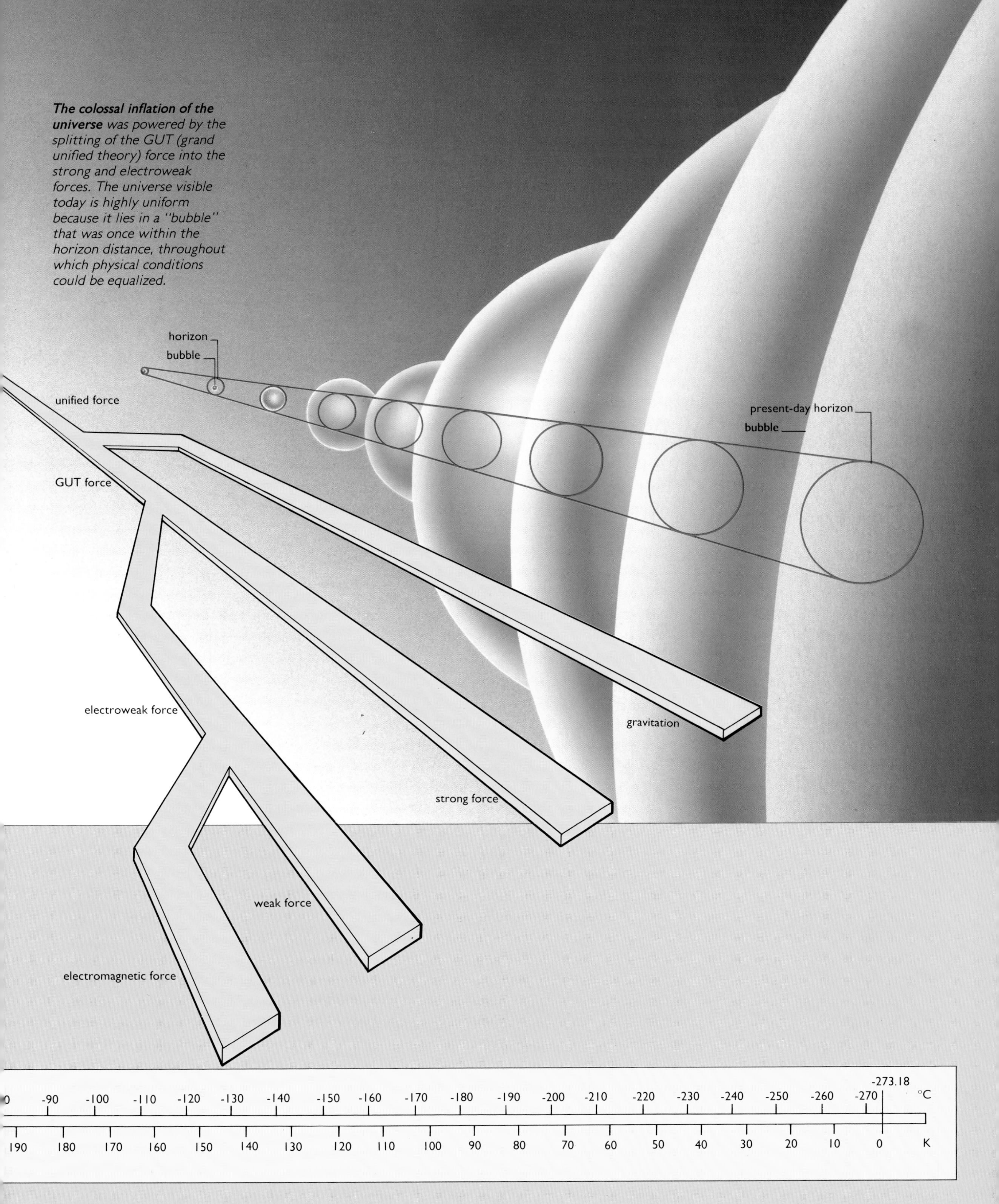


***The colossal inflation of the universe*** *was powered by the splitting of the GUT (grand unified theory) force into the strong and electroweak forces. The universe visible today is highly uniform because it lies in a "bubble" that was once within the horizon distance, throughout which physical conditions could be equalized.*

# BUILDING MATTER

## • *Stable atoms make their appearance*

The universe emerged from its burst of inflation at the colossal temperature of $10^{27}$ K. It was a seething mass of particles appearing fleetingly and disappearing again. In this "quark soup", quarks and antiquarks were constantly colliding with each other and being annihilated, giving out a flash of radiation. At first there was sufficient energy to produce quarks and antiquarks to balance this destruction, but as the universe expanded and cooled, production of these particles ceased. But because there had been a slight preponderance of quarks over antiquarks, these survived to form the basis of matter today.

Later the temperature dropped far enough to ensure that quarks and leptons could not turn into each other. There was an abundance of W and Z particles, so the weak force was still united with the electromagnetic force (p. 28). But when the universe was a hundred-millionth of a second old, and its temperature had fallen to some $10^{14}$ K, the electroweak force separated into the weak and electromagnetic forces as they are known today. This change did not, however, have such dramatic effects as the separation of the strong force, which powered inflation.

Then, when the universe was still only a millionth of a second old, the quarks began to bind together to produce hadrons, particles sensitive to the strong nuclear force. These included protons and neutrons, consisting of triplets of quarks, and mesons, consisting of quark pairs. Theory shows that some of the quarks may still exist singly, and could occasionally reach the Earth in cosmic rays. However, none has definitely been observed so far.

The hadrons present at this time included equal numbers of protons and neutrons. They were constantly being converted into each other by high-energy reactions. Protons and electrons combined to form neutrons, also giving out neutrinos. At the same time, neutrons were colliding with positrons to form protons and antineutrinos.

However, these reactions required the presence of a great number of electrons and positrons. These were produced in pairs by the annihilation of high-energy photons, but when the universe was about one second old such pair production ceased, and with it the equilibrium between the numbers of neutrons and protons. There were fewer neutrons than protons, because slightly more energy is required to produce a neutron from a proton than vice versa.

Nevertheless there were still a good number of neutrons in existence – one to every six protons. By themselves neutrons are unstable: they have a 50 per cent chance of decaying after 15 minutes. So now the free neutrons present began to decay into protons and two kinds of lepton: electrons and neutrinos. The leptonic era had commenced.

Yet there was time for neutrons to combine with other particles. In other words, they took part in the creation of the elements by nuclear fusion, the process that releases energy in a hydrogen bomb and in the stars.

Nuclear fusion became significant when the universe was one minute old. To begin with, one neutron and one proton combined to produce nuclei of heavy hydrogen, or deuterium. Each deuterium nucleus had one proton, which determined its identity as a type, or isotope, of hydrogen. But unlike ordinary hydrogen nuclei, it also contained a neutron. Some deuterium nuclei next collected a second neutron to produce nuclei of tritium, an even heavier isotope of hydrogen.

Reacting with another proton, the tritium then built a nucleus of a new element, helium, containing two protons and two neutrons. Virtually all the neutrons were incorporated into helium nuclei, so it is possible to work out that at this time there were 10 hydrogen nuclei to every helium nucleus. Other nuclei were also built up in small proportions by a variety of reactions. They included helium 3 (two protons, one neutron), beryllium 7 (four protons and three neutrons), and lithium 7 (three protons and four neutrons).

Radiation still played the dominant role in the universe when it was a few minutes old. This radiation era lasted from about one minute to some 10,000 years. At first much of the radiation consisted of high-energy gamma rays produced during the breakdown of deuterium and build-up of light nuclei. The radiation lost energy as the universe continued to expand, although photons could still knock particles about and be scattered in turn. It would have been impossible to see far in any direction.

As the universe continued to expand and cool, the photons lost energy. The masses of the nuclear particles did not change, and 10,000 years after the Big Bang their mass came to dominate the energy of the radiation, even though there were still about 10 billion photons to every proton. The universe had entered the matter era.

Even so, the photons were still moving at high enough speeds to break up any atoms that might briefly form when a nucleus and an electron came together. But after 300,000 years the energy of even the most energetic photons had dropped sufficiently to allow atoms to form and survive. Atoms and photons could now coexist, pursuing their respective paths, since the electrically neutral atoms did not significantly

free quarks

***In the first millionth of a second*** *of its life the universe was populated by photons and free quarks. Then quarks combined to form hadrons – protons and neutrons, consisting of triplets of quarks – and mesons, consisting of pairs of quarks.*

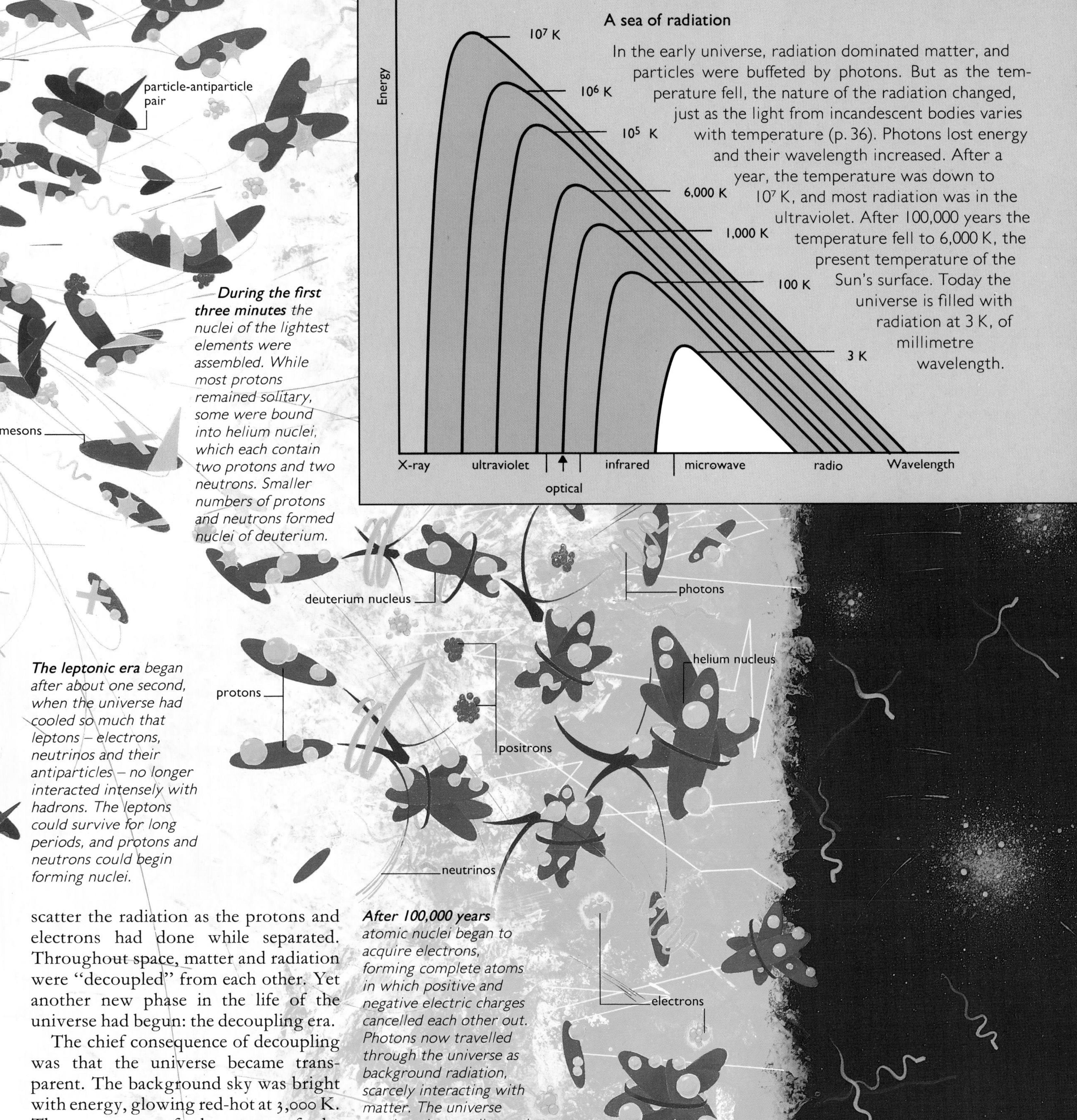


### A sea of radiation

In the early universe, radiation dominated matter, and particles were buffeted by photons. But as the temperature fell, the nature of the radiation changed, just as the light from incandescent bodies varies with temperature (p. 36). Photons lost energy and their wavelength increased. After a year, the temperature was down to $10^7$ K, and most radiation was in the ultraviolet. After 100,000 years the temperature fell to 6,000 K, the present temperature of the Sun's surface. Today the universe is filled with radiation at 3 K, of millimetre wavelength.

***During the first three minutes*** *the nuclei of the lightest elements were assembled. While most protons remained solitary, some were bound into helium nuclei, which each contain two protons and two neutrons. Smaller numbers of protons and neutrons formed nuclei of deuterium.*

***The leptonic era*** *began after about one second, when the universe had cooled so much that leptons – electrons, neutrinos and their antiparticles – no longer interacted intensely with hadrons. The leptons could survive for long periods, and protons and neutrons could begin forming nuclei.*

***After 100,000 years*** *atomic nuclei began to acquire electrons, forming complete atoms in which positive and negative electric charges cancelled each other out. Photons now travelled through the universe as background radiation, scarcely interacting with matter. The universe continued expanding and cooling.*

scatter the radiation as the protons and electrons had done while separated. Throughout space, matter and radiation were "decoupled" from each other. Yet another new phase in the life of the universe had begun: the decoupling era.

The chief consequence of decoupling was that the universe became transparent. The background sky was bright with energy, glowing red-hot at 3,000 K. Three-quarters of the mass of the universe was in the form of hydrogen, the rest was nearly all helium. These were the substances from which galaxies were destined to be made.

# THE COOLING UNIVERSE

## • *Big Bang's lingering glow*

The temperature of the universe has played a significant role throughout its development. The electromagnetic radiation that bathed matter from the moment of the Big Bang has cooled along with matter itself. It survives today, providing perhaps the most powerful evidence for the occurrence of the Big Bang.

A basic concept used in studies of radiation is that of a black body. This is an ideal object that, by definition, absorbs all electromagnetic radiation falling on it and reflects none whatsoever, which is why it is called a black body. Nevertheless, such a body does radiate energy, in a manner that varies depending on its temperature. When it is cool, it emits most energy at long (radio) wavelengths, less at others. As energy is fed into it and its temperature rises, its total emission increases, while its peak emission shifts progressively to shorter and shorter wavelengths.

The peak moves first through the infrared region, and then into visible wavelengths. The object gradually becomes bright red and then yellowish. At about 6,000°C, it becomes white-hot and its light resembles sunlight. As its temperature rises further, its radiation peaks successively at blue, violet, ultraviolet and eventually X-ray and gamma-ray wavelengths.

The universe can be regarded as a black body, cooling as it expands from the Big Bang. At each moment most of its radiation is concentrated at the wavelength corresponding to its temperature, but a drop in temperature is to be expected as its energy spreads out ever more thinly. Space should be filled today with radiation at the average temperature of the present universe.

This fact was first recognized by the theoretical astronomers Ralph Alpher, Robert Herman and George Gamow, working in the United States as early as 1948. But their conclusion was largely forgotten since there was then no equipment to detect such radiation.

Only in 1965 did the situation change, when Arno Penzias and Robert Wilson at the Bell Telephone Laboratories in Holmdel, New Jersey, constructed a radio telescope to observe radiation from the Galaxy, and picked up a constant background "noise". This did not change in strength as they pointed their horn-shaped antenna to various parts of the sky, nor as the Earth rotated, as it would have been expected to do if its source were celestial.

It was at this juncture that they were contacted by Robert Dicke of Princeton University, who was himself trying to detect radiation left over from the Big Bang. It immediately became evident that this was just what Penzias and Wilson had observed. The wavelength of the noise detected by their equipment was 7.3 centimetres. The radiation has since been detected over a range of wavelengths, and it has been found that it peaks at about 1 millimetre, corresponding to a temperature of 2.7 K.

In 1965 there were alternative theories of the origins of the universe besides that of a hot Big Bang, but none of them had predicted the existence of such radiation and none could account for it. Now almost all astronomers have been persuaded that the background radiation is the strongest evidence for a high-temperature Big Bang some 15 billion years ago or more.

***More remote "slices" of space*** *contain more stars, though they are fainter. If the slices continue to infinity, they send us an infinite amount of light.*

## Olbers' paradox

The expansion of the universe and its finite age are relevant to a simply stated, yet profound, puzzle: Why is the sky dark at night? In 1826 Heinrich Olbers, a German physician and astronomer, discussed this question, and his argument is now called Olbers' paradox, although the problem had been mentioned long before.

Olbers started by assuming that the true brightness of all stars is the same. This is an approximation, but it is satisfactory as a start. He also assumed that the stars are evenly distributed through space and that the universe is static.

Consider, then, a sphere with a radius of, say, 10 arbitrary units and a thickness of 1 unit, centred on the Earth. It is easy to calculate the total brightness of the stars lying within this shell. In another sphere of twice the radius, but the same thickness, the individual stars will be of only a quarter the brightness (since the brightness of an object falls to a quarter if its distance doubles). On the other hand, there will be four times as many stars to observe, since the volume of such a spherical shell will be four times as great. So the total amount of light coming from this shell will be the same as that from the first one. The argument can be continued by assuming further shells, each contributing the same amount of light. The brightness of the sky will be given by the total of the light from all these shells, which, by assumption, can be continued to infinity.

However, modern research has shown that the universe is not static, but expanding. As a result of the expansion, the radiation emitted from a distant object loses energy – the farther away it is, the greater the extra dimming. So in the argument above, the more distant spherical shells contribute less light.

More importantly, the universe is not infinitely old: it started at the time of the Big Bang. Even in an infinite universe most light could not yet have reached the Earth. The observable universe is only a tiny fraction of what would be needed to make the sky bright.

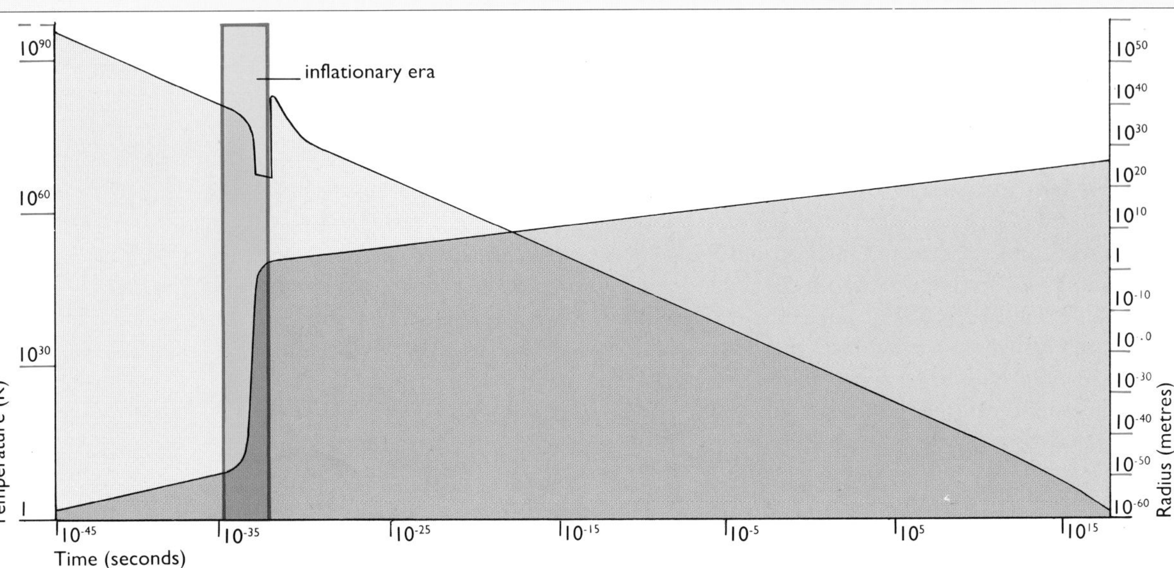


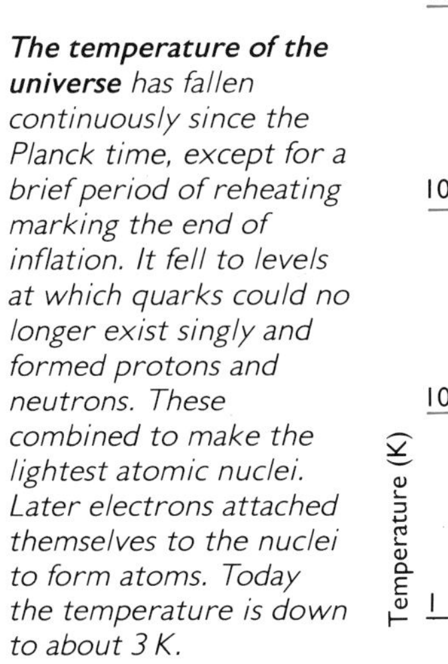

***The temperature of the universe*** *has fallen continuously since the Planck time, except for a brief period of reheating marking the end of inflation. It fell to levels at which quarks could no longer exist singly and formed protons and neutrons. These combined to make the lightest atomic nuclei. Later electrons attached themselves to the nuclei to form atoms. Today the temperature is down to about 3 K.*

# PROTOGALAXIES

## • *The first celestial bodies*

The primordial gas seems to have expanded for fully two billion years before the next great event in the evolution of the universe occurred: the appearance of embryonic galaxies. While the gas was thinning and cooling everywhere, in some places it was slightly denser than elsewhere. Here the gas expanded slightly more slowly because of its own gravitational attraction, until the expansion stopped and was turned into contraction.

In such areas the material of future galaxies slowly concentrated, probably condensing to form discs of material, each much larger than the galaxy, or cluster of galaxies, that was to be born from it. The amount of material in these protogalaxies varied from one place to another, and observation of the universe today gives every reason to suppose that often a single cloud gave birth to one of the clusters of galaxies now visible.

There is a slight problem in understanding how the process of "seeding" galaxies could commence. For gas to begin to gather at one place rather than another, there had initially to be some minute differences in density from place to place. But after the "smoothing" of the universe by the process of inflation, it would seem that the distribution of matter throughout all space would have been too uniform to allow this process to begin. However, recent research indicates that "fossil" traces of inflation left their mark in the young universe. One of the most widely discussed concepts is that of cosmic strings.

Cosmic strings, if they exist, consist of tremendously dense matter and energy – in fact, surviving samples of the false vacuum that initiated inflation (p. 30) – trapped in long, closed loops. Their masses would be measured in thousands of billions of tonnes per centimetre, yet any that now survive will have dimensions of millions of light-years. But formerly loops of all sizes would have existed, and would have vibrated very fast indeed – almost at the speed of light – emitting energy in the form of gravitational waves as they did so. As it lost energy, each loop would shrink until it became a concentrated lump of very dense material, ready to act as a seed around which a galaxy or, perhaps, a cluster of galaxies could form.

Attractive though this suggestion is, it presents certain theoretical difficulties, and its originator, Neil Turok in the United States, has modified and developed the idea into another stimulating theory. He now suggests that after inflation space was left with a "knot" texture. Each knot was a tiny region of space where energy was concentrated, distorting the geometry of space-time. When the supercooling of the universe that occurred during inflation had ended, the knots in space gradually became unravelled. As each one disappeared, a spherical blast wave of energy blew out from it. Where this met particles, it pressed them together, and such concentrated matter gave rise later to embryonic galaxies.

In this theory, the ending of inflation may also have given rise to strings of energy, which would leave behind a wake as they moved, just as a speeding ship leaves a wake in the water. Such wakes would be an additional cause of clumping in the universe.

Whatever the processes by which density differences were created, only after some two billion years did they become marked. Where density was greater, protogalaxies formed, so that by the time the universe was seven billion years old, the process of expansion had made it possible for galaxies to form in considerable numbers.

It seems that relatively concentrated regions of gas may first have condensed into large clouds of cold, dark matter, which later broke up to give rise to galaxies. The resulting galaxies would have an immense range of sizes – from one hundred times the mass of our galaxy down to a hundred-thousandth of it. Clusters of galaxies would also grow by attracting new galaxies to them, and the largest clusters would even be attracted to each other to give rise to superclusters.

The possibility of galaxy formation from hot rather than cold matter has been considered, using computers to investigate the results. The outcome was not at first encouraging because a vast number of very large clusters of galaxies would be created, though observations now seem to favour this scenario.

What is more, a computer simulation of formation from cold, dark matter shows that on larger and larger scales, the clustering of galaxies decreases: there would be fewer extremely large clusters than might be expected if they formed from matter that was hot, but observation indicates otherwise. In short, material is not evenly spread through the universe on large scales and is also unevenly distributed on smaller and smaller scales. Astronomers refer to this as hierarchical clustering.

Computer studies also demonstrate not only that the first condensations occurred about two billion years after the Big Bang but also that smaller-scale condensations took place as the larger condensations merged. They make it

***A million years after the Big Bang,*** *the universe was a featureless sea of hydrogen and helium gas. It became differentiated into regions of higher and lower density that ultimately gave birth to the present clusters and superclusters of galaxies. Computer simulations attempt to reproduce the details of this process. The images shown here were produced by a simulation based on the idea that neutrinos, generally thought to be massless, may have a non-zero mass. The distribution of matter in the universe is shown at one-eighth and one-third of its present age, and finally at the present day. The organization of matter into strings separated by voids is a fair representation of the observed grouping of clusters of galaxies.*

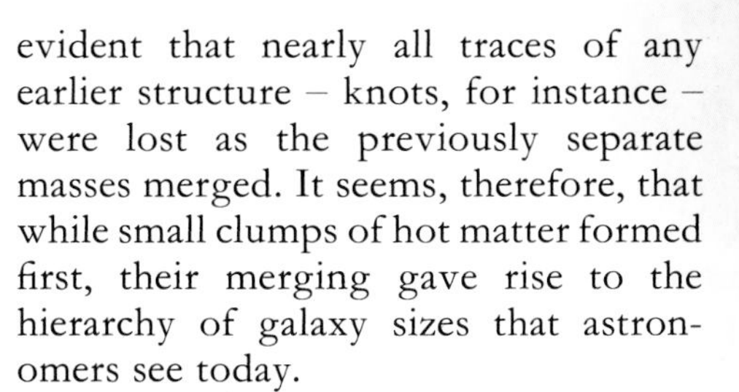

evident that nearly all traces of any earlier structure – knots, for instance – were lost as the previously separate masses merged. It seems, therefore, that while small clumps of hot matter formed first, their merging gave rise to the hierarchy of galaxy sizes that astronomers see today.

There are still problems over this picture of the formation of galaxies. When astronomers study the way the brightness of galaxies is distributed among the different types, they find that there are now more faint galaxies than bright ones. The reason may be that as embryonic galaxies gently collided and merged, energy was released and gas was driven off, limiting the sizes and hence brightnesses of the galaxies that formed. Only in a small proportion of cases would a sufficient number of gas clouds collide in a short enough time to build up those very large galaxies that we observe.

The universe had been dark for two billion years. Then, while the gas clouds were still contracting, they began to give birth to the first generation of stars. The scene was set for the evolution of the universe as it is known today.

# THE GRAND DESIGN

The universe is an enormous hierarchy, ranging from vast clusters of galaxies to minute fragments of interplanetary debris. What is more, everything in it obeys the same laws. Even at the farthest distances that can be observed, there is the same kind of material, behaving just as it does in the nearer reaches of space.

The description of the universe that follows deals with those celestial objects that formed first and goes on to describe the others in the order in which they may well have come into being. It starts with clusters of galaxies, and even clusters of clusters, and then moves on to the galaxies themselves, then the stars and clouds of gas and dust of which the galaxies are composed, and so on to the planets and their moons.

Galaxies were once known as island universes because they are conglomerations of stars and gas, and sometimes much dust as well, separated by great distances. They can be observed because they emit light and other radiation, which travels over vast distances of space before it reaches Earth. This radiation takes time to travel. Even though it covers almost 300,000 kilometres each second, it takes billions of years to reach Earth from the most distant galaxies. So here astronomical distances on the largest scale are being considered.

The next class of objects in the hierarchy of the universe comprises the gas clouds, or nebulae, within the galaxies. These are the birthplaces of the stars, which form the main visible content of galaxies. Billions of stars go to make up even the smallest galaxies. They shine by their own light, which is generated within them by nuclear fusion. Our Sun is just such a body, its central regions resembling a vast collection of exploding nuclear bombs.

Moving still farther down the hierarchical order is cool matter associated with the stars and derived from the material from which the stars themselves formed. Little is known of such material except in the case of our own solar system, where much of it has condensed to form lumps that orbit the Sun. These lumps are planets, some of which have smaller condensations – their moons – orbiting them.

The Sun's family comprises nine major planets and scores of moons, as well as thousands of fragments known variously as minor planets, planetoids or, more usually, asteroids – "little stars". Other material has condensed into small pieces of matter, the comets and meteors. The Sun's planetary system and interplanetary debris are all cold and are visible only because they reflect the Sun's light. There is good reason to believe that this system is not unique in our galaxy.

***The matter and energy** that poured forth from the Big Bang gave rise to the hierarchy of the present-day universe. Minute irregularities in the infant universe led to the growth of structures on all scales. Largest of all were the superclusters – groupings of galaxy clusters. A map of the superclusters is the basis of this computer-generated image, expressing the dynamism of the universe revealed by modern cosmology.*

# OBSERVING THE UNIVERSE

## • *New windows on the cosmos*

From the earliest times until a little less than 400 years ago, the only way of observing the universe was simply to look at it. There were no optical instruments that could be used – only the eyes, and this brought with it certain restrictions. Above all it meant that there was a limit to the faintness of the objects that could be detected, because the pupil of the eye is small and so allows only a small amount of light to fall on the light-sensitive retina.

The situation had changed by the early 17th century, as a result of the application of the telescope to astronomy by Galileo and others. Because the telescope gathers a greater amount of light than the human eye, thousands of faint stars suddenly became visible for the first time. Astronomers could now probe farther into space. Moreover, the telescope also magnified the objects that it viewed. While this made no difference to the images of stars, which remained pointlike, the telescope revealed to the observer detail that the naked eye could not detect in extended objects, notably the Moon and planets.

With the new instrument Galileo made discoveries that helped overthrow the theory of an Earth-centred universe. For example, he detected the phases of Venus, showing that the planet revolves around the Sun, and he discovered moons revolving around Jupiter. Moreover, he found that the Moon was a mountainous world, not unlike the Earth.

As the centuries passed, larger and more efficient telescopes were designed and built, and the bounds of the known universe were pushed still farther into space. In the 19th century the power of the telescope to penetrate space was much enhanced by the introduction of photography. Whereas the eye's sensitivity is not increased by staring for a long time at something too dim for it to detect in the first place, a photographic plate will allow images to build up while it is exposed, and so enable dimmer objects to be detected.

Another vitally important tool of the astronomer is the spectroscope. It was in the mid-17th century that Isaac Newton showed that "white" light is a combination of light of all colours. This he did by passing sunlight through a prism to split it into its component parts, forming a coloured band – a spectrum. The later spectroscope was a more sophisticated device designed to disperse the components more thoroughly.

Light was first passed through a narrow slit to produce a thin beam, before going to a prism or other light-scattering device. Each colour in the light formed a separate line – an image of the slit. Although with many types of source the lines merged to form a continuous band of colours, gases at low pressure emitted certain very specific colours when incandescent, and absorbed the same colours when cool. These colours appeared as characteristic bright or dark lines crossing the spectrum.

Thousands of dark lines appear in the spectrum of the Sun. Investigations, notably by Gustav Kirchhoff and Robert Bunsen in Germany during the 1860s, proved that these were the "fingerprints" of chemical elements in the Sun's outer layers. Later still it was found that all stars display such spectra, and that detailed examination of the lines could not only indicate the elements present but also show much about temperatures and pressures in the stars.

It was found that visible light does not constitute the whole of the spectrum of sunlight. There is also a component lying beyond the violet end of the visible spectrum, and hence called ultraviolet radiation. It affects human skins by producing a tanning reaction. There is another component, which lies beyond the red end of the spectrum, called infrared radiation. It is also described as heat radiation, since it warms objects that absorb it and is the main type of radiation that is given out by objects at the normal temperatures of surroundings on Earth.

Although a few astronomical observations were made in the infrared and ultraviolet bands, they were not pursued, primarily because their significance was not appreciated. The work of the fourth Earl of Rosse, who as early as 1877 measured the Moon's temperature by observing its infrared radiation, was followed only sporadically for decades. Yet James Clerk Maxwell had shown (pp. 14–15) that even ultraviolet, infrared and visible light make up only part of a wider-ranging spectrum of electromagnetic radiation. New fields of observation were awaiting the entry of astronomical pioneers using new techniques.

***X-ray sources in the skies*** *(right, above) were detected by the HEAO-1 X-ray observatory. They show a concentration of bright sources toward the centre of the Galaxy, but away from the obscuring material in the galactic plane, a more uniform distribution of such remote objects is visible.*

***The form of the Galaxy*** *at visible wavelengths is revealed (above) by the patterns of light and dark nebulae and the distribution of stars down to the 10th magnitude – 40 times fainter than the dimmest stars the naked eye can see. The best optical telescopes can see much fainter stars, but to penetrate the heart of the Galaxy other wavelengths are required.*

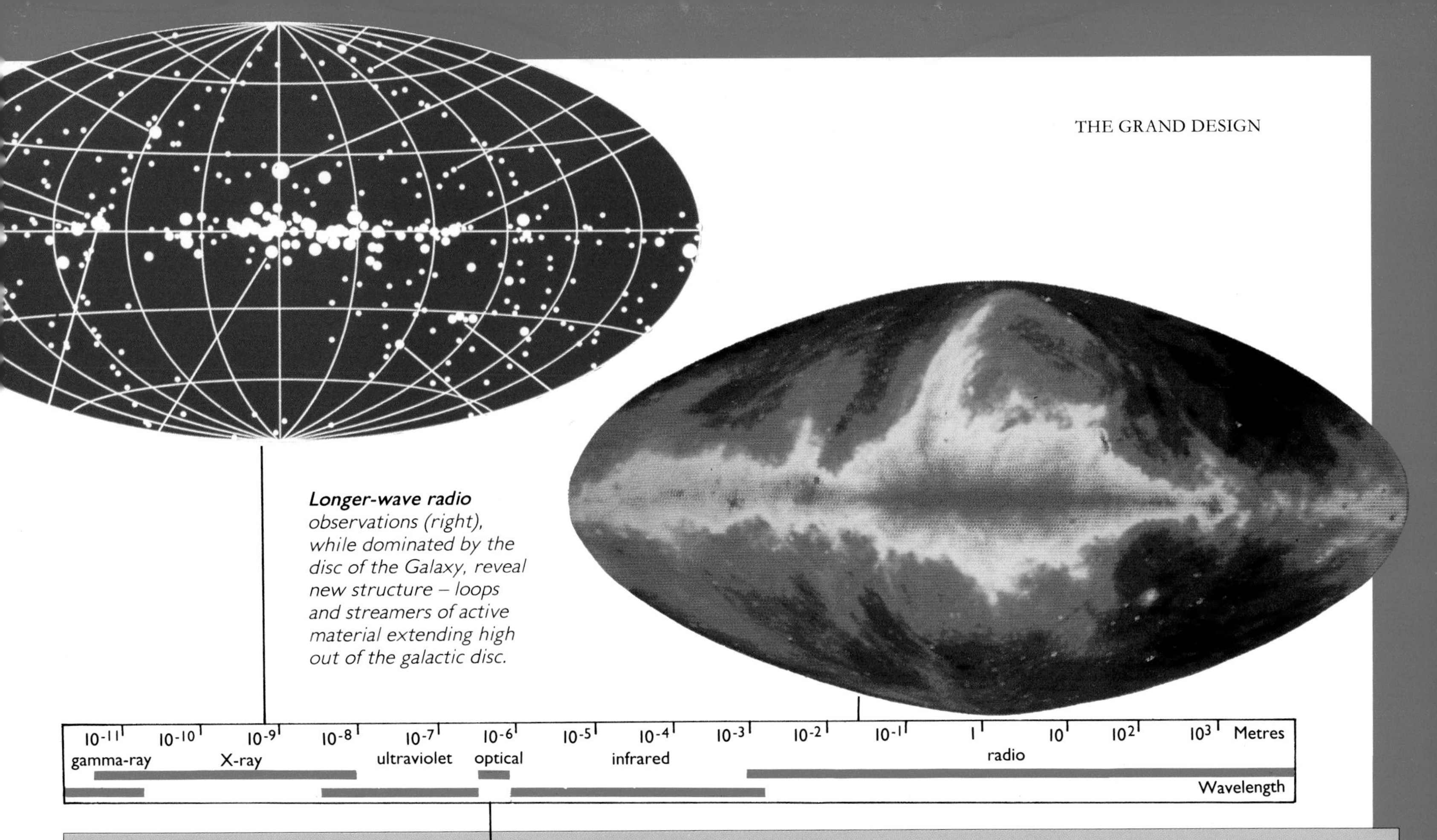


***Longer-wave radio*** *observations (right), while dominated by the disc of the Galaxy, reveal new structure – loops and streamers of active material extending high out of the galactic disc.*

**The barrier of the atmosphere**

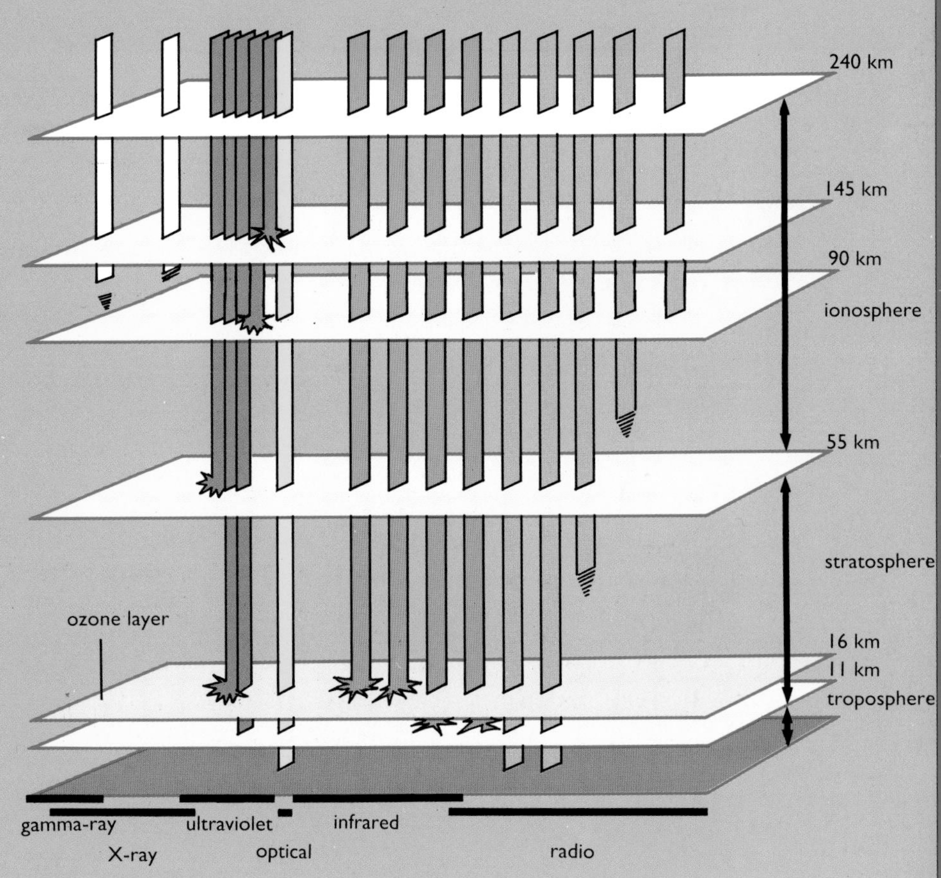


The thin skin of air that sustains life on Earth shields us from harmful radiations, while blocking the astronomer's view of the universe at most wavelengths. There are two major "windows": one allows visible light to reach the ground. The other lets through most radio wavelengths, but the ionosphere contains relatively high concentrations of ions which act like mirrors to some radio waves from above and below. Other wavelengths are absorbed at various heights. The ozone layer absorbs damaging ultraviolet rays.

# TELESCOPES OF TODAY

• *New technology for the astronomer*

***Optical astronomy*** *continues to advance, despite the explosive growth of "invisible" astronomy. The Multiple Mirror Telescope (above) on Mount Hopkins, Arizona, uses six main mirrors, each 1.8 m wide, which work together to produce a single image. They are the equivalent of a single mirror 4.5 m wide.*

***The dishes of the Very Large Array*** *(far right) make up the world's largest single radio telescope. There are 27 receivers, moving on railway tracks that form a Y shape. Each arm of the Y is about 20 km long. The VLA is located at Socorro, New Mexico.*

The first major breakthrough in extending the wavelength range of astronomical observations came in 1932, when an American radio engineer, Karl Jansky, made the serendipitous discovery that radio waves were emanating from the Milky Way. His work was extended by another American, an amateur radio ham, Grote Reber. Yet only after World War II did radio astronomy really get under way.

The results have proved of vital importance in extending our understanding of the universe. For instance, radio telescopes have made it possible to detect molecules in space that are invisible at optical wavelengths; they have made it clear that galaxies are not the peaceful islands of gas, dust and stars that they were once thought to be; and, above all, they have shown that the universe is bathed in background radiation that survives from soon after the Big Bang.

The desire to extend the observable spectrum farther led, in the 1960s, to the construction of special infrared telescopes, such as the United Kingdom Infrared Telescope (UKIRT) set up on a mountain peak in Hawaii in order to observe in very clear, dry air. But even instruments placed as high as this are severely limited by water vapour in the Earth's atmosphere, which acts like a blanket, absorbing much of the infrared radiation from space. This limitation can only be overcome by Earth-orbiting space probes.

In 1983 the joint European and United States Infrared Astronomy Satellite, IRAS, was launched. It had many triumphs, discovering places where starbirth is proceeding in our galaxy, detecting cool stars that are invisible optically, and producing evidence of the formation of planetary systems around other stars.

The equipment on an infrared telescope, including that on IRAS, has to be refrigerated, for at normal temperatures the instrument and its surroundings emit vast quantities of infrared radiation that would swamp the images of faint celestial sources. To view the shorter infrared wavelengths, the component that detects the focused infrared radiation is cooled with liquid nitrogen to −196°C; to view longer wavelengths it must be cooled to 2°C above absolute zero with liquid helium.

Space technology has, of course, proved of immense importance at all wavelengths. Its most obvious applications have been the close-up examination of planets and moons, but of even wider significance has been the extension of observation into the extreme ranges of the electromagnetic spectrum. Although some ultraviolet radiation reaches the surface of the Earth, most is blocked by the atmosphere. Space satellites can detect this radiation, and also X-rays and the even shorter-wavelength gamma rays originating from space.

Great ingenuity is often used in designing modern astronomical equipment. For instance, since X-rays are not deflected when they penetrate a material, it might seem impossible to construct an X-ray telescope. However, if the X-rays graze the surface of a suitable metal they can be reflected, and so can be made to form an image. Such grazing-incidence telescopes are now commonplace in X-ray satellites.

The optical telescope has gained a new lease of life with the advent of computer-controlled optics. This has permitted the building of large instruments that exploit the interference of light waves falling on separated mirrors, a technique derived from radio astronomy. Such instruments include special telescopes that can measure the diameters of large nearby stars.

Another revolution has occurred in the processing of the information from telescopes, satellites and space probes. Until the advent of the microchip and the computers based on it, the measurement of the positions of images on photographic plates and their analysis was a time-consuming process. For every night spent at the telescope, an astronomer might spend five days or more processing his results. Now automatic computer-controlled plate-measuring machines do the task in a few hours. Another computerized electronic technique that has proved extremely useful is the replacement of the photographic plate by more sensitive electronic detectors, known as charge-coupled devices (CCDs).

Extremely valuable also is the electronic process of image enhancement. Computer processing of the data from spacecraft – for example, the two Voyager interplanetary probes – is vital to obtain really good and detailed pictures. The astronomer can program enhancements of certain features to help the analysis of observations in each wavelength range. The resulting images make interpretation easier and more exact.

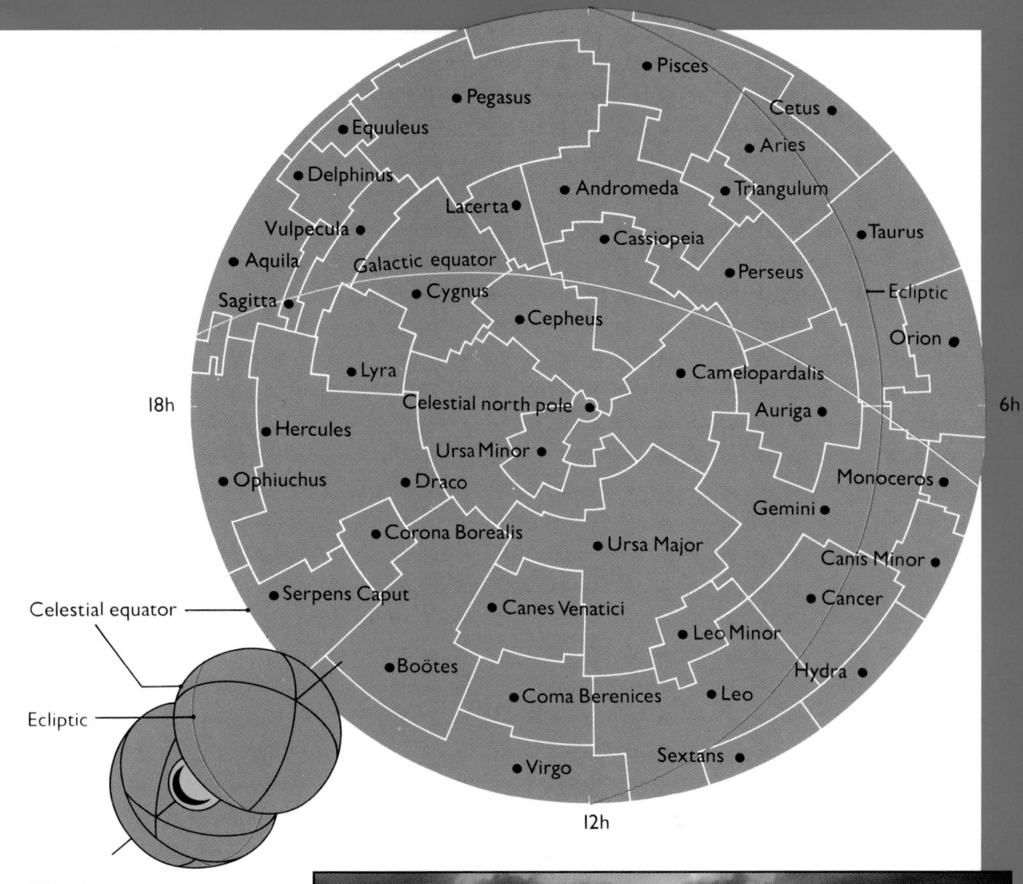


*The sky is mapped by imagining it to be a huge sphere centred on the Earth (above). It is divided into areas, each representing a single constellation (top and below).*

*The celestial poles and equator lie above their equivalents on Earth (below). The ecliptic is the Sun's apparent path around the sky – where the plane of the Earth's orbit cuts the celestial globe.*

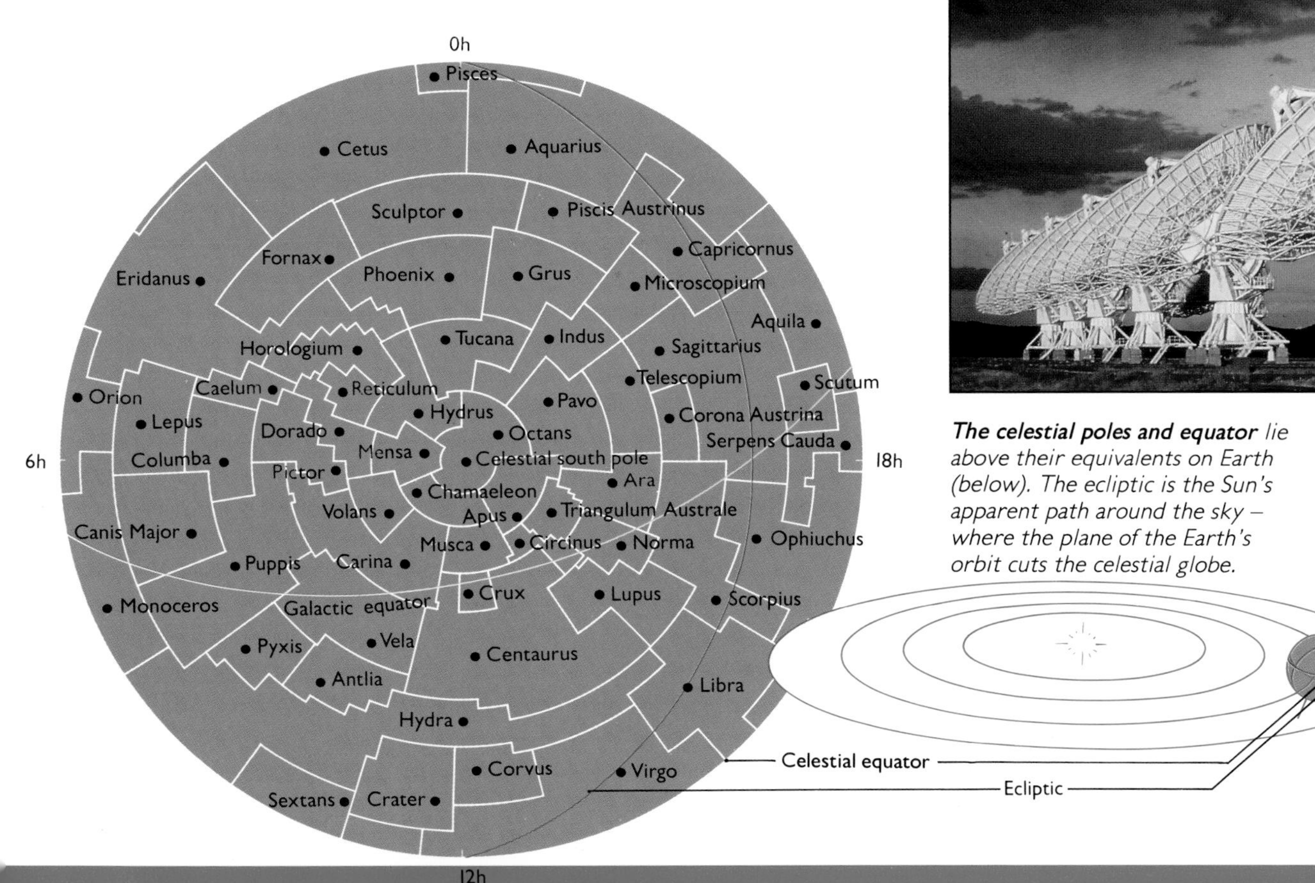

# YARDSTICKS OF THE COSMOS

## • *Techniques for measuring the universe*

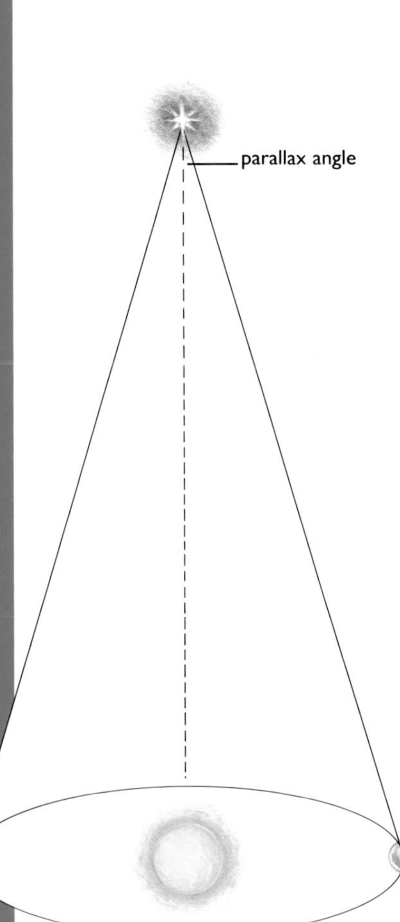


***The distance of a nearby star*** *can be found by observing its apparent yearly movement, caused by the Earth's motion around the Sun. Its position in relation to more distant stars is measured at intervals of six months, from opposite sides of the Earth's orbit. The star's parallax is defined as half the apparent angle of shift. Since the size of the Earth's orbit is known precisely, the star's distance can be calculated.*

Estimates of the distances of the celestial bodies by the earliest astronomers were woefully small. But it is not surprising that they should have erred: the difficulty of measuring the distances of the stars, in particular, is such that the feat was not accomplished until 1839. Today astronomers have a large repertoire of methods of measuring distance; they are of varying precision and each is limited to a certain range of distances.

Since World War II the distances of the Moon, Sun and planets, long established by older methods, have been obtained with unprecedented accuracy by means of three new techniques. Laser beams are bounced from the surface of the Moon and their return is carefully timed to yield the distance with such precision that it can be calculated to the nearest centimetre; radar pulses are used for determining the distance of the Sun; and distances for all the planets except Pluto have been determined by means of radio signals from visiting space probes. Finding the distances of the stars, however, is not quite as simple because of their sheer immensity.

The method used for the nearer stars is triangulation, which is also used by terrestrial surveyors. The astronomer selects a star that, because of its apparent brightness or comparatively large motion, is believed to be near the Earth, and measures its position in relation to other stars that are more distant. Six months later, when the Earth is at the other side of its orbit around the Sun, the star's apparent position is measured again. The star will seem to have moved in relation to the background. The star's annual parallax is defined as equal to half the angle through which it has moved (see illustration). Using trigonometry, the actual distance of the star can be calculated, since the diameter of the Earth's orbit is known accurately.

When such an angle has thus been translated into a distance, the result is huge. To make the figures understandable, the speed of light provides a convenient yardstick. On this scale the Moon is 1.3 light-seconds away from us and the Sun is 8.3 light-minutes away. But the nearest star, Proxima Centauri in the southern sky, is 4.3 light-years from us. The light-year is equal to the colossal distance of approximately 9.5 thousand billion kilometres.

Astronomers have found other, less direct, ways of determining distances beyond the range of parallax measurements. The most important means of taking larger strides into space is to use a form of celestial "lighthouse" – the Cepheid variable stars (pp. 86–87). These stars vary in brightness in a regular way, each repeating a pattern of variations over a fixed period measured in days or months. Their importance lies in the fact that their true brightnesses and their variation periods are intimately linked.

If the period of a Cepheid variable star is measured, its true brightness is known. From its apparent brightness the distance of the star can be determined. Cepheids have been detected in some of the nearer galaxies, thereby showing them to be millions of light-years distant. Yet what of those galaxies that are too far off for the Cepheids they contain to be detected?

One of the consequences of the Big Bang is that all the galaxies are moving away from each other. The farther a galaxy is from us, the faster its velocity of recession. Astronomers have detected this by the red shift of the tell-tale lines in a galaxy's spectrum: all the lines in the visible part of the spectrum are shifted toward the red, and lines in the red part of the spectrum are shifted into the infrared, and so on.

This happens because the crests and troughs of the electromagnetic radiation reach Earth less frequently the larger a galaxy's speed of recession. Consider, for example, the conspicuous twin yellow lines produced by sodium in a stationary source of light: the wave crests reach Earth 500 thousand billion times a second. But in the same light coming from a galaxy receding at one-tenth the speed of light, each crest has farther to travel than the one preceding it. They therefore reach Earth only about 450 thousand billion times per second, and the light seems to us to have a lower frequency and longer wavelength – to be reddened, in fact. Conversely, if the light source were moving toward Earth at this velocity, the crests would arrive 550 thousand billion times per second, and a shift of the line toward the blue end of the spectrum would be observed.

By measuring red shifts, astronomers can

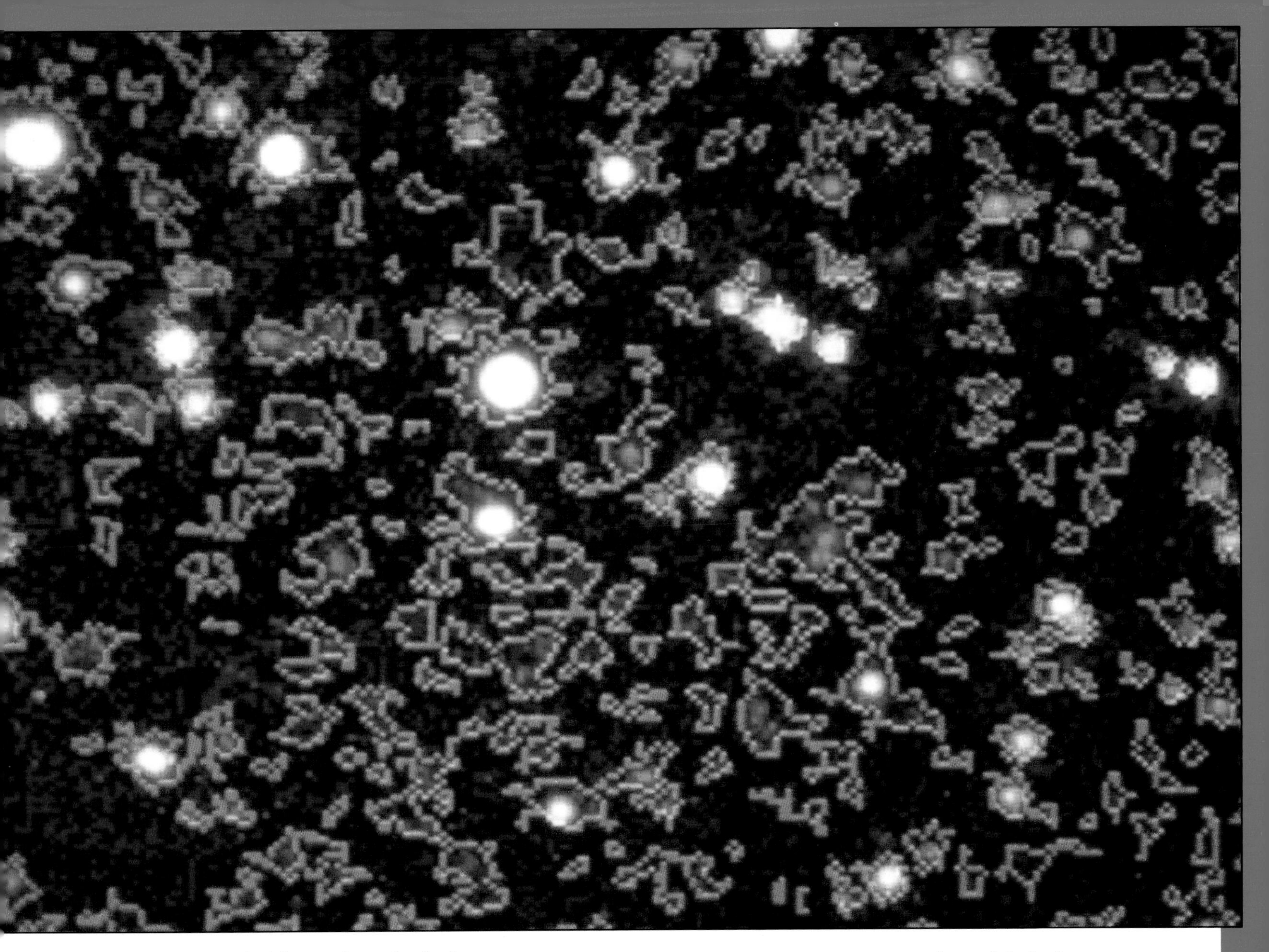

determine the speeds of recession of galaxies, and since they know that the farther off they are the faster they move, they can calculate their distances. Of course, this is possible only if they know exactly how velocity increases with distance, and at the moment there is uncertainty about this. The ratio of speed to distance – called the Hubble constant – could be as low as 15 kilometres per second per million light-years or as high as 30.

The value of the Hubble constant affects astronomers' ideas about the size of the universe: the smaller the constant is, the farther away must be a galaxy of a given speed of recession, and the larger the universe must be. And the larger the universe is, the longer it has taken to reach that size. At present the upper value of the age of the universe is taken to be 20 billion years, but this value is highly dependent on the accuracy of the distance scale that astronomers have painstakingly constructed.

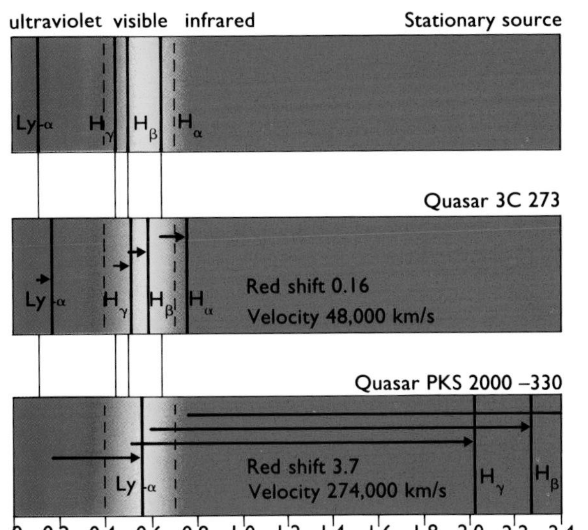


***Some of the farthest galaxies yet observed*** *appear in this image, obtained with a sensitive electronic detector attached to a telescope. The original image, obtained with blue light, has here been computer-processed. Galaxies of magnitude 26.5 are shown – some 100 million times fainter than the faintest star the naked eye can see.*

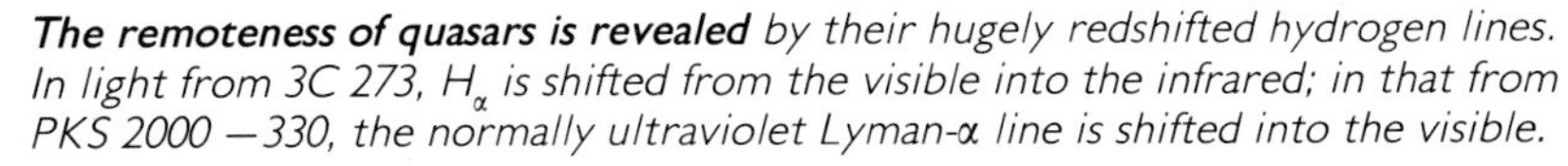

***The remoteness of quasars is revealed*** *by their hugely redshifted hydrogen lines. In light from 3C 273, $H_\alpha$ is shifted from the visible into the infrared; in that from PKS 2000 –330, the normally ultraviolet Lyman-α line is shifted into the visible.*

# SUPERCLUSTERS

## • *The largest structures in the universe*

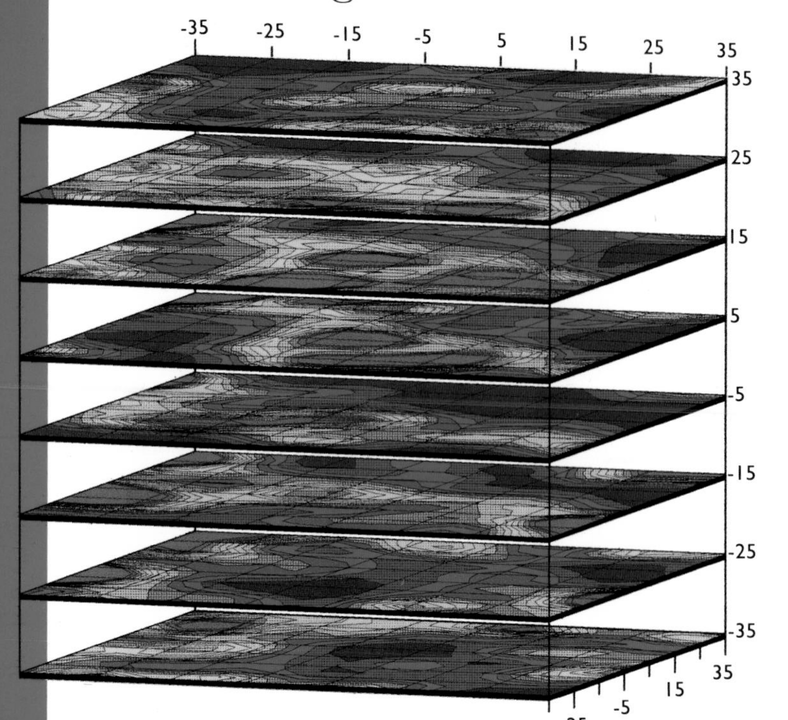


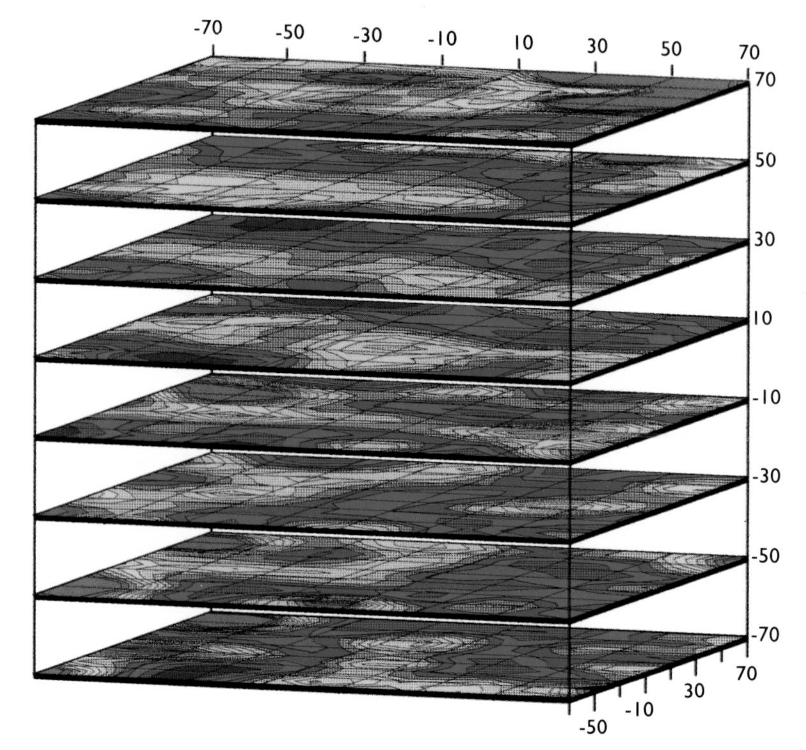


***Theories of cluster formation** from cold matter do not appear to match newly made observations. The blocks show the distribution of clusters of galaxies, with our galaxy at the centre of each block. The scales are in megaparsecs (a megaparsec is equal to just over three million light-years). The first three blocks show data from the Infrared Astronomical Satellite (IRAS), while the fourth is a computer simulation based on the theory that the clusters formed from cold dark matter (CDM). In the observations at the largest scale (third block), very high and very low densities of galaxies are more apparent than in the CDM simulation and galaxy formation from hot matter must therefore be considered.*

In the evolution of the hierarchy of the present universe, the largest structures appeared first: protogalaxies were formed and then shrank to become the galaxies that can now be observed. These, however, were probably grouped from the first into clusters and superclusters, the largest units that can be discerned and the highest stages of the cosmic hierarchy.

In the early years of this century it was already known from the work of the English astronomer William Huggins that nebulae (so called from the Latin word for "clouds") were of two kinds – hazy patches of gas, and spiral or elliptical shapes. Vesto Slipher at the Lowell Observatory in Arizona and Francis Pease at the Mount Wilson Observatory in California began an intensive study of the spectra of spirals and found that almost all displayed red shifts (pp. 46–47): in other words, virtually all were moving away from the Earth at great velocities. No such general recession was displayed by other kinds of nebulae. The question arose of whether the spirals were even part of our "star-island", the Galaxy.

In the 1920s Edwin Hubble at Mount Wilson was able to detect variable stars, of the type known as Cepheids, in a few nearby spiral nebulae. Since the true brightnesses of Cepheids can be determined (pp. 86–87), he could thus obtain definite distances for the spirals, showing that they lay far beyond the confines of our own galaxy. They came to be called spiral galaxies, for they are the same kind of system as the Galaxy. It became clear, too, that some other types of nebulae are also galaxies beyond our own.

By 1930 it had been established not only that most galaxies are receding from Earth but that their velocities of recession are greater the fainter, and therefore farther away, they are. This was the first evidence to indicate an expanding universe.

In the 1930s the Swiss astronomer Fritz Zwicky, who had settled in the United States, discovered that galaxies are grouped into clusters, some of them thousands strong and spreading tens of millions of light-years across space. Later it became clear that our own galaxy and the famous spiral in Andromeda are members, together with a number of other galaxies, of a small cluster now known as the Local Group. Members of a cluster move under the influence of their mutual gravitational attractions, and this provided a simple explanation for those few galaxies that display blue shifts in their spectra. They are all

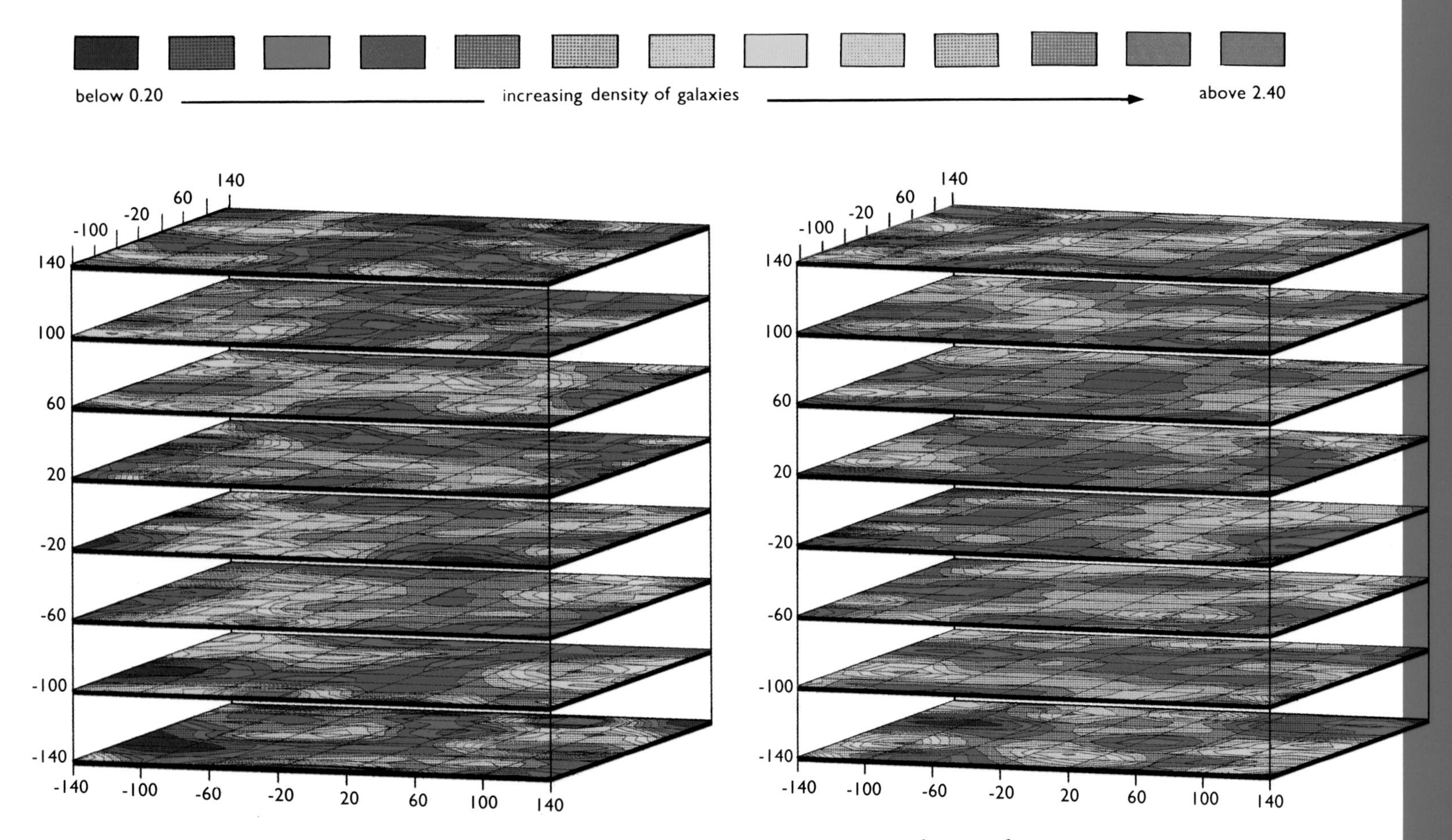


members of the Local Group that happen to be moving toward Earth at present.

Further work revealed a concentration of clusters of galaxies in a belt – a veritable Milky Way – around the northern and southern skies. Viewed from Earth, the centre of this string appears to lie in a vast cluster of thousands of galaxies, the Virgo cluster – so called because its centre appears to lie in the constellation Virgo. The string itself is therefore called the Virgo supercluster, or Local Supercluster. The Virgo supercluster is not a unique phenomenon; other superclusters also take the form of lines or strings. There seem to be no observations to lead astronomers to assume any structure larger than superclusters.

Astronomers are vigorously mapping out the superclusters. The Local Supercluster comprises not only the vast Virgo cluster, as already mentioned, but also our Local Group and dozens of smaller groups that can be seen in the constellations of Canes Venatici, Leo and Crater. The supercluster's shape is a flattened ellipsoid, with one or two protruding sections, so that the whole requires a cube of 100 million light-years on a side to contain it. All the member clusters are attracted toward the central Virgo cluster: the Local Group is moving at some 250 kilometres per second toward a point some 20 to 30 degrees from its centre.

There are other huge conglomerations – the Coma, Hercules and Perseus superclusters, for example. Each has a large cluster at its heart, the cluster at the centre of the Coma supercluster containing well over a thousand galaxies. The Hercules supercluster is centred on a much less dense cluster, which lies at a mean distance of 650 million light-years, and the Perseus supercluster is centred on a large cluster at a mean distance of about 235 million light-years. Another supercluster that deserves mention is the Southern Supergalaxy, centred on clusters in the constellations Fornax and Dorado.

Some astronomers believe that motions noted in the Virgo supercluster indicate the presence of an even larger supercluster farther out in space. Possibly some 300 million light-years from Earth, in the direction of the constellations Hydra and Centaurus, it has been named the Great Attractor. However, some doubt has now been cast on its existence, on the grounds that it would lead to an unevenness in the intensity of the microwave background radiation of the universe – an unevenness that has not been observed. But the question is still open.

# CLUSTERS OF GALAXIES

## • *Galaxies that travel in concert*

***Tenuous gas** (in blue, below left) surrounds and links the galaxies in the cluster called Pavo 5. Where there is a significant amount of gas and dust between galaxies in a cluster, scientists reason that the cluster is relatively old as the gas will have needed time to escape from the gravitation of the individual galaxies. Typical of the type of galaxy in the outer regions of a cluster is a spiral galaxy, such as NGC 6744 (below right), in Pavo (taken by the Anglo-Australian telescope). Toward the centre of a cluster, large elliptical galaxies tend to predominate.*

The size of clusters of galaxies ranges from mere pairs or triplets of galaxies to giants with thousands of members. Although there appear to be a few isolated galaxies drifting in space, they are probably former members of nearby clusters that have escaped.

All clusters consist of a variety of galaxies, of which there are three main types. Spirals have spiral arms with central bulges; ellipticals are approximately the same shape as an American football, although they approach a spherical shape in varying degrees; irregulars, although fairly formless as their name implies, seem to be more closely allied to spirals than to ellipticals. Ellipticals and spirals usually predominate in a cluster. The galaxies vary greatly in size, from thousands to hundreds of thousands of light-years in diameter. The smallest have recently been termed dwarf galaxies.

The cluster to which our own galaxy belongs, the Local Group, is a small one. How many galaxies it contains depends on whether certain outlying members are included. Certainly there are at least 26 chief members, and the figure is extended by general agreement to include a few other small, faint galaxies in the vicinity. These bring the total to at least 28, and doubtless there are further faint galaxies yet to be discovered.

All these galaxies are in motion: our own travels at a velocity of about 200 kilometres per second in relation to the group as a whole. The entire group is now known to have a diameter of almost four million light-years. No other galaxies are known within a distance of about 10 million light-years from our galaxy.

The nearest rich cluster to our Local Group lies in the constellation Virgo. This is a vast conglomeration of thousands of galaxies of every kind, of which 2,500 are fairly bright. Of these, 30 percent are elliptical galaxies. Prominent at the centre of the Virgo cluster is a huge elliptical galaxy, M87. A similar feature is evident in many clusters.

The average distance of the Virgo cluster is 52 million light-years. It has a somewhat irregular shape, the longest dimension of which is about 8.8 million light-years. The members of the cluster are embedded in gas. Recent observations show that M87 is also a source of strong X-ray emissions.

The large cluster in Coma Berenices, the chief component of the Coma supercluster, can be examined in some detail because it lies above the gas and dust associated with the Milky Way. Containing well over a thousand galaxies, it lies at an average distance of 326 million light-years. It is approximately spherical and its diameter is estimated as at least 10 million light-years. It has a central core that is three times as dense as the rest of the cluster, containing a number of massive spiral and elliptical galaxies. Possibly this is because the cluster is old, with

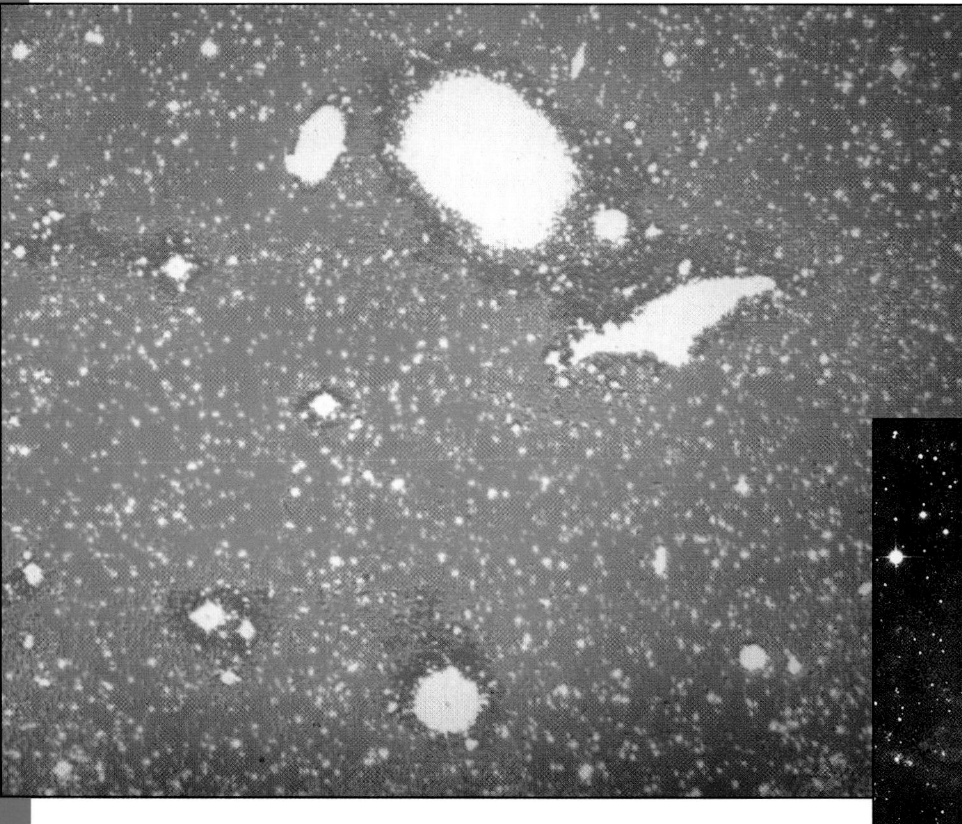

500,000 light-years

Galaxy

the result that its gravitational field has had time to attract these large galaxies toward its centre. A cluster that has evolved to such a state is said by astronomers to be "relaxed".

As is typical of large clusters, vast elliptical galaxies are present in Coma. The two brightest galaxies in the cluster are a supergiant elliptical and another large elliptical, which, because of the angle at which it is seen from Earth, appears to be spherical.

The Hercules cluster is a much less dense conglomeration than that in Coma. Irregular in shape, it contains galaxies of all types, but with a preponderance of spirals. Many of its members are associated in pairs or small groups. An unusual feature of the cluster is that many of the spirals seem to have retained more gas than normal, so that the intergalactic gas in the cluster is rather sparse. Since gas is expected to escape from the galaxies as time passes, this observation, and the general way in which the galaxies are moving, make it seem likely that the Hercules cluster is younger than, for instance, that in Coma.

The large Perseus cluster, 235 million light-years from Earth, has a supergiant elliptical galaxy at its centre. It also contains a number of radio sources – evidence of eruptive galaxies – and is rich in high-energy intracluster gas. Member galaxies have high velocities in relation to each other, and this, coupled with the evidence of the gas, may indicate that, like the Coma supercluster, the Perseus cluster is not particularly young.

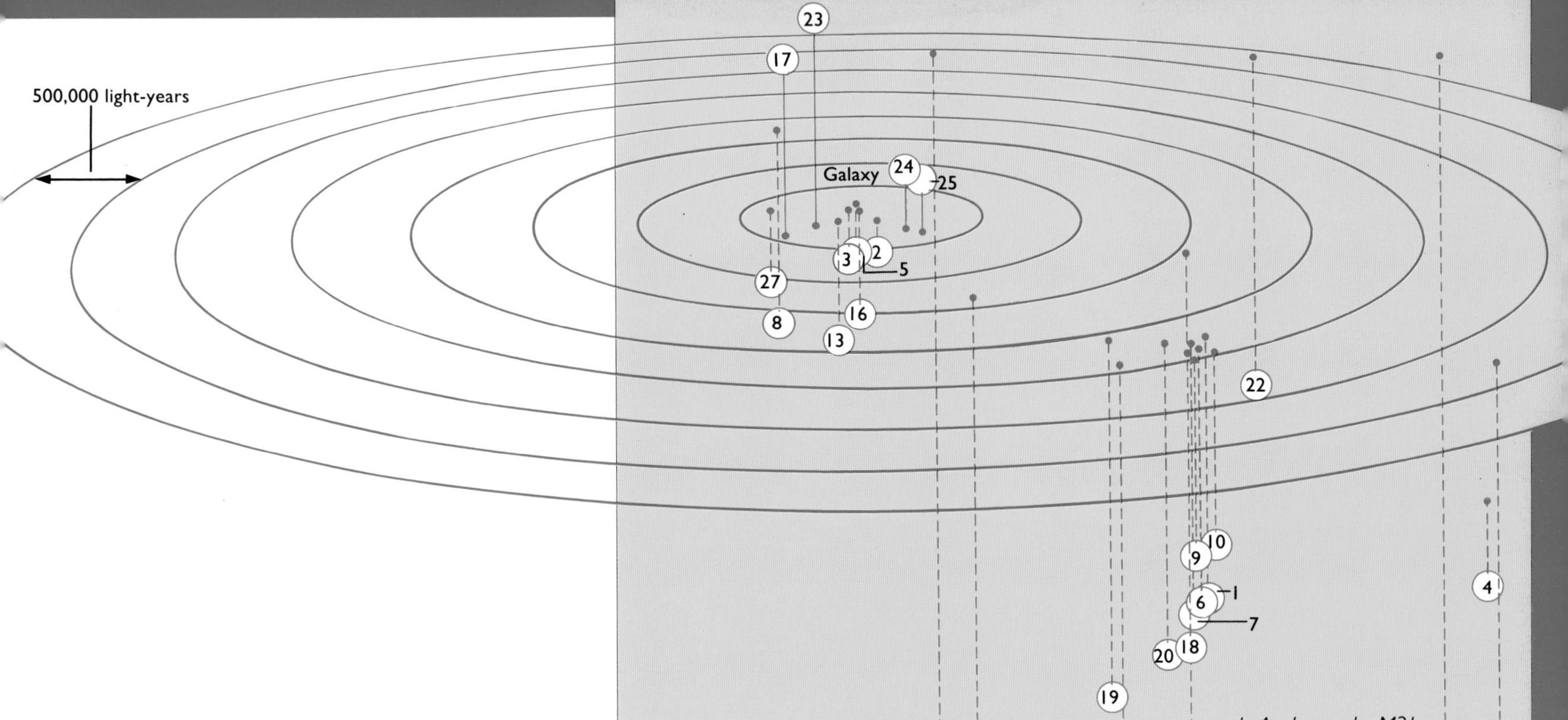

## The Local Group

Our cluster of galaxies, the Local Group, is dominated by two great spirals: the Milky Way and the Andromeda spiral, which is probably the larger of the two. The remaining galaxies are grouped around each of these. The most conspicuous companions of the Milky Way are the Large and Small Magellanic Clouds. Each of these satellite galaxies seems to have a central bar of densely packed faint stars, as do some spirals. This leads some astronomers to classify the Magellanic Clouds as rudimentary spirals.

There is a third large spiral in the Local Group, M33. It possesses no companion galaxies. The remaining galaxies are dwarf ellipticals and small irregulars. It is not certain exactly how many faint, relatively nearby galaxies belong to the Local Group: the main members are shown here.

*1 Andromeda, M31 (NGC 224) Sb*
*2 Triangulum, M33 (NGC 598) Sc*
*3 Large Magellanic Cloud SB (?)*
*4 IC 10 SB*
*5 Small Magellanic Cloud SB (?)*
*6 M110 (NGC 205) E6*
*7 M32 (NGC 221) E2*
*8 Barnard's galaxy (NGC 6822) Irr*
*9 NGC 185 dE*
*10 NGC 147 dE*
*11 IC 1613 Irr*
*12 Wolf-Lundmark Irr*
*13 Fornax dE*
*14 IC 5152 Irr*
*15 Pegasus Irr*
*16 Sculptor dE*
*17 Leo I dE*
*18 Andromeda I dE*
*19 Andromeda II dE*
*20 Andromeda III dE*
*21 Aquarius Irr*
*22 Sagittarius Irr*
*23 Leo II (Leo B) dE*
*24 Ursa Minor dE*
*25 Draco dE*
*26 LGS 3 Irr*
*27 Carina dE*

*dE = dwarf elliptical. Irr = irregular. For other symbols see pp. 52 and 60. Leo III (Leo A) is too remote to be shown.*

# SPIRAL GALAXIES

● *Pinwheels of the cosmos*

***A spiral galaxy,*** *NGC 2997, seen face-on. Old stars make up its yellowish core, which is virtually free of gas and dust. Bluish young stars, glowing gas and dark dust are visible in the arms.*

The pinwheel shapes of spiral galaxies are among the finest of cosmic spectacles. It is not surprising that most people think of a spiral whenever the word "galaxy" is mentioned.

The essential features of a spiral are a central bulge and a disc that spreads out from the bulge, far into space. The arms almost always lie in the plane of the disc. Spirals are classified according to how tightly their arms are wound: from So, for the most tightly wound, through Sa and Sb to Sc, which have arms that are extremely widely spread. Generally the central bulge is roughly spherical. However, in one class of spiral, the barred spirals, the central region is elongated to form a bar, from the ends of which the spiral arms protrude.

In addition to visible light, many spirals emit radio waves. These come from cold, dark hydrogen gas and from patches of glowing hydrogen. Furthermore, radio waves give evidence of haloes of star clusters and gas around spirals. Indeed, in recent years it has become evident that, on account of their haloes and gaseous envelopes, spirals are much larger and more massive than optical observations indicate. Some spiral galaxies are strong emitters of X-rays, which appear to emanate from the galaxies' central regions.

Spiral galaxies vary widely in size. The Andromeda spiral, M31, is a vast system, with a diameter of at least 124,000 light-years. Recent optical observations using new techniques for detecting faint radiation indicate gas extending much farther out than this. Our own galaxy is also very large, with a diameter of not less than 100,000 light-years.

However, the other substantial spiral in our Local Group, M33 in Triangulum, has a diameter considerably less than either of these. Its main disc is only 52,000 light-years across, yet it is still large compared with most spirals. Two examples of smaller spirals are the Magellanic Clouds, which are companions to the Milky Way. They are now classed as barred spirals, although formerly they were regarded as being irregular in shape.

Spiral galaxies contain myriads of stars. There are about 100 billion in our own galaxy, and the Andromeda galaxy may have as many as 1,000 billion. But these stars are not evenly distributed. The main concentration is in the central bulge. These are comparatively old stars, which collectively appear yellowish. Those in the disc and spiral arms are far younger: many are in fact being born out of the gas and dust of these regions at the present time, and they give both the disc and the arms a bluish colour.

How did spiral galaxies originate? They could not form until there were collections of stars to act as nuclei that could attract other material gravitationally. The same processes that gave rise to the protogalaxies (pp. 38–39) were also favourable to star formation.

*The appearance of a spiral galaxy* depends on the angle at which it is tilted toward us. Some are edge-on to us, or nearly so (top). Others display their coiled arms to varying degrees.

Collisions between atoms and molecules in the early universe meant a loss of energy, with the result that they slowed down and gravitation took over. The material gathered into small massive concentrations, the protogalaxies, which then began to contract rapidly under their own gravity to form central cores. At the same time, these broke into separate lumps of material, which contracted to become stars. Stars are still forming out of such concentrations of gas and dust today.

The material from which this central conglomeration of stars formed was in motion around the central core. Thus as the stars formed they were concentrated into a rotating mass. The galaxy's rotation had the effect of throwing gaseous material outward to form a flattened disc surrounding the central regions. The disc would display differential rotation, with the slowest-moving material being closest to the rim. Initially, such a disc did not have spiral arms.

However, present theories suggest that density waves appeared. Owing to odd disturbances in the gravitational field in the disc – caused, presumably, by the precise way in which material was distributed – stars would travel at different speeds. When stars passed each other, they caused mutual alterations in their orbits. These set up gravitational disturbances that travelled like waves through the material and gave rise to spiral arms.

The density waves were influenced by a further factor: the rotation of the gaseous material itself. This rotation and the formation of arms together generated further shock waves. These caused gas molecules to crowd together into denser clouds. In due course, gravitational attraction between the molecules in each of these clouds caused stars to be born; it is these later-forming, bright young stars that give spiral arms their blue colour.

The shock waves generated by these processes die away after a billion years at most. Nevertheless, if stars are to be born continually out of the disc material, as they seem to be, new shock waves must be set up all the time within spiral galaxies. Computer simulations of the conditions in spiral arms have shown astronomers how this could happen, and they also lead to other interesting results.

*NGC 4565* (below), a spiral galaxy that is edge-on to us. The dust in the disc is visible as a dark band crossing the central bulge. Areas of bright gas can be made out in the disc.

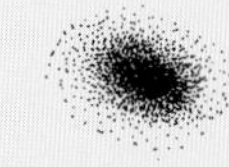


### How spirals are classified

Spiral galaxies are classified according to whether their arms are tightly or loosely wound, and whether they have normal or "barred" cores. They can be arranged in a double sequence, as in the diagram shown here, first drawn by the American astronomer Edwin Hubble. Normal spirals range from S0, which have almost no arms, through Sa, Sb, and so on, which have increasingly well developed and loosely wound arms. Intermediate classifications may be added. Barred spirals are similarly classified: SB0, SBa, and so on. It was once believed that these were evolutionary sequences – that spiral galaxies evolve from more open to more tightly wound forms – but this view is no longer held.

*__NGC 1365,__ an SBb barred spiral, is shown below. The bar is probably a feature that develops late in the life of a spiral galaxy.*

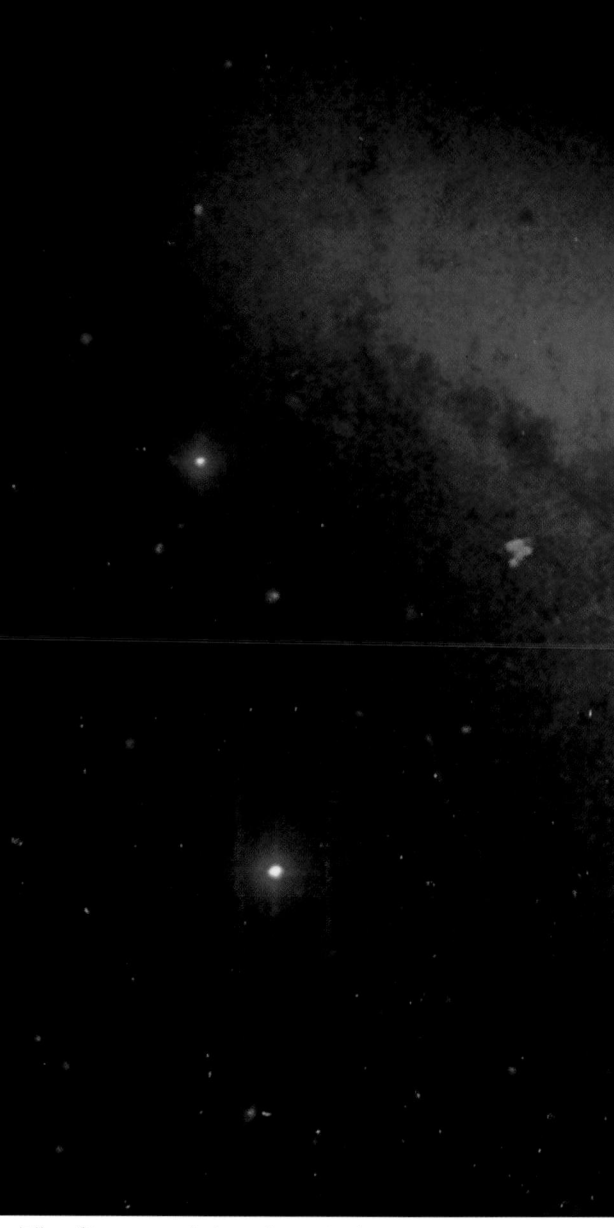

The disc containing the spiral arms is always thin, its thickness being about one-fifteenth of its diameter. Calculations show that such discs are inherently unstable and, furthermore, that the individual random motions of stars will become larger as time passes.

The result of such an increase is to destabilize the disc even more, setting up new shock waves. The principal waves will not be spiral in form, but long and narrow. This means that a uniform disc will first form spiral arms and then, after the galaxy has rotated a few times, the arms will largely disappear and be replaced by a bar. Such bars may contain as much as 40 percent of the optically visible material of the galaxy.

This is how barred spirals form, and clearly they are a late stage in the evolution of spiral galaxies. However, the age at which a spiral galaxy changes into a barred spiral depends on its enveloping halo of material. The more

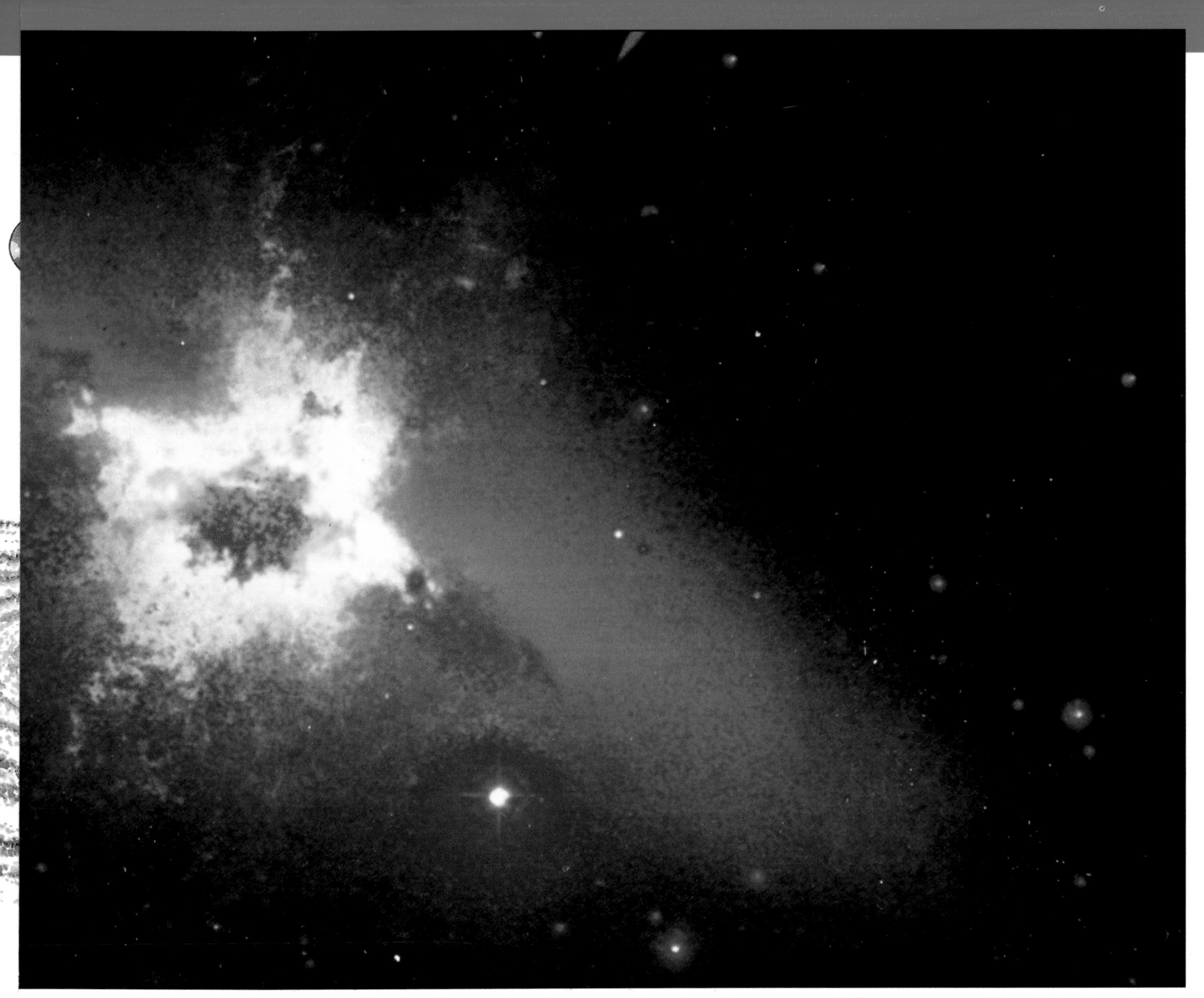

material there is, the more stable the disc, and the more slowly the galaxy will evolve.

This general scenario of the formation of spiral galaxies has been supported by the observation of a supergiant spiral in the Virgo constellation. There seems to be no doubt that this galaxy, called Malin 1, is a lens-shaped spiral at an immense distance. The galaxy is actually fainter than the night-sky background. That we can see it at all is thanks to a powerful technique of photography which was developed by David Malin – for whom the galaxy is named – of the Anglo-Australian Observatory in New South Wales.

What is so surprising about the Malin 1 galaxy is its vastness. Its central region is dominated by a huge bulge resembling an elliptical galaxy. It shows bright lines in its spectrum, indicating the presence of a considerable amount of glowing gas. This region is surrounded by a huge disc of gas, with a diameter of not less than 490,000 light-years, or approximately five times that of our own, not inconsiderable, galaxy.

The whole vast disc of neutral hydrogen is rotating more slowly than our own galaxy. If Malin 1 later contracts, its speed of rotation will increase. Indeed, its present velocity is to be expected if, as its discoverers suggest, this supergiant galaxy is a spiral in the early stages of formation.

If Malin 1 is a very young spiral it fits in well with the development of spirals sketched above. It is just what we should expect to find if galaxies contract from more diffuse protogalaxies: a core of older stars and a disc of gaseous material that will in due course split into spiral arms, owing to shock waves within it. We see it as it was 800 million years ago, long after the first spiral galaxies were born, and it seems to be evidence for the continuing production of galaxies in the universe.

***The spiral galaxy M82,*** *10 million light-years from the Earth, is seen in computer-enhanced colours in this photograph. It was long believed to be an exploding irregular. Now, however, it is thought to be a normal spiral. The reddish filaments that appear to emerge from the core probably belong to a cloud of gas and dust through which the galaxy is moving.*

# COLLIDING GALAXIES

• *Catastrophes in deep space*

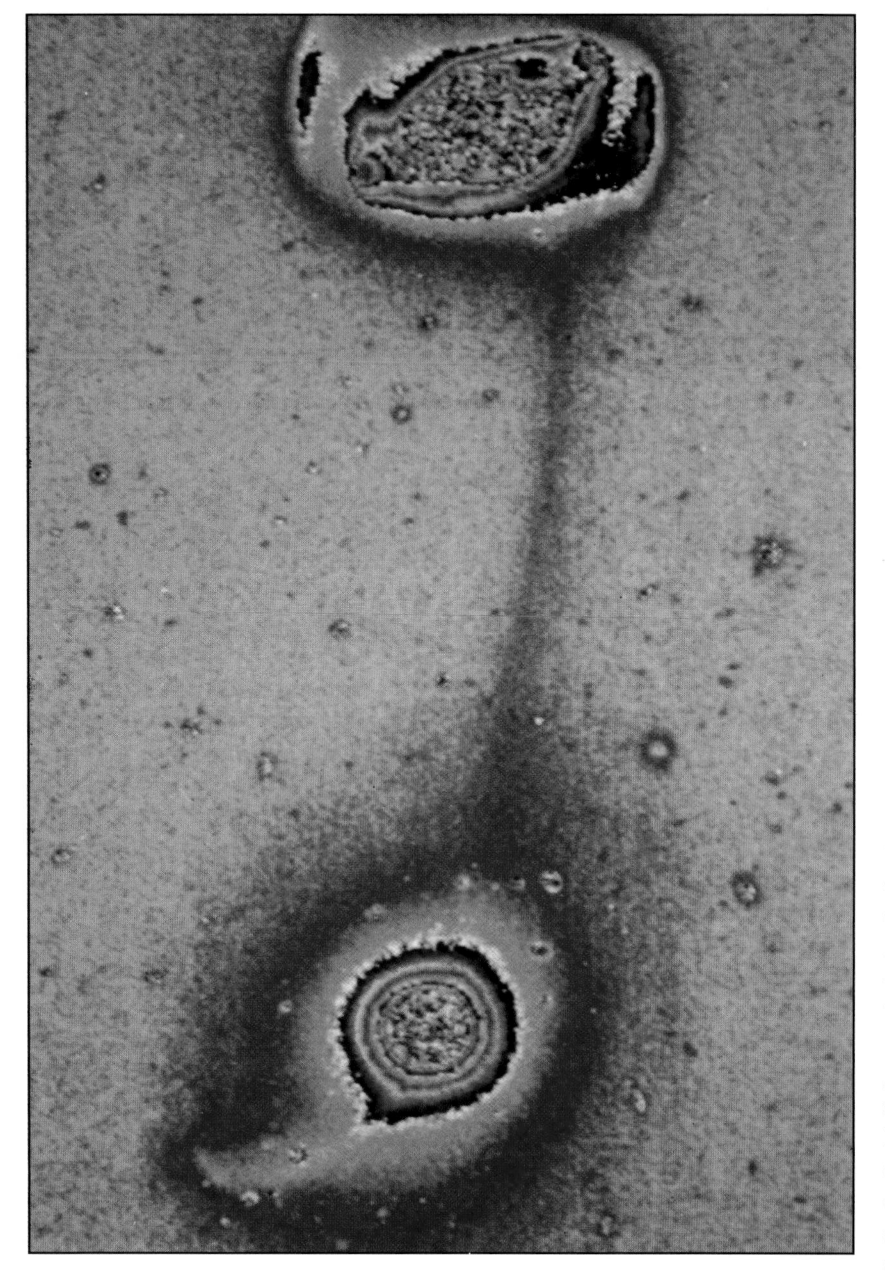

***A "rope" of stars** links the two galaxies NGC 5216 and NGC 5218. The rope was torn out of the two galaxies by their mutual gravitational pull. This image was obtained by the Isaac Newton telescope at La Palma Observatory, Tenerife, using a sensitive electronic detector in place of film.*

Most galaxies are a long way apart: even within clusters the average separation of the larger ones is about 1,500,000 light-years. Nevertheless, approaches sufficiently close to cause gravitational interactions and even collisions should occasionally occur, and observations confirm this.

The two Magellanic Clouds are in orbit around our own galaxy, completing a circuit in something like 500 million years. They have noticeable gravitational effects on the Galaxy: they distort its disc of stars, gas and dust, pulling one side upward and the other downward. In one part of the disc, in a region that is a little farther from the centre than our Sun, the disc is bent northward, while at an equal distance from the centre on the opposite side it is bent southward. This warp is not fixed in position: it revolves slowly around the disc, taking perhaps a billion years to complete a circuit. For its part, the Galaxy has tidal effects on the Magellanic Clouds, and an extended gas trail – the Magellanic Stream – has been found to circulate among the three galaxies.

Such mutual tidal effects are to be found in 8 out of 10 spiral galaxies, whose bent or distorted discs can be observed with modern techniques. The Andromeda galaxy suffers a similar distortion, although the cause in this case is not clear since its satellite galaxies seem to be too far away to do the warping. However, the haloes of spiral galaxies very probably contain dark matter (pp. 168–69), and such material could react on the discs.

This material may take various forms: some may consist of WIMPs (weakly interacting massive particles), a hypothetical type of matter that is still being sought in the laboratory, and some may consist of MACHOs (massive compact halo objects). The latter would be the remnants of stars that have collapsed long ago, or else are brown dwarfs or jupiters – objects of too low a mass to have formed into stars.

Many interesting interactions between galaxies have now been observed. One example is the interaction betwen the two spirals NGC 5426 and NGC 5427, in Virgo, whose spiral arms are actually linked. Moreover, their discs are distorted by mutual tidal action, and seem forced to oscillate with a period of the order of 100 million years.

Sometimes more than two galaxies are concerned in an interaction, as in the case of the Magellanic Clouds and our galaxy. In the constellation of Serpens Caput there is a group originally named Stephan's Sextet after its discoverer. It had to be renamed when detailed studies showed that what appears to be a sextet of interacting galaxies is not quite what it seems. Only five of the objects are interacting; the sixth object, a spiral galaxy, merely happens to be in the same field of vision: it is in fact five times as far away as the others.

The remaining five objects are certainly connected: they consist of four spirals and a large cloud of gas that has been ejected by one of them. Their masses and mutual separations show that they must be interacting gravitationally. Lanes of gas can actually be seen running between all five objects.

It is only a step from this type of situation to an actual collision between galaxies. A pair of spirals in Corvus seem to have collided and as a result to have ejected two vast streams of glowing gas, looking like the antennae of a giant insect. Computer simulations have shown just how this would occur if the disc of one spiral were to meet the disc of the other almost at right angles. The centres of these galaxies now lie some 65,000 light-years apart: the magnitude of the eruption that was generated may be judged from the fact that the ends of the antennae are now almost 500,000 light-years away from each other.

There are other examples of colliding galaxies that are even more spectacular in their effects. Two spirals called IG 29 and IG 30 have evidently collided. Again the discs seem to have met at right angles; what may seem surprising is that the two galaxies have passed right through each other and are still recognizably spirals. Nevertheless, they are still gravitationally bound, and there is a luminous bridge of gas joining their centres. One of the galaxies is surrounded by a ring of gas and bright stars; galaxies such as this are given the name ring galaxies.

Perhaps even more spectacular is the Cartwheel galaxy, lying at a distance of some 650 million light-years from us. This seems originally to have been a spiral galaxy, which developed into a ring galaxy in the aftermath of a head-on collision with a smaller galaxy that passed through it completely. The two are now over 250,000 light-years apart. The rim of the Cartwheel galaxy is an expanding shock wave that is triggering the generation of a vast number of bright, massive stars, which lead short lives before they explode as supernovae (pp. 90–91). The rate of production of such exploding stars is 100 times greater than in a normal spiral, which underlines the catastrophic scale of the collision.

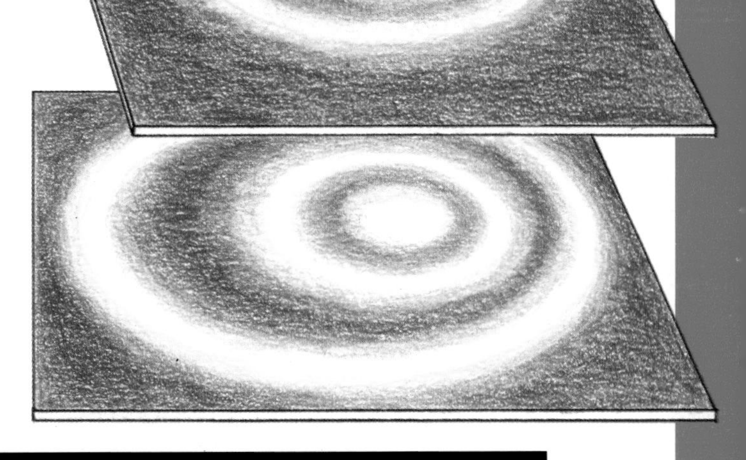

***A small elliptical galaxy*** *passing through a spiral galaxy (similar to our own) attracts stars, gas and dust from the larger galaxy toward the point of impact. After the elliptical galaxy has passed through, the material rebounds, forming rings rather like the ripples on a pond when a stone is thrown in. The rings spread out and trigger star formation as they compress gas and dust.*

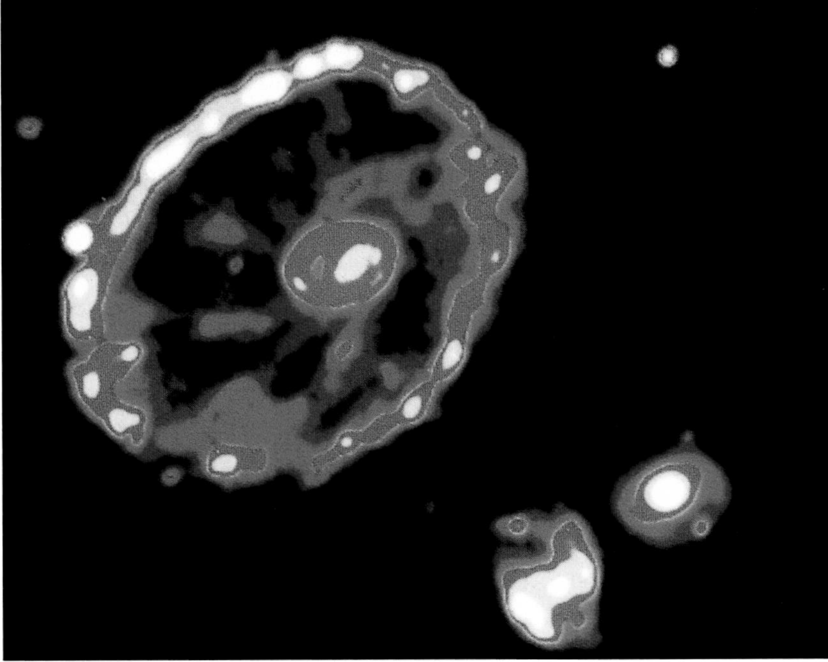

***The Cartwheel galaxy*** *(below) shows an outer and an inner ring, thought to be the result of an encounter with a small galaxy about 300 million years ago.*

# ELLIPTICAL GALAXIES

## • *Remnants of galactic collisions?*

***Lenticular galaxy NGC 5102,*** *about 13 million light-years from Earth, is visible in the southern sky. Lenticular, or lens-shaped, galaxies are intermediate between elliptical galaxies and spirals. NGC 5102 has a bright central elliptical bulge, about 3,400 light-years wide by 8,000 light-years long. The elongated disc is 33,000 light-years across.*

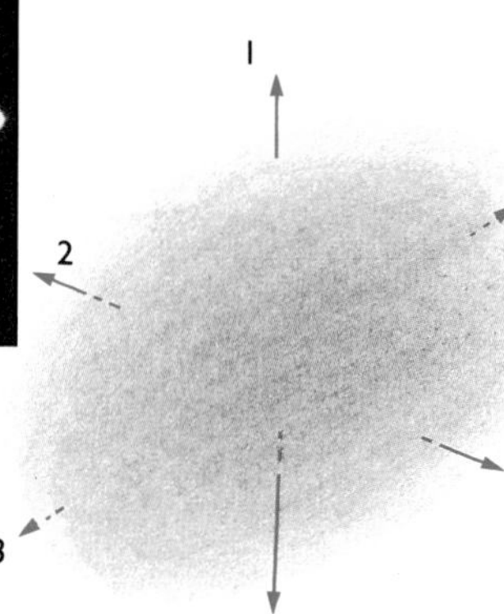


***Elliptical galaxies have complex shapes.*** *In general, the three main axes, which pass through the centre at right angles to each other, are all of unequal lengths.*

Elliptical galaxies seem at optical wavelengths to be composed entirely of old, reddish stars. They do not possess the discs, consisting of gas, dust and young stars, that characterize spiral galaxies. Their shapes are classified from E0, spherical, to E7, the most elongated. These last merge into the S0 galaxies, sometimes referred to as lenticular galaxies, which seem to form an intermediate class of galaxy between ellipticals and spirals.

It was once thought that the ellipticals had the shape of a rugby ball or American football (apart from some that happened to be spherical). Such a galaxy would look like an E7 if it presented a side view to Earth, but like an E0 galaxy if it were end-on to Earth. However, since the 1970s observational techniques have developed to a stage where astronomers can closely examine the orbital speeds of stars in galaxies. The results have been something of a surprise, for they have shown that in general all three of the axes of an elliptical are unequal.

It was also once thought that elliptical galaxies evolve into spirals, but the standard view now is that all galaxies were formed at much the same time. Whether a protogalaxy (pp. 38–39) gave rise to a spiral or an elliptical was a consequence of the speed of its rotation. A high rotation velocity resulted in condensation into a spiral galaxy; a slower one led to the formation of an elliptical galaxy.

In elliptical galaxies the formation of stars was largely completed as the protogalaxy contracted, and there is now an absence of young stars. This is the more striking since radio observations have shown that ellipticals contain a considerable amount of neutral hydrogen, which gives no sign of its presence at optical wavelengths but is very evident at the radio wavelength of 21 centimetres. Furthermore, modern optical detectors have made it clear that ellipticals contain considerable quantities of dust. Despite these facts, it remains true that elliptical galaxies consist predominantly of older stars. Spirals also contain ageing stars in their cores and haloes; it is in their discs that the process of starbirth from gas clouds continues.

However, very massive ellipticals are found in the central regions of clusters of galaxies (pp. 50–51), and it now appears that these may have been formed more recently in a different manner. They may indeed have been formed from spirals, but not by a process of evolution, as was once supposed.

One clue is provided by the study of the way the brightness falls off toward the edge of an elliptical galaxy. The form of this brightness distribution is called de Vaucouleurs' law. Since 1987 it has gradually become clear to astronomers that this law is also applicable to another class of objects known as starburst galaxies. These are irregular galaxies that contain much dust spread throughout them in a seemingly chaotic way. In these galaxies stars are being formed in abundance.

Few starburst galaxies are found in isolation. Most are interacting with other galaxies, and there is also an extensive class formed by the actual collision of two or even more galaxies.

These are now observed as single objects, surrounded by trails of gas and dust and other remnants. Optical observations, supplemented by infrared studies, often show that the movements of the older stars as they were before the collision is of the kind seen in elliptical galaxies.

In addition, computer studies have shown that the violent disturbances that would occur during collisions of two similar spirals with large dark haloes would produce two ordinary ellipticals to which de Vaucouleurs' law would apply. Recent observations of the galaxy NGC 7252 in Aquarius seem to confirm this.

The study of colliding galaxies has been taken further by Gillian Wright and her colleagues, using the United Kingdom Infrared Telescope in Hawaii. They have examined galaxies that emit vast amounts of infrared radiation, with very long wavelengths up to one-tenth of a millimetre. These include the starburst galaxy Arp 220, one of the most luminous known sources at these wavelengths.

Such galaxies display the wreckage to be expected if two spiral galaxies had collided. Optical pictures of Arp 220 show it to be accompanied by much dust, some of it in the form of a lane running across the image. This is the kind of dust associated with the formation of new stars. The infrared observations show not only large amounts of hot dust but also the presence of young stars and complex gas molecules.

These observations prove that elliptical galaxies with active centres are still forming. At earlier times in the history of the universe, when galaxies were closer together, collisions must have been more frequent. It is therefore possible that, in contrast to the predominant view, most or even all ellipticals are in fact the result of collisions between spirals.

*__Centaurus A is an intense radio source__, the third strongest in the sky. The dark band seen here is a disc of gas and dust, lying at the centre of a much more extensive elliptical galaxy, which needs special photographic techniques to reveal it. The disc is a site of star formation.*

# BLACK HOLES

## • *Cut off from the universe*

***Time slows to a halt** at the boundary of a black hole, a region of space-time from which neither matter nor energy can escape. The watches above are initially synchronized. As viewed from the outside universe, time passes normally for the watches far from the hole – for example, they record 45 minutes as having passed between an arbitrary start time and end time. But the passage of time is slower for watches that are closer to the black hole. For an observer with each watch, however, time passes as normal. Close to the black hole space is affected as drastically as time: intense gravitational forces would deform and finally destroy any object approaching too close.*

As long ago as 1783 John Michell, an astronomer and geologist, suggested that gravity could act on light as well as on matter. At that time Newton's theory that light was composed of small particles or "corpuscles" was still current, so Michell's idea did not seem too far-fetched. He drew the interesting conclusion that some stars might be so large as to have escape velocities greater than the speed of light, thus preventing any light corpuscles that they radiated from escaping into space. To an observer such massive stars would therefore be black, and invisible against the night sky.

Pierre Laplace, the great French mathematical physicist, put forward similar ideas independently. Yet to most scientists they appeared to be no more than a curiosity. Moreover, they were suspect because there was no reason to suppose that light could have one and only one velocity in all circumstances. In the early 19th century, when Newton's corpuscular theory gave way to the idea that light is a wave disturbance, it seemed even more unlikely that gravity would be able to trap light.

Only in 1915, with the advent of Einstein's general theory of relativity, was the situation radically altered. The bending of light was one consequence of the theory, and once the observations made at the total eclipse of 1919 had confirmed that light is bent by the Sun, the whole subject of the effects of gravity on light was revived.

The English physicist Oliver Lodge calculated that a body of "reasonable" mass could trap light completely only if it were of a density that could not be attained by matter as it was conceived of in his day. But later the concept of "degenerate" matter was proposed (p. 88). This is extraordinarily dense, and put a different complexion on the idea of a "black star".

But interest in the subject did not really awaken until 1969. By then astronomers had evidence that some galaxies were strong emitters of X-rays, and they had to face the question of how the vast amounts of energy this implied could be generated. Donald Lynden-Bell at Cambridge University suggested that super-dense bodies could provide the answer.

Such a body would attract matter, accelerating it to huge velocities as it fell in. An accretion disc would form, consisting of matter revolving around the black hole before its final disappearance. The infalling material would move at immense velocities, emitting X-rays in huge quantities.

Relativity now enters the scene. The extreme density of such a body would create an intense gravitational field. This would mean that space-time around the body would be so strongly curved as to cause the interior to be completely closed off from the outside universe. Nothing could escape from it. This is why such objects are now called black holes.

These holes in space-time not only absorb whatever falls into them: they have other astounding effects. As a body approaches a black hole the immense gravitational field causes the passage of time for that body to slow down, as measured by an outside observer. The frequency of the light that it emits falls lower and lower – that is, it is redshifted – and it becomes weaker. The object approaches, but never quite reaches, the so-called event horizon surrounding the black hole, and finally seems to hover there. In consequence the outside observer never sees it actually fall into the hole.

In the frame of reference of the body itself, however, there will be no apparent change in the rate at which time passes. But the body will be subjected to immense forces that will tear apart any macroscopic body, and certainly make survival impossible for any living thing. Indeed, all structure is destroyed, so that matter inside the black hole loses all individuality – all "memory" of what it was before. It falls toward a centre known mathematically as a singularity, where the density of matter becomes infinite and space-time is reduced to a mere point.

To understand these and other facts about black holes, it is helpful to draw space-time diagrams. The vertical dimension is taken to represent not a space dimension but time. A vertical line directed upward represents the position of a stationary object as it moves into the future. A line that slants, or is curved, shows that the object is moving. Such a line is called its world line.

Waves of light emitted from some event spread out across space. In such a diagram it is represented by rings increasing in diameter up the page. We end up with a light-cone, its apex representing the original event.

Now imagine that a star falls into the black hole. Successive wavefronts of the star's light

start time

end time

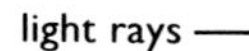


*Black holes are formed when matter reaches sufficiently high density. Light from a star of low density, such as a red giant, is scarcely deflected by the gravitational field. A white dwarf, which has a density of 1 tonne per cubic cm, also does not trap light, although the light rays are appreciably curved. They are even more strongly curved by the gravity of a neutron star, which is about a hundred million times more dense than a white dwarf. An even denser stellar remnant, left after the death of a very massive star, bends light rays so strongly that they cannot escape at all, and the body is a black hole.*

remain circular in the diagram but are emitted from a point that is falling inward with increasing speed. The result is that the light-cone appears to be tipped over toward the black hole. At the event horizon the light-cone is tilted so far that its outer edge is vertical, representing the fact that the light cannot escape from the event horizon, and we see the star apparently hovering there for ever.

This is probably an oversimplified picture. It seems likely that in the real universe black holes, being formed from material that is rotating, are themselves rotating. The British cosmologist Stephen Hawking and his colleagues have established that such black holes can be described by equations first discovered by the New Zealander Roy Kerr.

The area around the event horizon of a rotating black hole is known as the ergosphere. It represents that part of space-time in which the black hole interacts with the rest of the universe. For instance, the Kerr equations tell us that as matter falls through the ergosphere and into a rotating black hole the area of its surface increases. But strangely enough this infalling matter can cause a *decrease* in the mass, provided that the spin of the black hole is reduced as well.

Despite all that has been said above, Stephen Hawking has found that black holes can lose energy (see box). He has applied the laws of quantum theory and thermodynamics to black holes, with interesting results. The first is that black holes have a temperature, which is larger the smaller the mass of the hole. Even so, the temperatures are minute by everyday standards: even a comparatively small black hole with the mass of a star would be no hotter than a ten-millionth of a kelvin.

However, Hawking believes that minute black holes could have been formed from very dense matter crushed together at an early stage of the Big Bang. Being so small, their temperature would be very high, and some of them would be evaporating now, in explosions that would put an artificial nuclear explosion to shame. But no radiation from such mini-holes has been observed to date, and it seems probable that the Big Bang was too uniform to allow them to form.

But black holes of larger masses probably exist. Where are they? It seems likely that some reside in the central regions of spiral and elliptical galaxies. They also form after the catastrophic collapse of very large stars, and also, possibly, after the deterioration of weird supermassive objects called spinars. The existence of these objects is uncertain, but they could have a mass some 10 million times that of the Sun, crammed into a small space. Such an object – sometimes called a magnetoid because of its immensely powerful magnetic field – is believed to lurk at the centres of active galaxies. But for their rapid rotation they would collapse immediately. When the energy of their rotation is spent, they form black holes.

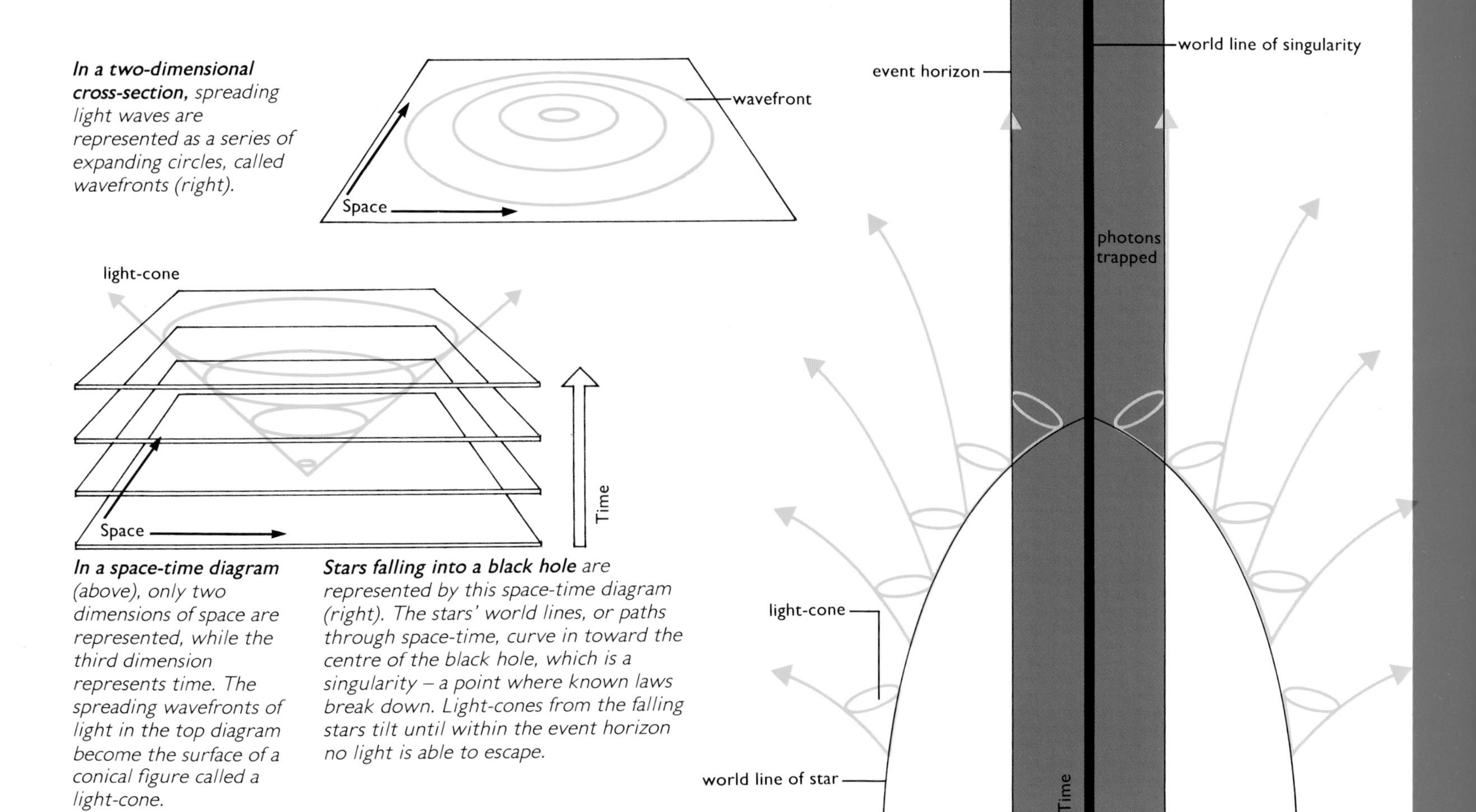


***In a two-dimensional cross-section,*** *spreading light waves are represented as a series of expanding circles, called wavefronts (right).*

***In a space-time diagram*** *(above), only two dimensions of space are represented, while the third dimension represents time. The spreading wavefronts of light in the top diagram become the surface of a conical figure called a light-cone.*

***Stars falling into a black hole*** *are represented by this space-time diagram (right). The stars' world lines, or paths through space-time, curve in toward the centre of the black hole, which is a singularity – a point where known laws break down. Light-cones from the falling stars tilt until within the event horizon no light is able to escape.*

## The decay of black holes

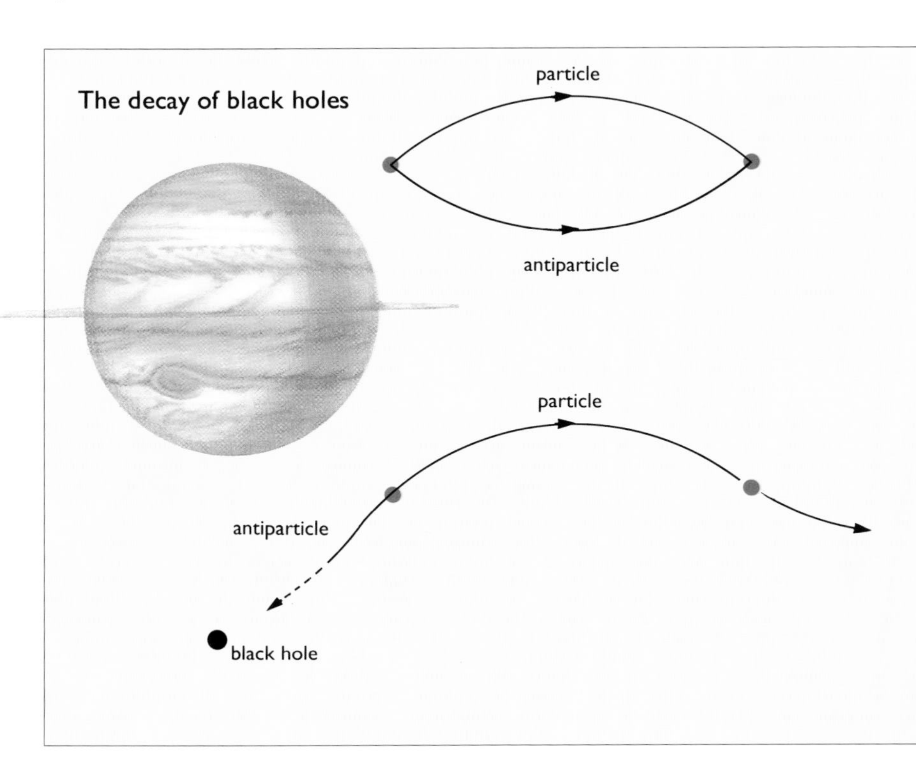


Throughout space virtual particles (p. 26) are constantly appearing fleetingly, and then disappearing. They appear in particle-antiparticle pairs, so that there is no net creation of electric charge or spin. Pair production is stimulated by the gravitational field of any massive body (upper diagram), but when it happens near a black hole, one member of the pair can fall into the black hole, while the other escapes. The one that escapes carries energy away, while the one that is captured carries negative energy into the black hole. The black hole appears to radiate energy with a spectrum resembling that of a body with a definite temperature. This process of decay is extraordinarily slow. A black hole with the mass of the Sun would take $10^{56}$ times the present age of the universe to evaporate.

# ACTIVE GALAXIES

## • *Star systems in turmoil*

Once galaxies were recognized as star systems lying far outside our own, astronomers generally gained the impression from photographs that they were staid, well-behaved objects. Only in the early 1950s did it suddenly become clear that this might not be true. Optical and radio astronomers working together established that two powerful radio sources, called Virgo A and Centaurus A, were actually galaxies. Virgo A turned out to be the E0-type elliptical galaxy M87 and Centaurus A the S0-type galaxy NGC 5128. Optically neither appeared remarkable except that Centaurus A was crossed by a band of dark material. The name "radio galaxies" was coined.

As the detail that could be "seen" with radio telescopes was improved, it was found that there were many radio galaxies. On optical photographs they appeared as galaxies might be expected to, but the radio observations gave a very different picture. Centaurus A was shown in detailed radio pictures to be a triple source. In the centre is a small intense radio source, and on each side of it are two large and strongly emitting radio "clouds".

At some time in the past, the central regions of the galaxy have suffered a vast explosion, causing two jets of matter to be thrown out in opposite directions. The jets are composed of very fast-moving particles, reaching velocities of about 60,000 kilometres per second (one-fifth of the speed of light). Their energy is so great that they burn their way through the surrounding intergalactic gas, setting up pressure waves rather like the bow waves of a ship.

The particles are ions (electrified atoms) and electrons. Their motion creates magnetic fields that restrict the volume of space through which the particles can spread. As a result, the particles pile up, forming lobes at the end of each jet. When the galaxy stops emitting jets, the two lobes of plasma continue to expand, forming the clouds we now observe. As time goes on their radio "brightness" diminishes.

In the case of Centaurus A there is also a radio halo, rather elongated in shape and about one million light-years in extent. The clouds lie within this, two at a distance of some 30,000 light-years from the galaxy's centre, one at a distance of 100,000 light-years. The two inner lobes form a ring around the galaxy, and were formed from two jets. The single more distant cloud is presumably the result of a single jet of atomic particles. The amounts of energy emitted by these clouds are prodigious. The radio emission is anything up to a hundred times the optical output of a typical galaxy, and lasts for between one million and 100 million years.

Not all radio galaxies display lobes. Some, such as M87, emit only a single jet of material as Centaurus A seems once to have done. The jet can be seen optically, as well as detected at radio wavelengths. Radio galaxies also emit at X-ray and gamma-ray wavelengths. Most of this very energetic radiation comes from a small region in the nucleus of the galaxy.

Clearly, radio galaxies are undergoing enormous eruptions. What is more, radio studies show that our own galaxy and M31 in Andromeda, which at first sight appear so quiet, also have active centres, although on nothing like the same scale.

Certain active galaxies have become known as BL Lac objects. The name is derived from the first object of this kind, which was originally thought to be a star and was designated as BL Lacertae (that is, variable star BL in the constellation Lacerta, the Lizard). It is now known to be a starlike object within the nucleus of a galaxy surrounded by very strong magnetic fields. These fields vary, like BL Lac's radiation output, occasionally becoming 100 times greater than usual. Moreover, the object itself is no larger than a few light-months in diameter. About 50 BL Lac objects are now known and it is evident that each is part of the radiating area of an active galactic nucleus.

What, then, is happening in the central regions of such explosive and active galaxies? Certainly atomic particles are being emitted at immensely high speeds. Indeed, sometimes the clouds appear to be travelling faster than light. Such superluminal velocities would contradict the theory of relativity if they were real (pp. 16–17), but they are in fact illusory. Such cases occur when the radiation-emitting jet is coming from the galactic nucleus in a direction very close to the line of sight.

The jet from an active galaxy comes from a source some 3 to 30 light-years away from its centre. Although we cannot directly observe this central powerhouse, we can come to some

***The structure of an active galaxy,*** *according to modern theory. The source of its energy is a black hole at its heart. Gas, dust and even stars are continually drawn into the black hole. As they spiral in, they form an accretion disc of orbiting matter. As the material is swallowed, energy equivalent to 10 percent or more of its mass may be emitted. This outpouring of energy creates jets of charged particles that escape along the black hole's axis of rotation.*

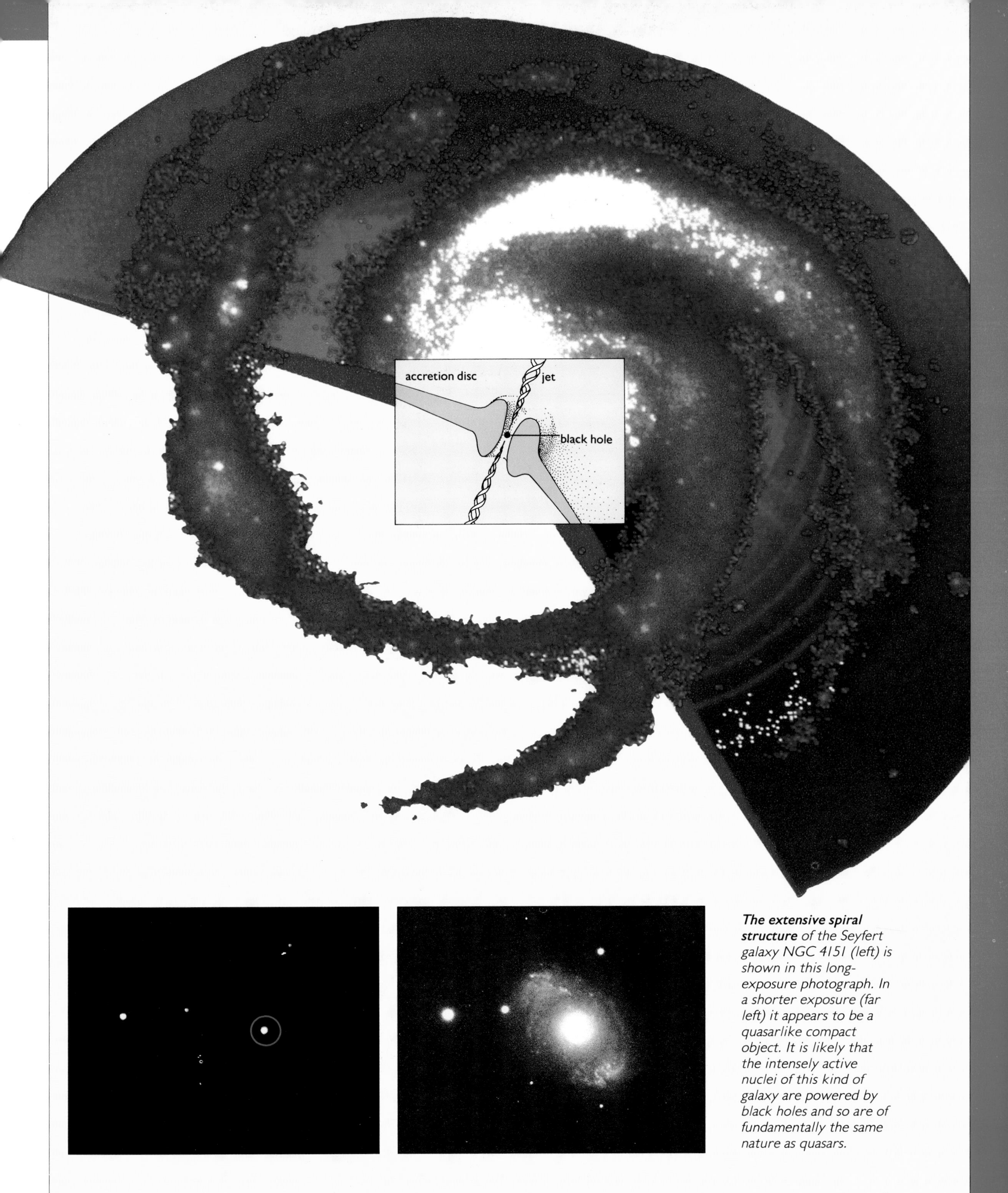


***The extensive spiral structure*** *of the Seyfert galaxy NGC 4151 (left) is shown in this long-exposure photograph. In a shorter exposure (far left) it appears to be a quasarlike compact object. It is likely that the intensely active nuclei of this kind of galaxy are powered by black holes and so are of fundamentally the same nature as quasars.*

*An intense jet of charged particles* streams from an active elliptical galaxy. The low density of interstellar gas and dust in the galaxy permits the jet to escape and create radio-bright lobes where it collides with intergalactic matter.

definite conclusions about it. Astronomers find that it must be associated with a disc of gas at the galaxy's centre. The disc is probably thinner at its centre than at the edge. In that event material driven outward by the violent energies at the centre would find it difficult to escape through the thickness of the disc; instead the material would stream away from the galaxy at right angles.

The disc is formed by accretion of material orbiting a tiny object at its centre; around this core the disc bulges. In addition to the beams of particles ejected, the disc itself would emit high-energy ultraviolet and X-radiation. If the disc is surrounded by very dense gas clouds, these would be heated by this bombardment and would give out radiation with the kind of spectrum observed in many active galaxies.

In such a scenario there has to be something very active at the centre of the galactic nucleus, something powerful enough to cause the accretion disc to blast away atomic particles at a sizable fraction of the speed of light and give rise to super-energetic synchrotron radiation, X-rays and short-wavelength ultraviolet.

The only viable candidate is a black hole. Black holes can be caused by the collapse of very large stars; at the centres of galaxies there are concentrations of stars, among which at least some massive examples are to be expected. Moreover, once a black hole – even a small one – has formed, it will gather other material into its embrace. This will revolve around the black hole, and will give rise to an accretion disc.

Matter near the centre will be heated intensely, emitting X-rays and ultraviolet radiation before plunging into the black hole. What is more, although formed from massive stars and much other material, the black hole would be very small by cosmic standards.

The presence of black holes at the centres of active galaxies makes it seem likely that black holes lurk at the centre of our own and perhaps in the centres of all galaxies, because all have dense central cores of old stars. What is more, active galaxies do

*The active galaxy Cygnus A* (below) is the brightest radio source in the sky. Most of its radio emission comes from two lobes of gas, 120,000 light-years from the galaxy. These are sending out a million times as much radio energy as our Milky Way galaxy.

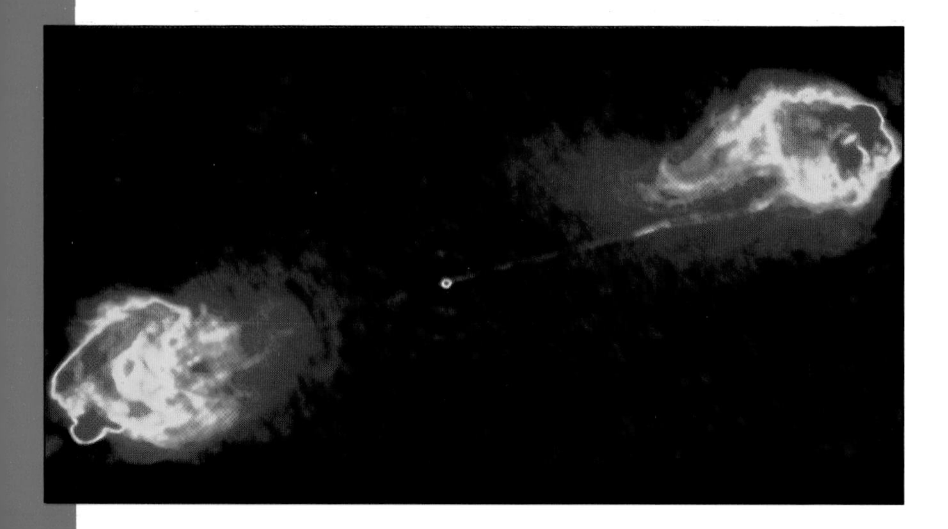

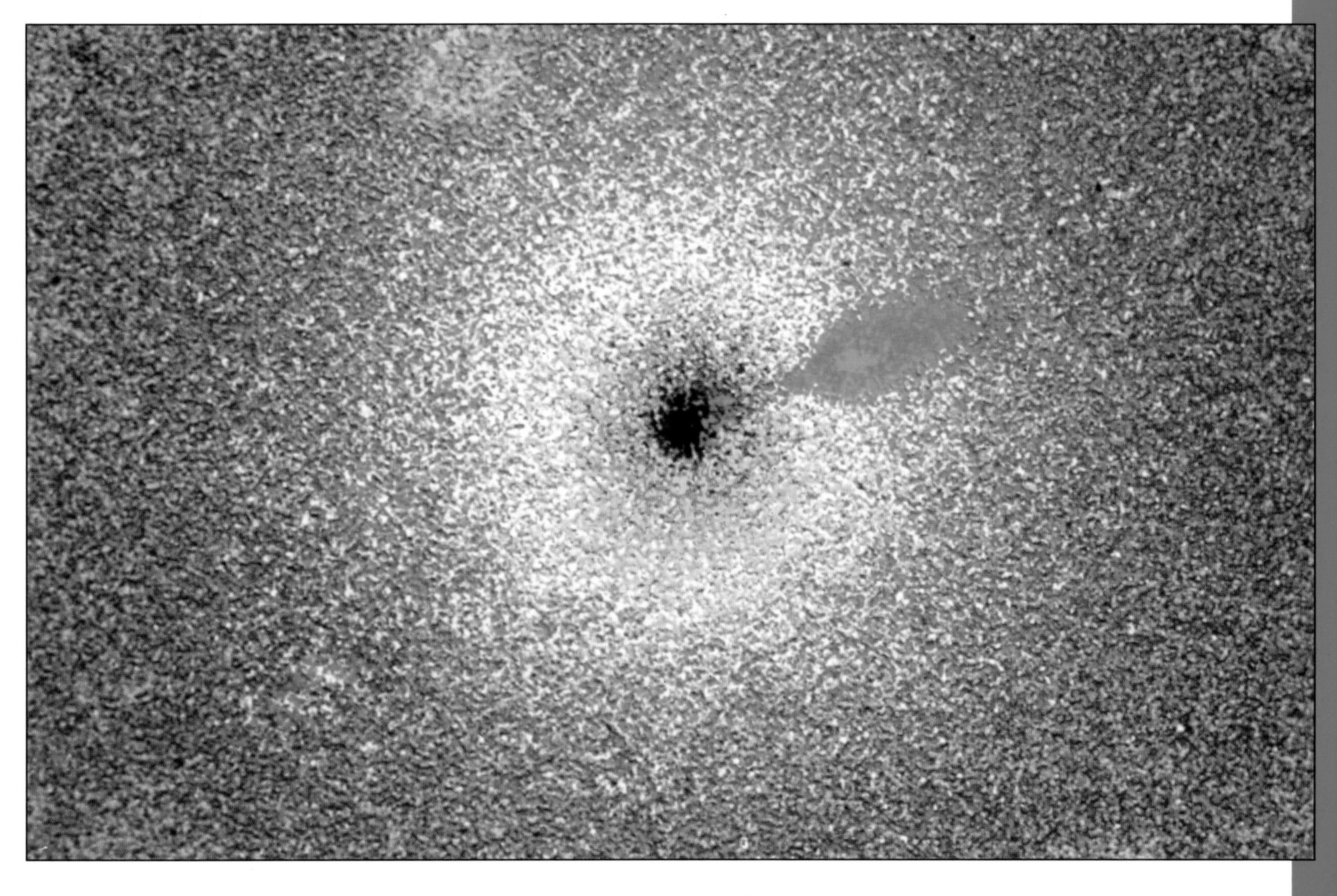

*A jet of gas* emanates from the giant elliptical galaxy M87, 50 million light-years from us in the Virgo cluster of galaxies. Radio waves and X-rays from fast-moving electrons come from the jet and the galaxy. The jet is probably emitted by a massive black hole at the centre of the galaxy. The black area at the centre of this false-colour computer-processed picture represents a region of closely packed, fast-moving stars, which are probably moving under the influence of the black hole's intense gravity.

not stay wildly active permanently, and may well settle down. The more sedate galaxies may once have been active themselves.

Even galaxies that do not possess violently energetic cores and do not puff out lobes of radio-emitting ions vary among themselves in their level of activity. They differ especially in the vigour with which they form stars. Some dwarf galaxies appear to be active in this way.

Dwarf galaxies are generally classified as dwarf ellipticals (dE), dwarf irregulars (dI) or blue compact dwarfs (BCD). Dwarf ellipticals possess very little unused hydrogen and show little evidence of current star formation. The light from their stars shows that they contain a comparatively high proportion of metals, which means that they have existed for many stellar life cycles. Dwarf ellipticals are old and some, perhaps most, were formed directly from the protogalaxy material. However, it also seems possible that some dE galaxies evolved later from dwarf irregulars.

Dwarf irregular and blue compact dwarf galaxies are very different. Both are roughly disc-shaped. The irregulars contain much hydrogen and give evidence that star formation is going on at present. When their hydrogen is depleted they could well become dwarf ellipticals. As far as the BCD galaxies are concerned, their colour alone shows that they are rich in young blue stars.

The chief difference between dI and BCD galaxies is that the dIs show only isolated patches of star formation, while the BCDs seem to have such activity everywhere. In consequence BCDs are brighter than dIs. However, it may well be that BCDs and dIs are really the same kind of dwarf galaxy. The apparent differences would be due to star formation proceeding in cycles. The BCDs would be those galaxies that happen to be at the more active parts of their cycles.

# QUASARS

## • *Celestial powerhouses*

In 1963 radio astronomers were watching intently as the Moon passed in front of the radio-bright source 3C 273. They were interested in this object because it had been tentatively identified with an optically dim blue starlike object with a peculiar spectrum. The identification remained uncertain, however, because radio telescopes could then fix the positions of sources only to an accuracy of five seconds of arc. When the Moon's edge blotted out the radio waves from 3C 273, it revealed the precise location of the radio source, and showed that it was indeed the suspected "star".

The occultation, as such events are called, also showed that 3C 273 was not a simple source but consisted of two parts. Detailed optical studies later showed that although one part was certainly a compact source, the other was associated with a faint jet.

The optical spectrum of the starlike object was puzzling because it seemed to display no trace of hydrogen, which is the main constituent of stars. However, these optical observations held an even greater surprise in store. It was realized late in 1963 that the lines characteristic of hydrogen were present: the lines normally seen in the optical region had been shifted into the infrared, while the lines that were now in the visible range were ones normally visible in the ultraviolet. In other words, the object had an unprecedented red shift, far larger than could be expected in a star.

What was true of 3C 273 proved to be true also of 3C 48, which had first been tentatively identified with a blue "star" in 1960. Because of their starlike appearance, the two objects became known as quasi-stellar radio sources, or quasars. Later it turned out that there were many more such objects, all showing strong red shifts, although most did not have their peak emission at radio wavelengths. They were given the general name quasi-stellar objects, or QSOs, but the name quasar has stuck.

But what were the quasars? If the immense red shifts indicated a recession, as those of galaxies were taken to do, then 3C 273 must lie at a distance of no less than 1,500 million light-years, and 3C 48 must be twice as far away. Yet if this were so, the optical radiation they were emitting must be more than 1,000 times greater than that of a galaxy such as our own.

There was another problem. The radiation from quasars is not only extremely intense but also varies in strength, often over a few weeks. This implies that the source can be no more than a few light-weeks across; otherwise radiation from the farthest parts of it would reach us with a delay sufficient to "smear" the variation over a longer period.

In 1963 no mechanism was known that could produce such immensely strong radiation in so tiny a region of space. Other explanations of the huge red shifts were therefore sought. Could the quasars be nuclei ejected from their parent galaxies, so that their high speeds had nothing to do with the expansion of the universe, and the objects were in fact relatively nearby? Or could their red shifts be caused by intense gravitational fields, which is possible according to relativity theory?

But neither of these explanations fitted all the observations, and consequently these suggestions had to be rejected. In fact there seemed to be only one satisfactory answer, and that was that the red shifts were the consequence of distance: they were truly "cosmological". This explanation is now generally accepted, with only a few dissenters (pp. 178–79).

A clue to their nature comes from the fact that quasars are so far away. We are observing objects formed in the comparatively early days of the universe, at the stage when the galaxies were coming into being. Their spectra display a continuous background on which are superimposed broad lines. The continuous part probably comes from a compact source, and the lines from clouds of hot gas surrounding this core.

Quasars' spectra are in fact very like those of galaxies of the kind known as Seyfert Type I, which have extremely active cores. Indeed, Seyfert galaxies and quasars seem to be the same kind of object, with Seyfert Type I being at a slightly later stage of evolution than quasars. Type II Seyferts, which have less active cores, are at a still later stage of development. Neither Seyfert galaxies nor quasars are permanently present in the universe, but rather mark different phases in its evolution.

Yet an explanation is still required for the intensity of these objects, since their emission of radiation is truly astounding. In 1969 it was

***Hot gas surrounds the core*** *of a quasar. At the centre enormous quantities of X-rays and ultraviolet radiation are released. This energy is absorbed and re-emitted by the surrounding matter. Within about one light-year of the core small, hot clouds move at almost the speed of light, both toward us (blue) and away from us (red). Cooler gas extends thousands of light-years beyond this (outer zone).*

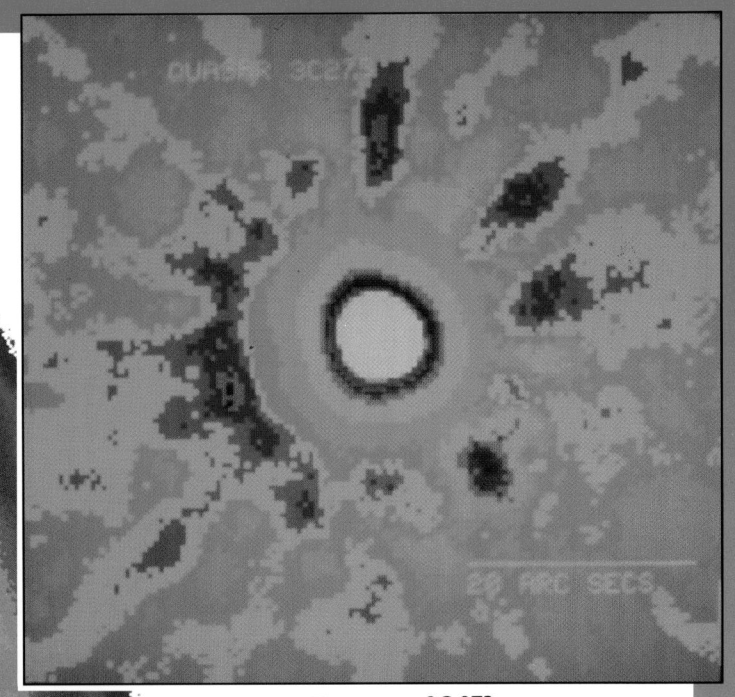


***The quasar 3C 273*** *appears in a computer-processed image (above), taken at X-ray wavelengths by the Einstein Observatory satellite. The region of most intense emission is the light blue patch at the centre. The quasar emits a conspicuous jet of particles, appearing as a dark patch at the bottom right of the quasar itself. 3C 273 is at the heart of a giant elliptical galaxy.*

## The spectrum of a quasar

A quasar's spectrum (below left) seems bafflingly complex. But it can be analysed into numerous lines superimposed on a continuum (below right). The continuum is emitted by the core; sharp lines come from relatively cool surrounding gas. Lines from fast-moving gas are strongly redshifted and blueshifted, and therefore broadened.

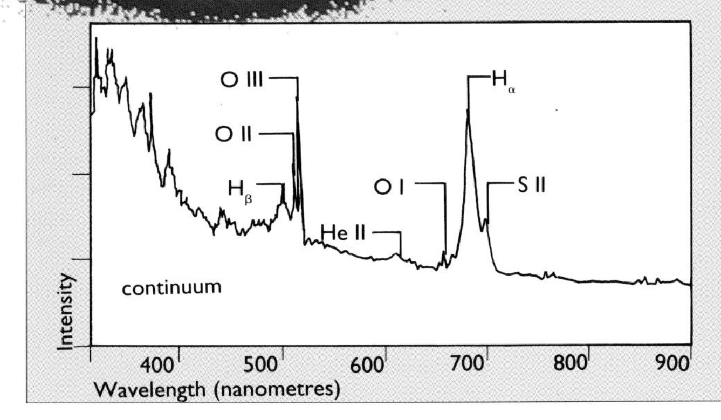


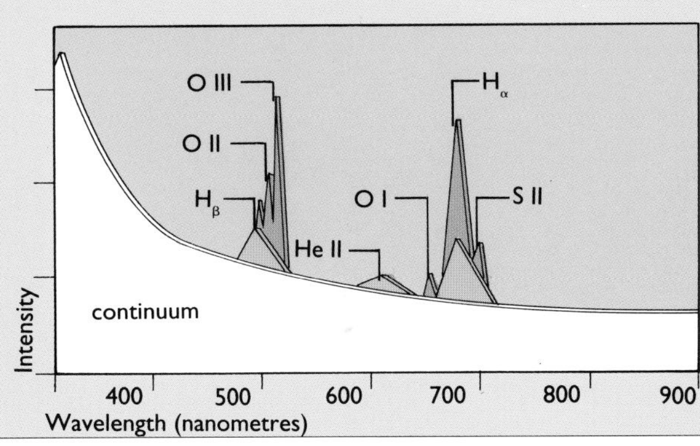


***Spectral lines are emitted*** *by atoms of particular elements, which have often lost some electrons. O III, for example, is doubly ionized oxygen, He II is singly ionized helium, and so on.*

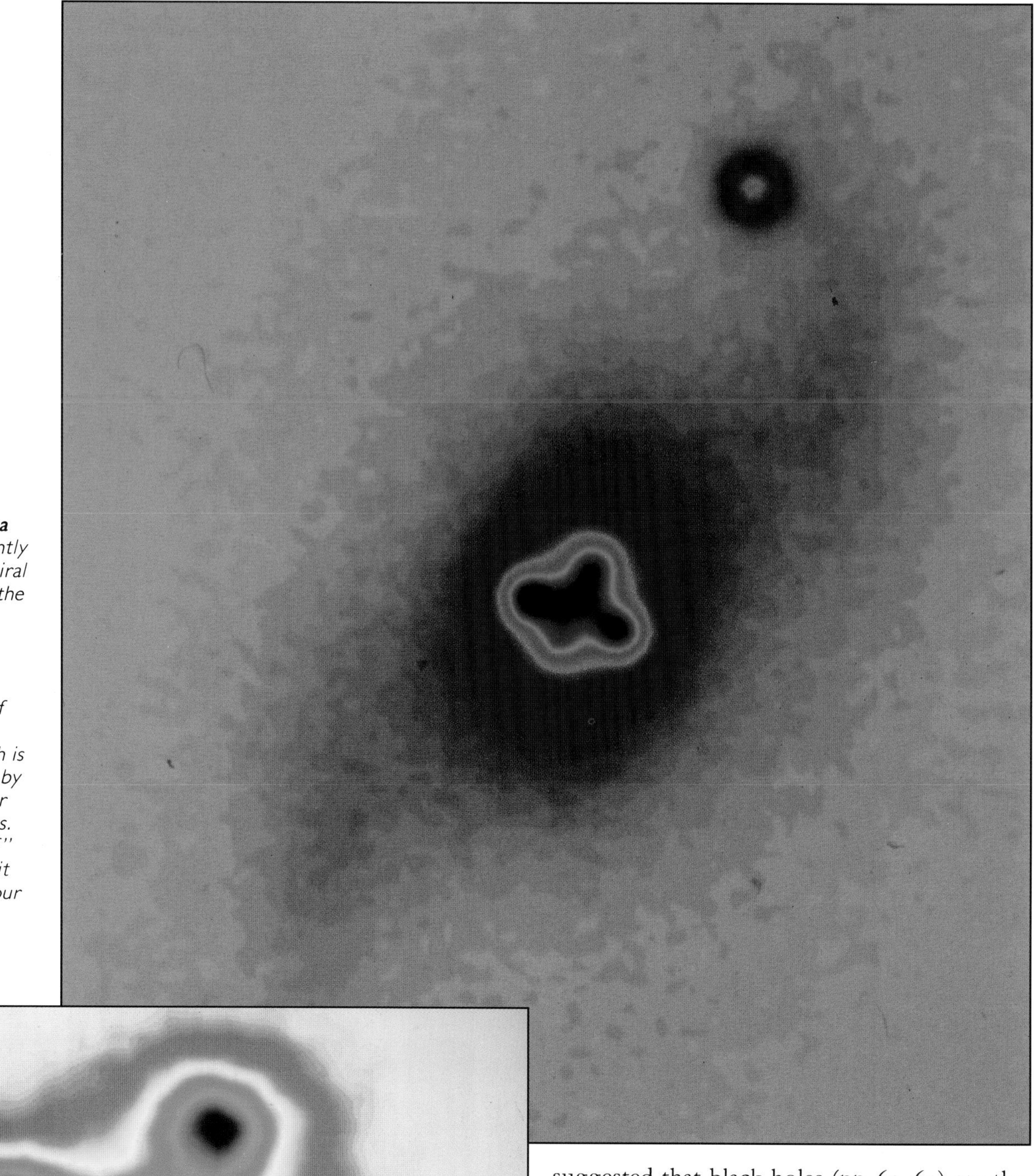

*__A cloverleaf image of a quasar__ is seen apparently superimposed on a spiral galaxy (right). In fact the galaxy is 500 million light-years away, whereas the quasar is billions of light-years distant. The cloverleaf effect is produced by "microlensing", which is gravitational focusing by individual stars, rather than by entire galaxies. When the "cloverleaf" is magnified (below), it proves to consist of four images of the quasar around the galaxy's bright core.*

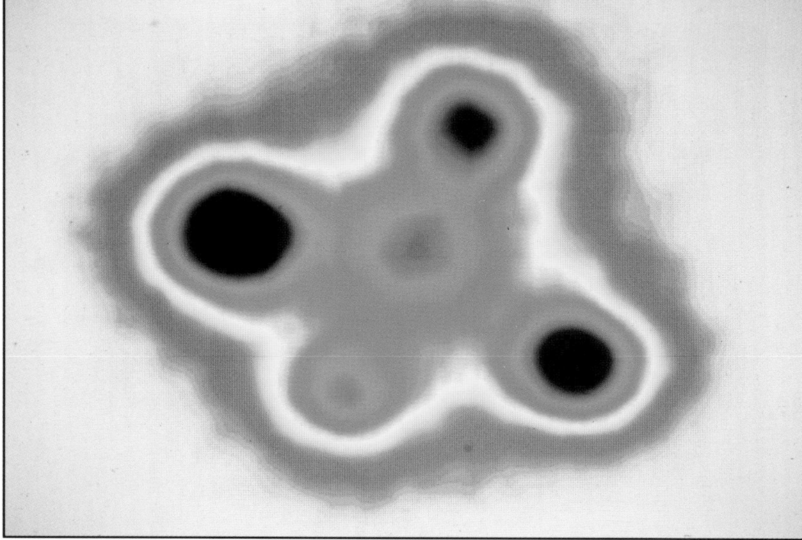

suggested that black holes (pp. 62–65) are the power sources for quasars. The amount of energy emitted when gas or other material falls into a black hole is a high proportion of that which would be produced by the complete annihilation of the matter, as given by Einstein's equation $E=mc^2$. This is about 1,000 times what the same amount of matter could yield if it fuelled nuclear reactions inside a star (pp. 82–83), and is of the right order of magnitude to account for a quasar's energy.

Such quantities of radiation would be

generated if the black hole at the centre of a quasar attracted material at relativistic speeds, producing intense synchrotron radiation (electro-magnetic waves emitted whenever charged particles are whirled round by magnetic fields). Much of the energy given out by quasars is indeed in the form of synchrotron radiation. What is more, strong "winds" of ionized particles would also blow outward. This is confirmed by observations showing that the central regions of quasars have been swept clear of dust. From every point of view black holes seem to be the cause of quasars.

Calculations based on the observational evidence indicate that the masses of black holes could span an enormous range. At the lower end of the scale, the holes might be the equivalent of no more than 100 million Suns, while at the top end they might be as massive as 1,000 billion Suns.

Because they are so bright, yet so remote, quasars are uniquely valuable as probes of the distant, early universe. That they are truly distant seems confirmed by additional observations, which show the gravitational deflection of their light by intervening galaxies.

From the general theory of relativity it can be calculated that the light from, say, a quasar or other distant source lying almost behind a nearer massive body should be deflected by the distortion of space-time. The image of the quasar should not only be shifted but should in general also be broken up into several images, or spread out into arcs or an "Einstein ring" apparently centred on the nearer body.

Such gravitational lensing was first detected by the Nuffield Radio Astronomy Observatory at Jodrell Bank, in what appeared to be a double source. In 1979, some years after the first radio observations, the pair was studied optically. It was found that the objects were quasars, and that they had identical spectra. This ruled out the possibility that they consisted of two clouds thrown out by an active galaxy lying between them. The two images were clearly of the same object, some 10 billion light-years from Earth. The cause of this double image is a giant elliptical galaxy that has been detected by radio and photographed. It is less than half as far away as the quasar whose light it affects.

A number of Einstein rings and arcs have been detected. Two arcs associated with the quasar MG 1654+1346 in Hercules are a fine example of gravitational lensing at radio wavelengths. The quasar possesses two lobes, in addition to the emission from its centre. The southern lobe is directly in line with a large intervening elliptical galaxy, which splits the image of the lobe into two arcs. The galaxy's mass can be calculated from the lensing it causes: it is equal to 89.8 billion solar masses, which makes it large as elliptical galaxies go.

Light can be focused by the gravitational fields of single stars, an effect known as "microlensing". There is little chance of observing this phenomenon in the case of stars in our own galaxy, but it has been detected in some external galaxies. Some astronomers claim that increased numbers of quasars can be seen close to the lines of sight to nearby galaxies; they believe this is due to the focusing, and hence brightening, of their light by stars in the galaxies.

Not all double quasars are the product of gravitational lensing. Some are true binary quasars. In 1989 it was reported that quasar PHL 1222 in the southern sky is double. Its two components are very close – no more than 100,000 light-years apart – but are not identical. The system is calculated to be 12 billion light-years away, so what we are probably observing here is a pair of quasars being born.

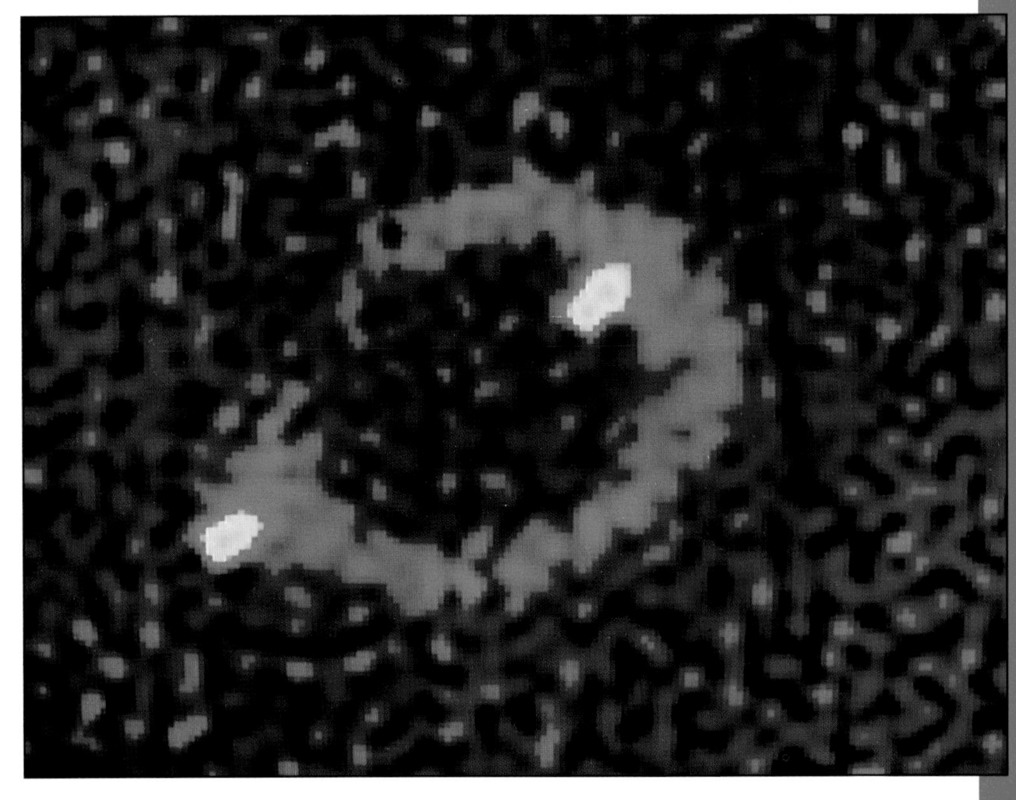

***A possible Einstein ring*** *is seen at radio wavelengths in the constellation of Leo. It may be produced by an elliptical galaxy (invisible here) lying almost exactly in the line of sight to a remote quasar. If so, the two bright patches are images of the quasar's core, while the ring is an image of a jet extending from the quasar.*

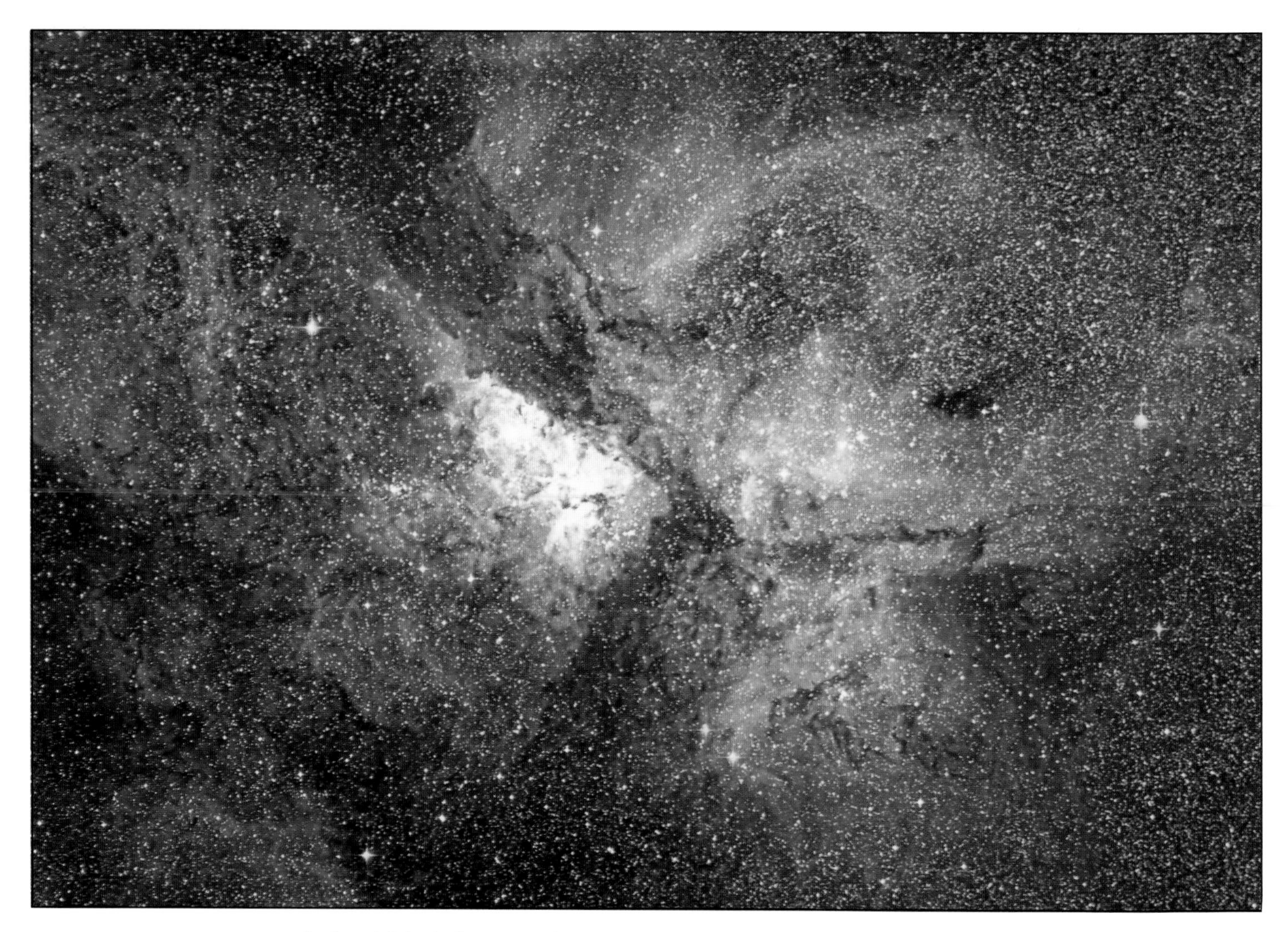

***The magnificent nebula in the constellation Carina*** *(the Keel of the Ship), in the southern skies. Dark lanes of dust cross the bright regions of gaseous matter. At the centre of the nebula lies a denser knot of gas and dust, swathing the remarkable variable star Eta Carinae. In the 1840s it suddenly brightened: having been a second-magnitude star, it became the second-brightest star in the sky. In a few decades it faded, until now it is invisible to the naked eye. It may have been 200 times as massive as the Sun when it was born. It has shed much of its matter into space, but it still ranks as a supermassive star.*

nebulae. This is because they cannot survive unless protected from the damaging effects of ultraviolet and other high-energy radiation, which would break up such complex molecules into their constituent atoms.

Molecules of water and hydroxyl are also present. In the Orion Nebula, M42, they are part of a vast molecular cloud, contracting under the influence of gravity. The nebula possesses two strong infrared sources. Almost certainly one of these is a star in the process of being born. The evidence for this comes from the water and hydroxyl molecules, which act as "masers". (This name is derived from the initials of the phrase "*m*icrowave *a*mplification by *s*timulated *e*mission of *r*adiation".) Radio waves emitted by the nascent star stimulate the molecules to radiate, providing a clue to its existence.

The material in a nebula is extraordinarily thinly spread. Generally there are no more than 10,000 atoms per cubic centimetre, even at the centre – billions of times less than the density of a puff of smoke. Some gas clouds are very much less dense even than this. As far as the dust is concerned, there is on average one grain per 100,000 cubic metres – the volume of an average concert hall. This, too, represents a very small quantity of matter, for typical dust grains are no larger than one-thousandth of a millimetre, and may be only one-tenth of this size. They are composed of carbon or, less frequently, silicon. However, they are sometimes coated with ice or with large, complex molecules.

It may seem astounding that such tenuous clouds can be observed at all, until one realizes their vast extent. As an example, the optically visible region of the Orion Nebula stretches 20 light-years across space. It would take over 10,000 solar systems in a line to span it. Such vastness makes it easier to understand how the incredibly thin dust of a nebula can dim and even block out the light of the stars beyond it.

Like many other objects in the universe, nebulae are in constant motion. They share in the general rotation of the Galaxy. And when new stars form within them, the newly emitted

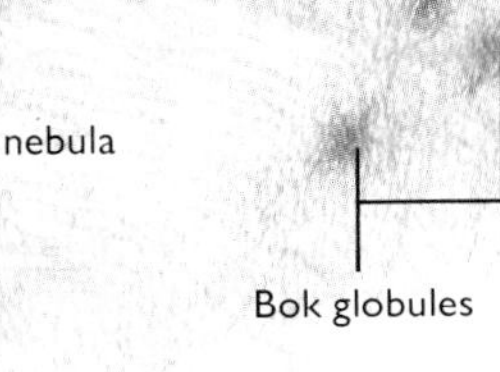


***A nebula that is sufficiently dense*** *is destined to form stars (left). It begins to contract, and concentrated regions called Bok globules form in places.*



***The collapse of the gas*** *heats the cores of the globules, forming protostars (left). These become true stars (below) when nuclear reactions are ignited. Shock waves spreading from the young stars trigger further starbirth.*

radiation disturbs the cloud, blowing away the gas and dust. It is estimated that the presently visible regions of M42 in Orion will be dispersed within the next 10,000 years.

Yet the disappearance of this material will not leave a gap, because new clouds of gas and dust will form in the Orion spiral arm, in which the nebula lies. This matter is at present spread out along the spiral arm, but it will gather into nebulae by the processes mentioned in the discussion of the formation of spiral galaxies (pp. 54–55). Thus new nebulae will appear, some of them lit up by stars that are still unborn.

The presence of nebulae and diffuse interstellar matter poses problems for the observational astronomer. They can dim the stars and even hide the most distant objects from view. The small size of the dust particles means that they scatter the shorter optical wavelengths and so produce an apparent reddening of distant objects. This makes it difficult to assess the true brightness and colour of distant stars and galaxies.

Astronomers can allow for the drop in intensity with distance, provided that they have some standard by which to assess the diminution. Some galaxies provide such a standard because they are of known brightness. The true brightness of certain stars can also be determined; for example, the brightness of a Cepheid variable can be determined with considerable precision from its period of variation (pp. 86–87). Such compensation regularly has to be applied when observing across interstellar space.

However, as was mentioned above, neutral hydrogen radiates at a radio wavelength of 21 centimetres. This allows radio astronomers to penetrate some of the dark clouds and to "see" farther into space in certain regions than is possible optically. This is one of the reasons why knowledge of the universe has expanded so greatly in recent decades.

***The Cone Nebula*** *(below), in Monoceros the Unicorn, is the scene of starbirth today. There are many young, bright T Tauri stars nearby, which vary erratically in brightness because they are still surrounded by swirling remnants of the gas and dust from which they formed.*

# STARBIRTH

## • *Lighting the nuclear fire*

If you look very closely at photographs of bright nebulae, you will find that many of them are pitted with tiny dark blobs. These are called Bok globules, after Bart J. Bok, the American astronomer who first observed them. The infrared and radio waves coming from them show that they are the birthplaces of stars.

Stars could not form in the very early universe, when matter was still at temperatures of hundreds of thousands of degrees. However, with the continuing expansion of the universe the hydrogen gas cooled. Some two billion years after the Big Bang the protogalaxies formed (pp. 38–39). Then local condensations of gas appeared within the protogalaxies, forming nebulae. In some places the density of gas in these reached billions of molecules per cubic metre. This density, though far thinner than the best laboratory vacuum, enabled gravity to pull the material into ever more concentrated conglomerations.

As this happened the central regions of each concentration heated up, rather as air heats up as it is pumped into a tyre. The temperature in each knot of matter rose until the molecules were broken up into individual atoms, and then still further, until the atoms were ionized, losing their outer electrons. Pressure rose as fresh material was pulled in, pressing down on the central regions.

Although nuclear reactions had not yet begun, this protostar was already producing copious amounts of energy. But no visible light could penetrate the gas and dust that enveloped it. Only infrared radiation could escape.

In due course the situation at the heart of the protostar became critical. The density had increased billions of times, and the temperature had risen to 10 million K, or even more. The positively charged hydrogen nuclei, now stripped of all their electrons, were pushed so close together that they began to collide, overcoming the strong repulsion of their electrical charges. Hydrogen nuclei began to fuse to form helium nuclei. The protostar became a true star.

Each helium nucleus had slightly less mass than the hydrogen nuclei that formed it. The mass that vanished was converted into huge amounts of energy, according to the relationship $E = mc^2$ discovered by Albert Einstein (pp. 14–15). Each gram of hydrogen that reacted produced as much energy as 600 million single-bar electric fires radiate every second. The temperature at the centre of the protostar rose further as these vast amounts of energy tried to escape. The new star would have exploded were it not for the huge weight of material of which it was composed.

As radiation made its way outward, it set up convection in the outer layers of the star. The radiation heated the deep-lying gas, which rose, cooled at the surface, and then sank, to begin the cycle all over again. (The same sort of movement can be seen in a pan of milk heated on a stove.) The light and heat from the young star blew the clouds of gas and dust away and it became visible to the universe beyond.

The same process of starbirth is going on today. Infrared astronomers think they can glimpse it in parts of some nebulae, such as the nebula in Orion. But not every concentration of gas and dust forms into stars. If there is insufficient material in a gas cloud, its gravitation will be too weak to make it condense sufficiently, and its temperature will not rise to the critical level at which nuclear fusion begins. The "star" that forms in this way will merely be a warm body, detectable mainly because of its infrared radiation.

Many such objects seem to exist in our own region of the Galaxy, particularly within the spiral arms. They have become known as brown dwarfs. Some failed stars are so small they are only about the size of the planet Jupiter, and accordingly are called jupiters.

Fully developed stars bear signs of the epoch in which they were born. The earliest stars – the first to be formed after the Big Bang – condensed from the primal material: hydrogen with a proportion of helium. Later generations of stars formed from material that had been mixed with the remnants of exploded stars. These included elements with nuclei heavier than that of helium, for these had been synthesized in the cores of the stars before they exploded (pp. 82–83). The Sun contains such elements, which shows that it is not a first-generation star. Furthermore, our theories of stellar life cycles tell us that the Sun is only five billion years old – making its origin more recent than the first period of star formation.

***A cloud of interstellar matter** begins to shrink, its central portions collapsing faster than its outer parts. The core warms up, while the cloud begins to rotate.*

***After hundreds of thousands of years of contraction,** the cloud is rotating faster and its globular shape has begun to flatten. Its central regions have formed a hot protostar, emitting several times as much radiation as today's Sun. But only infrared radiation can escape from the surrounding cloud.*

***The bulk of the circling gas and dust** has become concentrated into an accretion disc. Matter continues to fall on to the protostar, which is so hot that it violently ejects matter. This escapes at the poles, where there is least obstructing matter. This outflow sweeps away much of the gas and dust surrounding the protostar.*

nebula

***The sites of future stars are visible*** *(right) as dark condensations against a background of glowing red nebulosity, the Eagle Nebula. The bright stars, belonging to the cluster M16, were born from the gas cloud only about two million years ago. This stellar nursery lies in the equatorial constellation of Serpens, the Serpent.*

warm core

outflow

accretion disc

protostar

brown dwarf

***A protostar of small*** *mass – less than about 8 percent of the Sun's – becomes a brown dwarf. It is heated by gravitational energy released by its continuing slow shrinkage. How many of these faint "failed stars" exist in the Galaxy is unknown.*

nebulosity

main-sequence star

***In protostars of sufficient mass*** *– more than 8 percent of the Sun's – hydrogen begins to be converted into helium by nuclear reactions. The star is now on the main sequence. It is at first surrounded by residual gas and dust. Eventually this either disperses or forms a planetary system.*

# STELLAR COMPANIONS

## • *Star clusters and multiple stars*

***A globular cluster** beyond our galaxy (left). This object orbits the centre of the Large Magellanic Cloud, a companion galaxy to the Milky Way. Globular clusters are present in all galaxies. Although they contain hundreds of thousands of stars, these are widely spaced even near the centre of the cluster – but by light-months rather than the light-years that separate stars in our own stellar neighbourhood.*

Stars form not only singly but also in groups; this occurs when a number of protostars are born close together. They form a gravitationally linked cluster, with a collective motion on which their individual motions are superimposed.

Clusters with a few hundred members, loosely grouped, are termed open clusters. The conspicuous Pleiades, in the northern constellation Taurus, are of this kind. The nearest open cluster, however, is the group called the Hyades, also in Taurus, only some 130 light-years away. Its age is about two billion years.

In the southern skies the most spectacular open cluster is the Jewel Box, NGC 4755, some 7,800 light-years away. It derives its name from the fact that it consists of a group of about 50 bright young blue stars together with a bright red star, Kappa Crucis.

Even looser groupings of stars occur, but they are difficult to identify. They have only recently been recognized by astronomers and are called stellar associations.

On the other hand, it may happen that thousands or hundreds of thousands of stars form together – not enough to form an elliptical galaxy, but enough to make a roughly spherical grouping known as a globular cluster. Such clusters are always old, and it seems that they formed around the same time as the galaxies with which they are associated.

Our own galaxy has a halo of some 125 globular clusters evenly distributed around the galactic centre, and up to 60,000 light-years from the galactic plane. The most spectacular is Omega Centauri, in the southern skies, only 16,500 light-years away. Through a large telescope it is seen to cover more than a degree of arc – twice the apparent size of the full Moon. It is probably the most massive globular cluster in the Galaxy, spreading over some 65 light-years. It is not spherical but ellipsoidal, apparently because of its rapid rotation.

Single stars like our Sun account for only about half the stars we see: the other half are binary or multiple stars. Some apparent doubles do not consist of genuinely associated stars, but merely of ones that happen to lie in the same line of sight from the Earth. For example Mizar, the middle star of the handle of the Plough, or Big Dipper, has a fainter "companion", Alcor. Yet Mizar is some 58 light-years from the Earth, while Alcor is 81 light-years away, and the two are not physically connected, although each is itself a double.

However, most multiple star systems are genuinely associated stars. About 30 percent have more than two members; about one in nine have four or even more components. In the overwhelming majority of such systems, the components were formed at the same time, but in a few cases a massive star has captured a companion.

All the members of a multiple star system orbit around their common centre of gravity, and have a complex motion. The members of some systems like this cannot be separated in a telescope, but their combined spectrum shows distinctive behaviour. While one star is moving toward the Earth, its light will be blueshifted (pp. 46–47). At the same time its companion will be moving away from the Earth, and its light will be redshifted. The result will be that lines in the combined spectrum will periodically split and recombine, or at least become broader and narrower. Thus the multiple nature of the light source can be recognized.

***The brightest members of the Pleiades** in Taurus make a fine naked-eye cluster and form a spectacular grouping in a telescope. There may be 500 stars in all.*

*Cocooned in clouds of cold gas and dust, the cluster spreads across space for about 30 light-years, and is at a distance of 400 light-years. The cluster is young, having an age of no more than 50 million years, and its stars are bluish. The nebulosity shines by reflected starlight.*

# THE CHEMISTRY OF THE STARS

## • *Sources of stellar energy*

***In the nuclear powerhouse at the heart of a star*** *like the Sun, hydrogen nuclei are smashed together and welded into heavier nuclei. At the same time energy in the form of lighter particles and high-energy gamma rays is released. The hydrogen nucleus is a single particle, the proton, and the dominant process, shown here, is called the proton-proton cycle.*

*Protons are constantly colliding to form nuclei of deuterium, called deuterons. These in due course collide with protons to form nuclei of helium 3. The collision of two helium 3 nuclei forms one nucleus of helium 4, while releasing two protons. This ends the chain.*

The fundamental power source of all stars is the fusion of hydrogen into helium. But this conversion can be achieved by any of a variety of processes. However, in stars of the Sun's size it proceeds almost wholly by one process, the hydrogen, or proton–proton, cycle.

The sequence of events begins when two hydrogen nuclei combine. The nucleus of a hydrogen atom is the simplest of all nuclei. It consists of a single positively charged particle, the proton. Since protons have like electrical charges, they repel each other. Two of them can be forced together by the high temperatures and pressures of a stellar interior, but they can stay together only if one of them loses its charge. It can do this through reactions that turn it into a neutron, which has slightly more mass than a proton but is electrically neutral. The combination of proton and neutron thus formed is called a deuteron.

In the process a positron (positively charged electron) and a neutrino are given off. The neutrino has no electrical charge and no rest mass – or such a tiny mass that it has not yet been measured – and it has a tremendous penetrating power. It carries away a substantial amount of energy.

At Earthlike temperatures and pressures, a deuteron would soon pair up with a single electron, which would balance the deuteron's single proton. The combination would be an atom of deuterium. Deuterium is an isotope of hydrogen: since its atom has only one electron it has the chemical properties of a hydrogen atom, but roughly twice the weight.

In due course, the deuteron will collide with another proton, again giving out energy – this time in the form of a gamma-ray photon. The result will be a nucleus of two protons and one neutron – an isotope of helium. Two of these nuclei later collide and produce a nucleus of ordinary helium (two protons and two neutrons), together with two protons.

So the net result of the hydrogen cycle is the building up of a nucleus of helium from four protons. Part of the energy given out in the successive steps of the reaction goes into heating the star; part escapes from the star. Most of the neutrinos' energy is lost, because these ghostlike particles can slip through thousands of kilometres of matter as if it did not exist. The energy carried by positrons, on the other hand, is soon converted into radiation when the positrons are annihilated in collisions with electrons. The radiation is absorbed by the star, thus maintaining its temperature.

In stars much more massive than the Sun the temperature of the central regions is above 15 million degrees Celsius. At these temperatures heavier nuclei move at the speeds necessary to

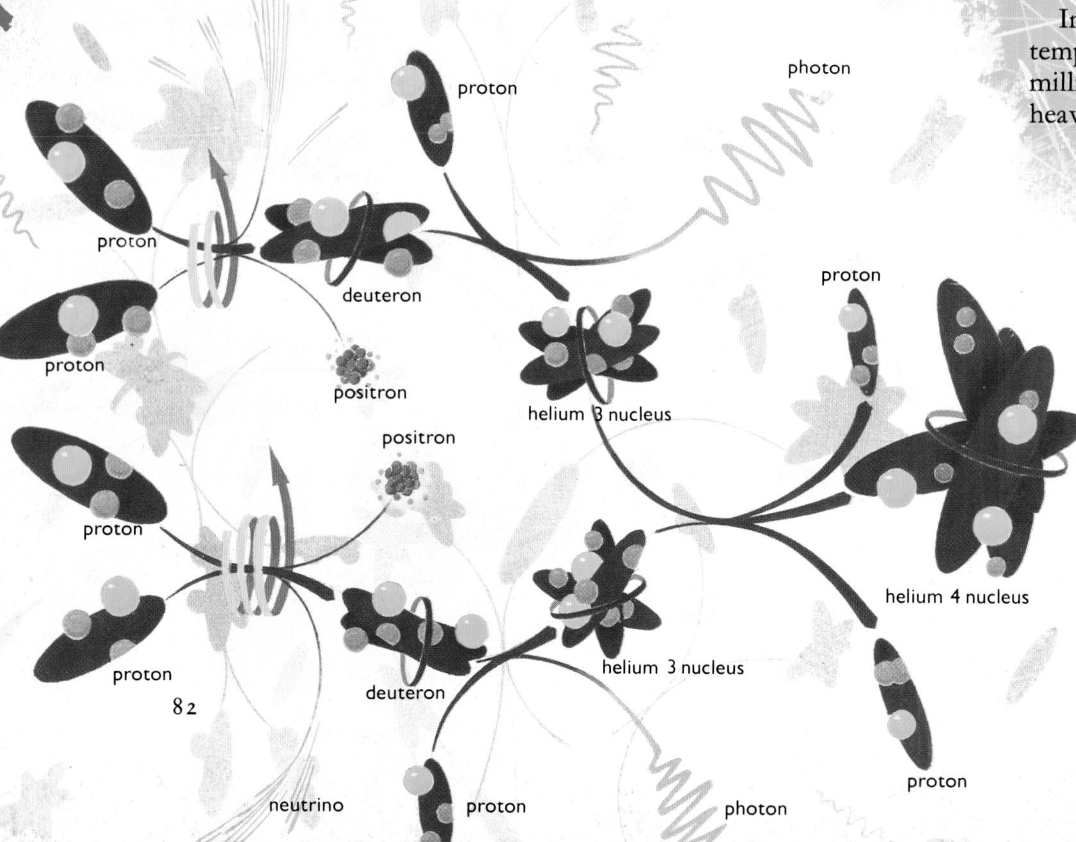

enable them to react. Carbon nuclei, in particular, take part in a sequence of reactions called the carbon cycle. The carbon nuclei emerge at the end of the carbon cycle unchanged – in the chemist's terminology they are catalysts:

1: A carbon nucleus (six protons and six neutrons) combines with a proton, forming a nitrogen nucleus (seven protons and six neutrons) and giving out a gamma-ray photon. This form of nitrogen is unstable, since there are not enough neutrons present; in all long-lived atomic nuclei the number of neutrons equals or exceeds the number of protons. Thus the nitrogen emits a positron (plus a neutrino), so that one of its protons turns into a neutron. The nucleus is now a carbon nucleus, although it has seven neutrons instead of six.

2: When this carbon nucleus encounters another proton, it forms a nitrogen nucleus again – more stable this time, since its seven protons are combined with seven neutrons. Another gamma-ray photon is given out as the nitrogen nucleus is formed.

3: When this nitrogen meets another proton, they combine to form an unstable isotope of oxygen, with eight protons and seven neutrons (eight neutrons are present in ordinary oxygen). At the same time another gamma is emitted. The oxygen soon "decays" by sending out a positron (with, for good measure, a neutrino as well), changing to yet another form of nitrogen (seven protons, eight neutrons).

4: The nitrogen nucleus now meets another proton, and they form a nucleus of helium (two protons, two neutrons) and one of carbon (six protons, six neutrons), identical to that which started the cycle.

The net result of the carbon cycle is the production of helium nuclei – one for each of the four protons that are used up – and the release of energy in the form of neutrinos, gamma-ray photons and positrons.

In massive stars the carbon cycle is faster than the proton-proton cycle at converting hydrogen into helium. It is also an effective producer of energy, since during each cycle three gamma-ray photons are brought into being. The carbon cycle is one of the main reasons for the much greater energy production of massive stars.

In giant red stars yet another process, the triple-alpha reaction, occurs. The process is so called because three helium nuclei are involved, and alpha-particle is another name for the helium nucleus. To begin with, two helium nuclei combine to make an isotope of the element beryllium (four protons, four neutrons), with the emission of a gamma-ray photon. Then another helium nucleus fuses with the beryllium; this produces a stable isotope of carbon, with six protons and six neutrons. Again, a gamma-ray photon is emitted. So the net result of this reaction is the conversion of helium into carbon.

There are still further reactions that lead to the synthesis of heavier chemical elements. For instance, two carbon nuclei can fuse to produce a magnesium nucleus. Under immensely hot conditions such heavier nuclei can themselves combine to make elements as heavy as iron. Such reactions occur only in the most massive stars, as they approach their deaths.

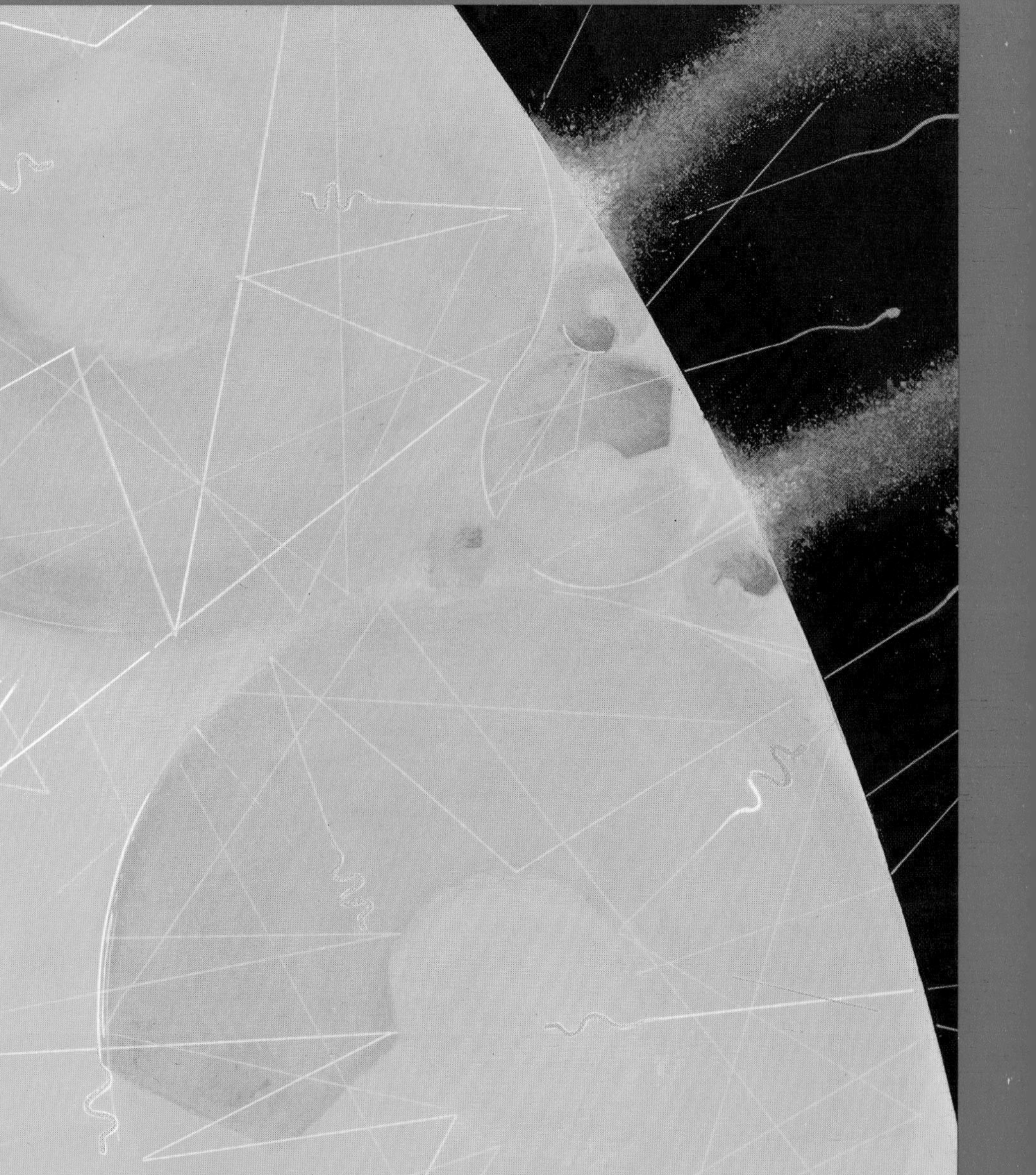

***Energy travels outward from the Sun's energy-producing core*** *in the form of high-energy radiation. Photons travel a zigzag path as they are absorbed and re-emitted by nuclei, taking thousands of years to battle their way to the surface. In the outermost layers, however, convection becomes more important in transferring heat. Here hot gas, being less dense, rises to the surface of the Sun. There it radiates light and heat, cools, becomes denser, and sinks. The gases circulate in convection cells (darker arrows above).*

# LIVES OF THE STARS

## • *Paths of stellar evolution*

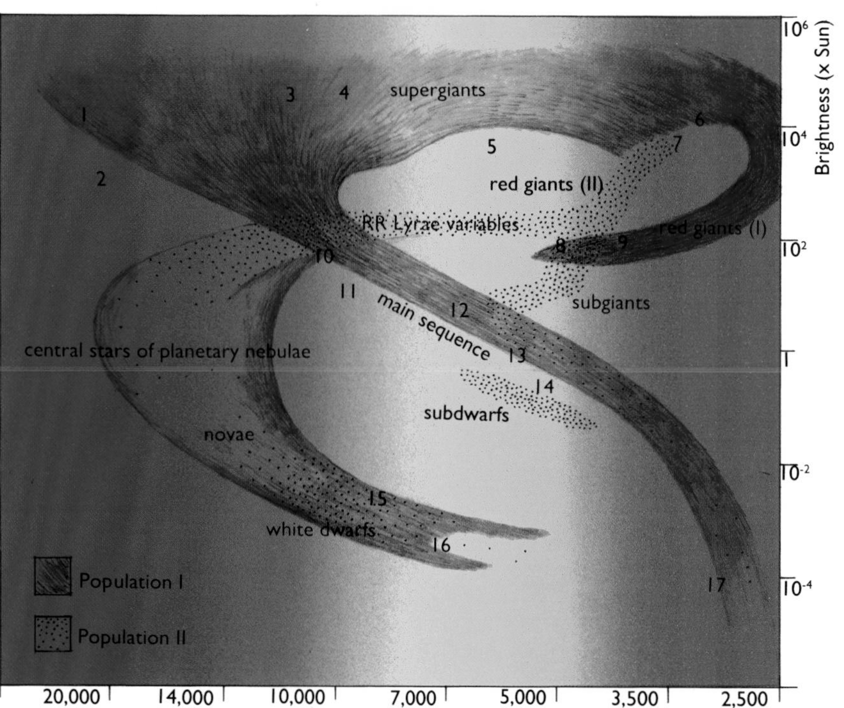


***In the Hertzsprung-Russell diagram*** *luminosity is plotted against temperature. Most stars fall in the central band called the main sequence. Population II stars are relatively old, and many have evolved away from the main sequence. Only the most massive and fastest-evolving of the younger Population I stars have yet done so.*

*1 δ Orionis*
*2 Spica*
*3 Rigel*
*4 Deneb*
*5 Polaris*
*6 Betelgeuse*
*7 Antares*
*8 Arcturus*
*9 Aldebaran*
*10 Vega*
*11 Sirius A*
*12 Procyon A*
*13 The Sun*
*14 τ Ceti*
*15 Sirius B*
*16 Procyon B*
*17 Proxima Centauri*

Nuclear reactions determine the complex life cycles of the stars. An essential tool in describing stellar evolution is called the Hertzsprung-Russell or H-R diagram. It was first devised during the second decade of this century by Ejnar Hertzsprung and Henry Norris Russell.

An H-R diagram carries a scale of brightness – the true brightness, or absolute magnitude, of a star – running upward, from faintest at the bottom to brightest at the top. Horizontally it carries a scale of temperature, or of types of spectra that correspond to temperatures. These classes bear the traditional labels O, B, A, F, G, K, M. (All English-speaking astronomers remember this sequence with the mnemonic "Oh Be A Fine Girl, Kiss Me!").

At the high-temperature end O-type stars are blue and have temperatures in excess of 25,000 K. B-type stars are a little cooler, having temperatures of 11,000–25,000 K, but are still bluish, and so on. The Sun is a G-type star, with a surface temperature of about 6,000 K, described by astronomers as yellowish. M-type stars are red and have temperatures below about 3,500 K.

When the brightnesses and temperatures of stars are plotted on the H-R diagram, it is found that some regions are occupied and others are not. The majority of stars lie on a band called the main sequence, running from bottom right to top left. It has become clear that the pattern of the H-R diagram is related to the evolution of stars: it is a plan of the way they spend their lives.

When a star has just been born in a Bok globule, and once nuclear reactions begin in its core, it appears close to the main sequence. Precisely where it enters the sequence depends on its mass.

Very small stars, of about a quarter the mass of the Sun, enter as M-type red dwarfs. The more massive Sun entered farther up, while even more massive stars join still farther along the sequence. All stars spend the greater parts of their lives on the main sequence, their positions changing very little, while the hydrogen in their central regions lasts.

A large central core of nonreactive helium "ash" gradually builds up. This core contracts and its temperature rises, but a surrounding shell of hydrogen still "burns". Now the star moves off the main sequence.

The length of life of a star depends on its mass, and thus on its position on the main sequence. A dim red dwarf evolves so slowly that it will take 200 billion years before it leaves the main sequence; the Sun will do this after about 10 billion years, whereas a star of five solar masses takes only 70 million years.

As a sunlike star evolves away from the main sequence it expands to something like 50 times its previous size. It becomes cooler and redder and therefore moves to the right of the H-R diagram. But because of its huge increase in size it also grows brighter, and so moves upward in the diagram, becoming a red giant.

But the core shrinks and its temperature rises still farther. When it reaches 100 million K, the helium begins to take part in nuclear reactions. The outer regions contract and the star ceases to be a red giant.

By this time the core of the star is mainly composed of carbon and oxygen which have been formed by the helium-burning (pp. 82–83). It now reaches the final stages of its life. At first the energy output becomes less and the star contracts. However, the interior of the star expands again and for a short time the star becomes a red giant once more, with helium as well as hydrogen burning around the core.

Suddenly there is a change: the energy generated around the core blows off the outer envelope and for a time the star is surrounded by a shell of gas. It has become a planetary nebula – so called because in a telescope it appears as a hazy disc, rather like that of a planet. After this the star contracts and all that remains is a superdense core, still burning nuclear fuel in its outer regions. The star ends its days as a white dwarf, cooling and fading (pp. 88–89).

After leaving the main sequence a Sunlike star will spend some 10 billion years evolving toward the red giant stage. More massive stars live their lives faster because their nuclear reactions proceed at a higher rate. Stars of five times the Sun's mass will take only 70 million years to evolve to the red giant stage; those 15 times as massive will last only 10 million years.

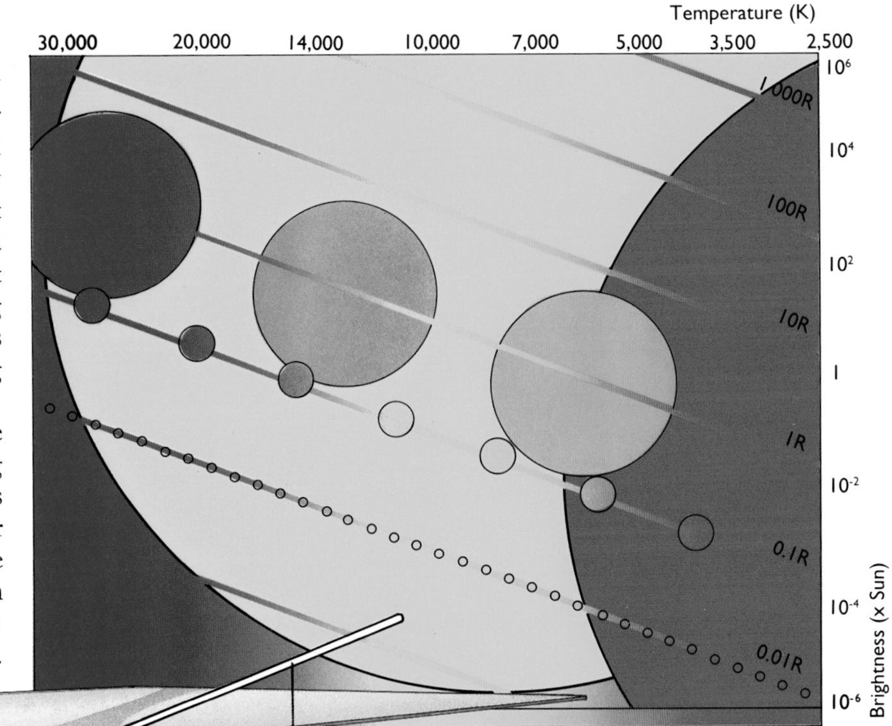


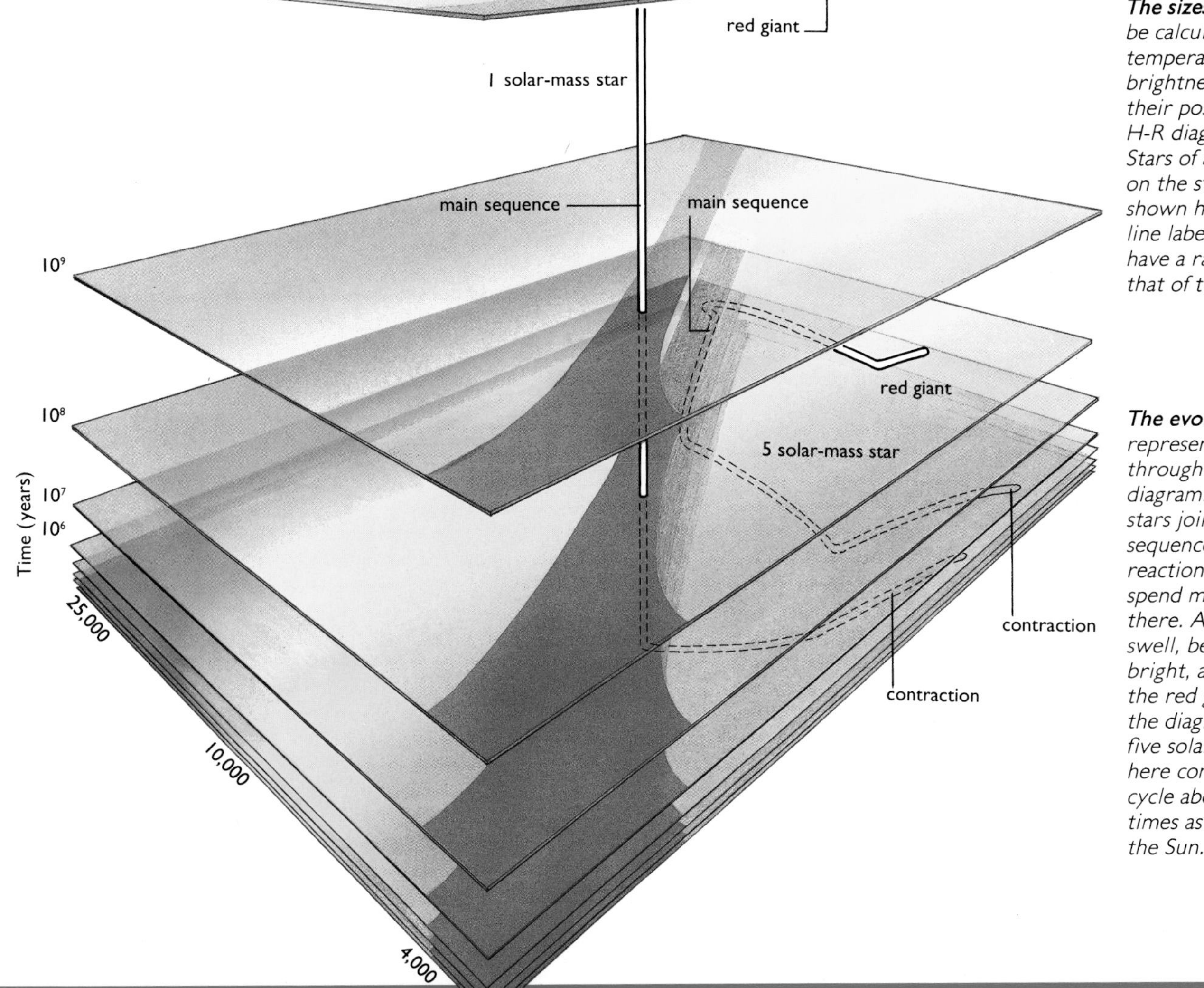


***The sizes of the stars*** *can be calculated from their temperature and brightness, as shown by their positions on the H-R diagram (above). Stars of a given size lie on the straight lines shown here; thus on the line labelled 10R they have a radius 10 times that of the Sun.*

***The evolution of a star*** *is represented by a track through the H-R diagram. Newly born stars join the main sequence as nuclear reactions begin, and spend most of their lives there. As they age they swell, become hot and bright, and move into the red giant region of the diagram. The star of five solar masses shown here completes its life cycle about a hundred times as fast as a star like the Sun.*

# VARIABLE STARS

## • *Shining with inconstant light*

Stars vary enormously in brightness in the course of their lives. But these changes occur over millions and even billions of years. Some stars, however, lead a less sedate existence in which their radiation output varies far more rapidly. Study of such variable stars brings new evidence about their internal processes.

One type of variable star does not truly vary at all. Such are binary stars with orbits that happen to be edge-on to us, so that we see each component eclipsing the other in turn. But in most variables there are intrinsic changes in the radiation emitted. This is the case with the irregular and eruptive variables. These are generally stars that are very young and still associated with the nebulosity from which they were born. They vary in brightness because from time to time huge "flares" explode on them, blowing masses of hot gas into space.

Then there are the periodic variables, which brighten and dim in a regular fashion. The surfaces of some are not uniformly hot but have brighter and dimmer spots. As a result their apparent brightness varies as they rotate, with periods from about 12 hours to a few hundred days. These are mainly hot stars, on or close to the main sequence. Their variations in surface temperature are probably caused partly by uneven mixing of the gases in their outer regions, and partly by an uneven magnetic field near the surface.

*One type of variable star has a surface of uneven temperature, and hence brightness. The changes in apparent brightness are solely due to the fact that the star presents different parts of its surface to us during each rotation.*

A quite different type of periodic variable is named after the star R Coronae Borealis; these are stars that suddenly dim by a factor of 10,000 times and then recover. They are old red supergiants, rich in carbon that periodically forms clouds of "soot" around the star.

A very important class of periodic variables comprises the Cepheids, named after the prototype, Delta Cephei. The outer gaseous envelopes of these stars pulsate outward and inward. This is caused by a complex process in which atoms in a layer below the star's atmosphere first lose some of their electrons and then, as the star contracts, lose even more, so that the layer becomes opaque to radiation from the interior. Pressure builds up, the star expands and cools, ionization is reduced, and radiation escapes. The cycle is then repeated.

A Cepheid's brightness varies by 10–20 percent of its maximum brightness, and this maximum is related to the period of the variations: the brighter the star, the longer its period. Most Cepheids are of Type I, massive young yellow supergiants. The pulsation periods of most of these range from about one day to about 50 days.

Type II Cepheids – sometimes called W Virginis variables – are old stars. They are found in globular clusters and toward the central regions of the Galaxy. The range of their variation periods is a little more restricted than those of Type I Cepheids, and they are some six times fainter.

Once the period (and the type) of a Cepheid is known, its true brightness can be worked out. By comparing this with the apparent brightness it is possible to calculate the star's distance (pp. 46–47).

There are other kinds of pulsating variables, among them the RR Lyrae stars, whose periods range from hours to about 1½ days. They are all old stars, originally deficient in carbon and oxygen, which later turn into Type II Cepheids. They are of about the same absolute

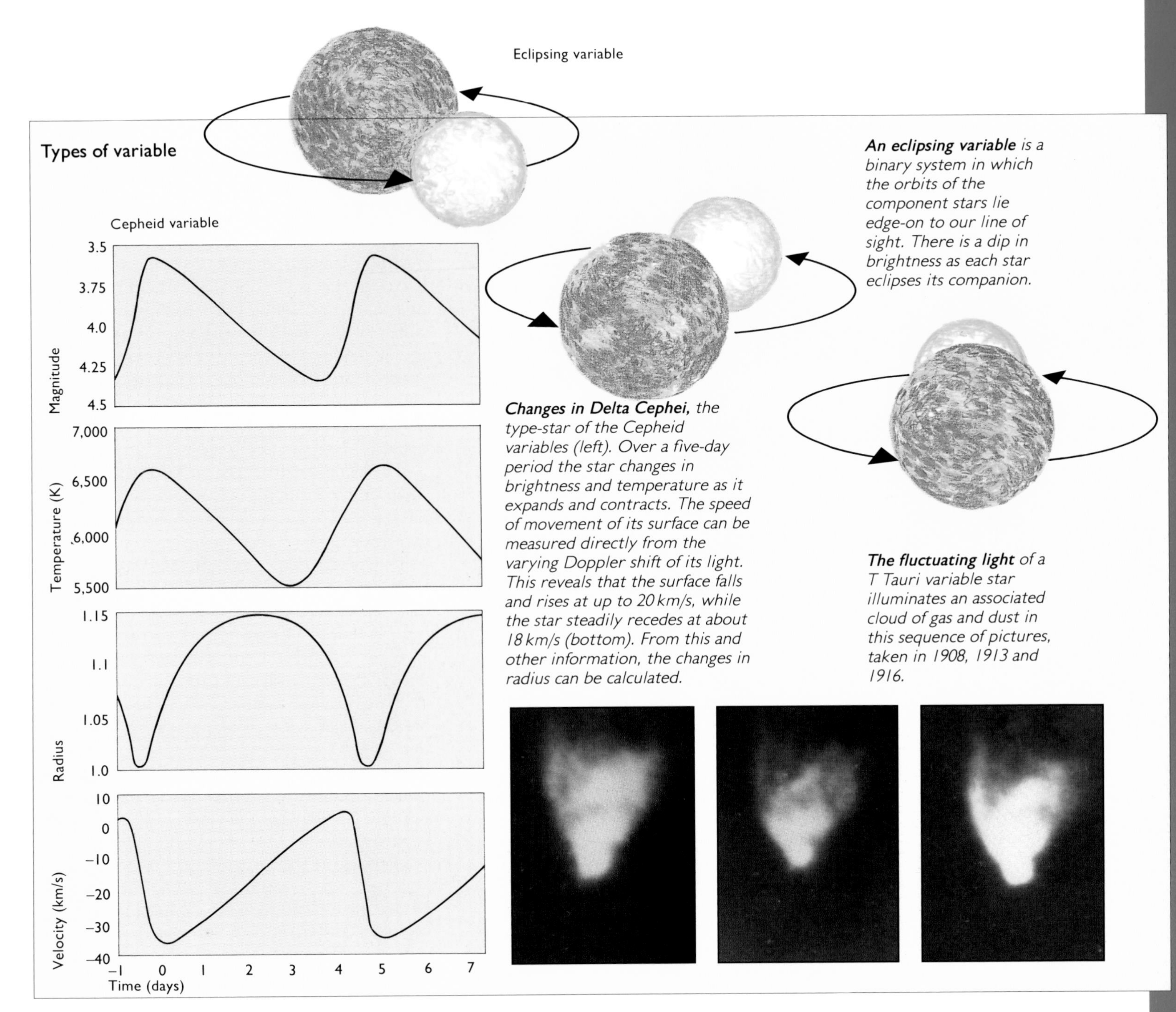


***An eclipsing variable*** *is a binary system in which the orbits of the component stars lie edge-on to our line of sight. There is a dip in brightness as each star eclipses its companion.*

***Changes in Delta Cephei,*** *the type-star of the Cepheid variables (left). Over a five-day period the star changes in brightness and temperature as it expands and contracts. The speed of movement of its surface can be measured directly from the varying Doppler shift of its light. This reveals that the surface falls and rises at up to 20 km/s, while the star steadily recedes at about 18 km/s (bottom). From this and other information, the changes in radius can be calculated.*

***The fluctuating light*** *of a T Tauri variable star illuminates an associated cloud of gas and dust in this sequence of pictures, taken in 1908, 1913 and 1916.*

brightness (about 100 times that of the Sun), which makes it easy to determine their distances from their apparent brightnesses.

Novae are yet another kind of variable star. "Nova" is the Latin word for "new"; in a few hours a nova brightens by some 10,000 to a million times. It then sinks back to its original luminosity in a couple of months. A star becomes a nova when it suddenly throws off a shell of matter, which may comprise about one-hundred-thousandth of its total mass. (Supernovae are even more extreme outbursts; they are much rarer and of quite a different nature, pp. 90–91.)

Many novae are observed to flare up repeatedly, at periods from 10 days to tens of years, and many more probably have longer periods. Such recurrent novae are close binary systems containing a white dwarf, which draws matter from its companion. This falls on to the dwarf and is violently burned in nuclear reactions, which create the outburst. A star that was invisible may become visible to the naked eye, truly appearing to be a new star.

# DEATH OF A STAR

## • *When the fuel runs low*

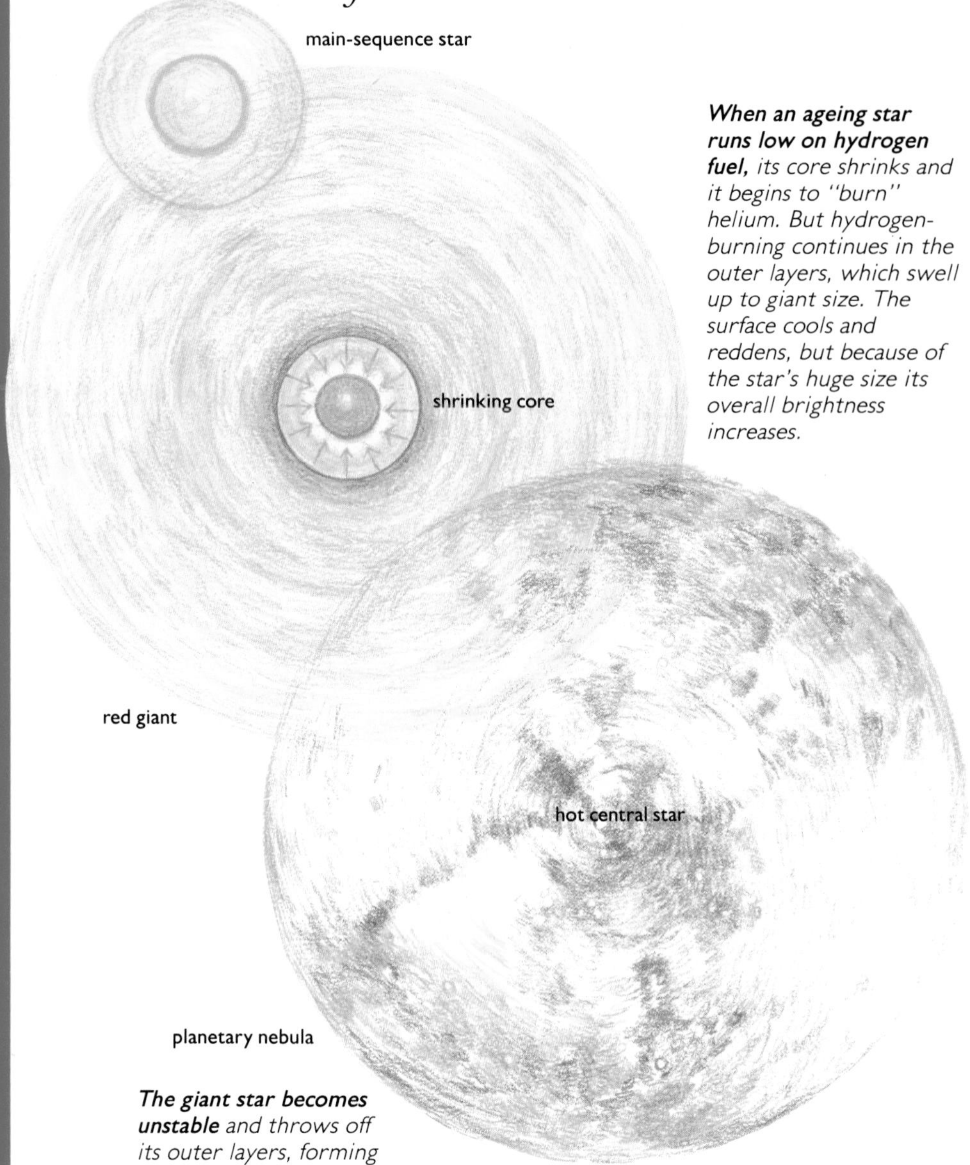

main-sequence star

shrinking core

red giant

hot central star

planetary nebula

white dwarf

***When an ageing star runs low on hydrogen fuel,*** *its core shrinks and it begins to "burn" helium. But hydrogen-burning continues in the outer layers, which swell up to giant size. The surface cools and reddens, but because of the star's huge size its overall brightness increases.*

***The giant star becomes unstable*** *and throws off its outer layers, forming an expanding planetary nebula. The core of the star, intensely hot and therefore bluish-white, is exposed. Its invisible ultraviolet radiation makes the nebula glow like a fluorescent lamp.*

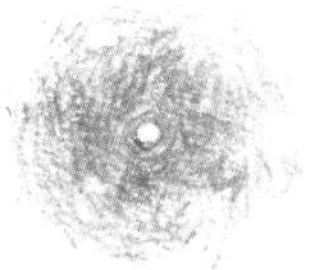

***Death is a slow fading away*** *for a star of about the Sun's mass. When all the helium in the core has been used up, the star shrinks to become a white dwarf, about the size of the Earth. It grows fainter as it radiates away its energy.*

Some stars end their lives spectacularly, torn apart by colossal explosions. Others, after suffering less violent disturbances, quietly fade from view over millions of years. What are the factors that determine which path a star takes at the end of its life?

A star leaves the main sequence when it has aged to the point at which helium-burning processes start up within it (p. 83). It begins to suffer from "middle-age spread" as it swells into a red giant or, if it is very massive, into a supergiant. In due course it becomes a variable, and throws off its outer layers, forming a planetary nebula around itself. This heralds the onset of death.

The simplest case is that of a star like the Sun, which is neither very large nor very small. Such a star is "well behaved" and lives its life at a modest rate. By the time it reaches the planetary nebula stage it has shrunk and become fairly hot, since it has used up its helium. After the nebula has formed, the central star cools and shrinks further.

What happens then was analysed in the 1920s by the Indian astrophysicist Subrahmanyan Chandrasekhar, when he developed his theory of white dwarf stars. He proposed that when a star shrinks the gravitational pressure at its centre grows so intense that the matter becomes far more closely packed than normal. It is forced into a state called degeneracy.

Degenerate matter was undreamed of before quantum theory. Normal matter is made of atoms, each consisting of a nucleus circled by one or more electrons. How many electrons there are depends on the chemical identity of the atom. The Pauli exclusion principle (p. 25) states that in a given region no two electrons can be in the same state – having the same energy, spin and so on. This holds the electrons apart by forcing them to occupy different energy levels. The principle is responsible for preventing atoms from collapsing. And it keeps the density of everyday matter below a certain maximum, about 90 times that of water.

Inside a star the extremely high temperature causes the atoms to be completely ionized – separated into atomic nuclei and electrons. In this state the matter can be crushed together, becoming much denser. It becomes denser still in the heart of a star that is shrinking as it ages. Nevertheless, the Pauli exclusion principle again sees to it that no two electrons can occupy the same state. As the electrons are squeezed closer together, they are forced to move faster and faster, and thus create a pressure that opposes the crushing force of gravitation.

In a star of moderate mass (up to 40 percent greater than the Sun's) the electron pressure becomes great enough to prevent further contraction when the density at the centre of the star is about one tonne per cubic centimetre – 10,000 times heavier than the densest material met with on Earth. This is the first stage of degeneracy, and the pressure supporting the star is termed degeneracy pressure.

By this stage the envelope of the star has been thrown off into space. The core of the star is exposed and is so hot that it shines with a white light. The star is now a white dwarf. However, its internal temperature is not high enough for more complex nuclear reactions to occur, and so it shines merely by radiating its internal energy. It cools as it does so and gradually fades to become a black dwarf.

Observation has shown that white dwarfs do exist. In 1844 it was realized that wobbles in the path of Sirius across the sky reveal the presence of an unseen companion. The companion was observed in 1862. From the strength of its pull on Sirius its mass was found to be about equal to the Sun's. Yet analysis of its light showed that it could be no more than five times as big as the Earth. Sirius B, as it is called, is a white dwarf. Several hundred white dwarfs have since been discovered, and thousands more "suspects" await confirmation of their nature.

The above account applies, as we have seen, to any star with a mass up to 40 percent greater than the Sun's. (This figure is called the Chandrasekhar limit.) If the star is above this limit, the temperatures reached in the core will be higher, and new, more complex, nuclear reactions will occur. In these reactions nuclei as heavy as those of iron are synthesized.

But once the core has been converted to iron no further nuclear reactions are available to release energy. The pressure needed to prevent further collapse cannot be maintained. The force of gravity in such a star overwhelms even the electron degeneracy pressure and the core collapses catastrophically, to a density much greater than that at the centre of a white dwarf. Electrons crash into protons and combine with them to form neutrons. As a result a neutron gas is generated.

Collapse continues until the neutrons are moving so fast that they exert enough degeneracy pressure to prevent any further collapse. Since neutrons are 2,000 times heavier than electrons, the neutron gas can exert much greater pressures than an electron gas. The core is now in a superdense state.

The collapse of the core is the trigger for a supernova explosion (pp. 90–91). But there is a further stage of collapse that can be reached by the most massive stars – those that are more than five times as massive as the Sun. When these stars collapse, the gravitational force of their overlying material crushes even the dense core of neutrons. Although the outer parts of the star are torn apart by the supernova explosion, a dark remnant of the core is left.

Since matter and energy cannot escape from the star, it becomes a black hole (pp. 62–65). Such objects are what Chandrasekhar called "the most perfect macroscopic bodies in the universe" because, as he put it, "the only elements in their construction are our concepts of space and time".

***The Helix, or Sunflower, Nebula*** *lies in the zodiacal constellation of Aquarius, the Water-carrier. It is the largest planetary nebula in apparent size, but it is invisible to the naked eye. Estimates of its distance conflict: according to some, it is about 450 light-years from us, and 4 light-years in diameter. The central star is a white dwarf.*

Nearly a thousand years ago, in AD 1054, Chinese astronomers noticed that a new star had suddenly appeared in the constellation we now know as Taurus. It was so bright that it could be seen even in daylight. This was the first supernova of which we have records, and the wreckage of its explosion is visible today as the Crab Nebula. Such extremely bright objects are rare in our galaxy – none has been seen since 1604. Hundreds have been observed in other galaxies, however.

Some of these immense catastrophes occur in binary systems of which one member is a white dwarf, the dying remnant of a star (pp. 88–89). Like all white dwarfs, it has consumed almost all its hydrogen, but does contain elements built up from hydrogen in its vigorous youth: carbon, calcium, magnesium, oxygen, silicon and sulphur. The white dwarf attracts material from the outer layers of its companion star, which "burns" when it falls on to the dwarf's outer regions.

At the same time the dwarf's core becomes extremely hot, reaching a temperature of at least 100 million K. In consequence, carbon-burning takes over in the core and a runaway reaction occurs; the white dwarf is consumed and its remains are thrown out in a vast cloud of gas containing elements with heavy nuclei, which have been synthesized in the great heat.

This is a Type 1 supernova. On average, one occurs in a particular galaxy every 140 years. Because such supernovae are always of much the same brightness, observing them allows astronomers to determine the distances of quite remote galaxies.

A Type 2 supernova is very different. It is the result of the sudden collapse of a lone star. Such an event occurs on average once every 91 years in a particular galaxy and, by a stroke of good fortune for science, one happened comparatively close to Earth in 1987, in the Large Magellanic Cloud.

Before its explosion, this supernova – known as SN 1987A – was a star some 15 times as massive as the Sun. It had passed through its short life of about 10 million years on the main sequence, followed by less than a million as a red giant, before entering the blue-supergiant stage. Calculations show that at the end of its life iron was synthesized in its core. The core was tiny – about half the diameter of the Earth – yet so dense that it accounted for one-tenth of the mass of the entire star.

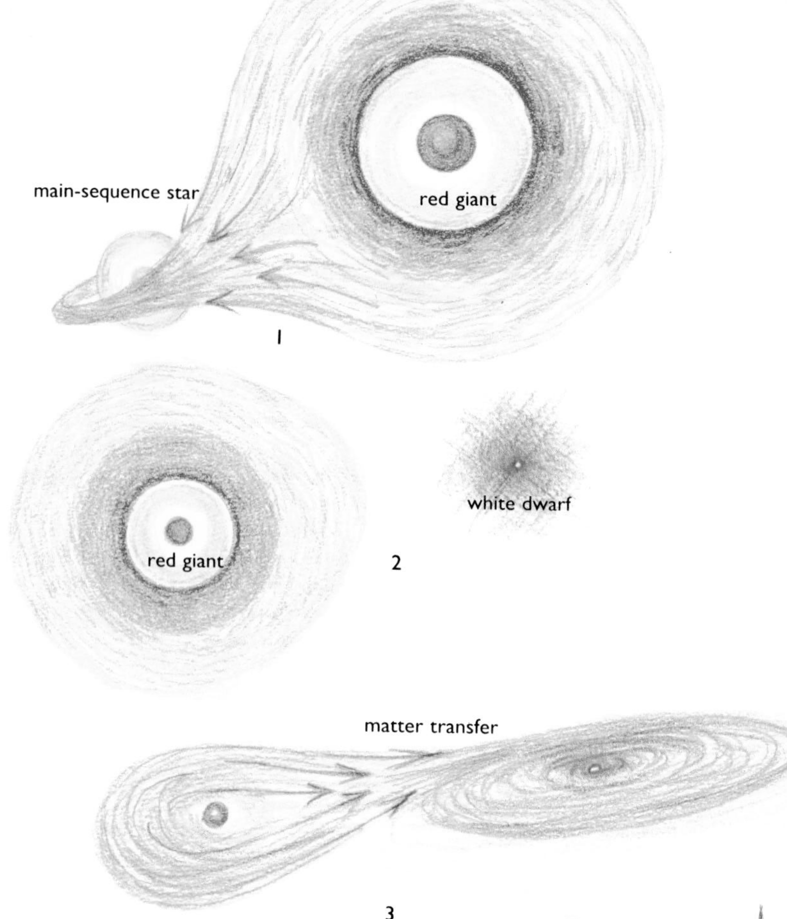


The supernova explosion occurred when this core shrank catastrophically. Because the core was massive, the collapse did not stop at the density levels characteristic of a white dwarf, but went farther. Two things therefore happened at the same time: the newly made iron nuclei were broken down again into lighter elements; and electrons and protons combined to form neutrons. The core turned into a ball of neutrons – a neutron star.

The rest of the star's material then fell in on to the core under its own weight. In doing so, it reached about one-tenth of the speed of light, and hit the core with such force that it bounced back outward again. It spread out into space, emitting vast amounts of energy as it did so. This gaseous material was moving at a prodigious velocity: within 10 hours it had swollen to the diameter of the Earth's orbit about the Sun, about 300 million kilometres.

The core of SN 1987A remains as a tiny neutron star, its diameter no more than 20 kilometres. Yet its mass is comparable with that of the Sun. That means its density is huge – some 300 million tonnes per cubic centimetre. Its gases continue to spread outward into space, just as those that now form the Crab Nebula have been doing for over 900 years.

**Type I supernova**

4

***A Type I supernova*** *results from a star belonging to a binary system. When it becomes a red giant, it loses matter to its less massive, slower-evolving companion (1) and becomes a white dwarf. The other star evolves to the giant stage (2), sheds matter on to the dwarf (3), and triggers an enormous explosion (4).*

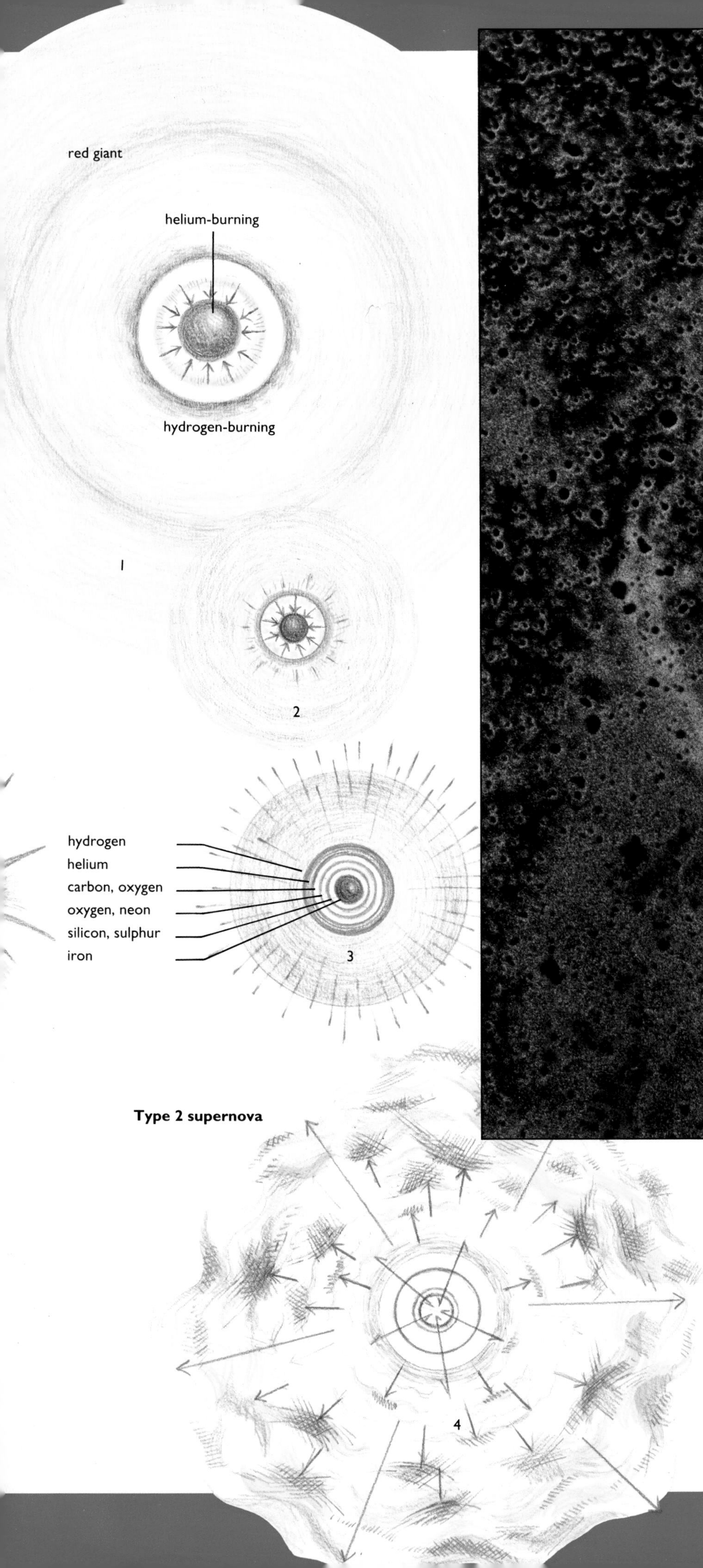

red giant

helium-burning

hydrogen-burning

1

2

hydrogen
helium
carbon, oxygen
oxygen, neon
silicon, sulphur
iron

3

**Type 2 supernova**

4

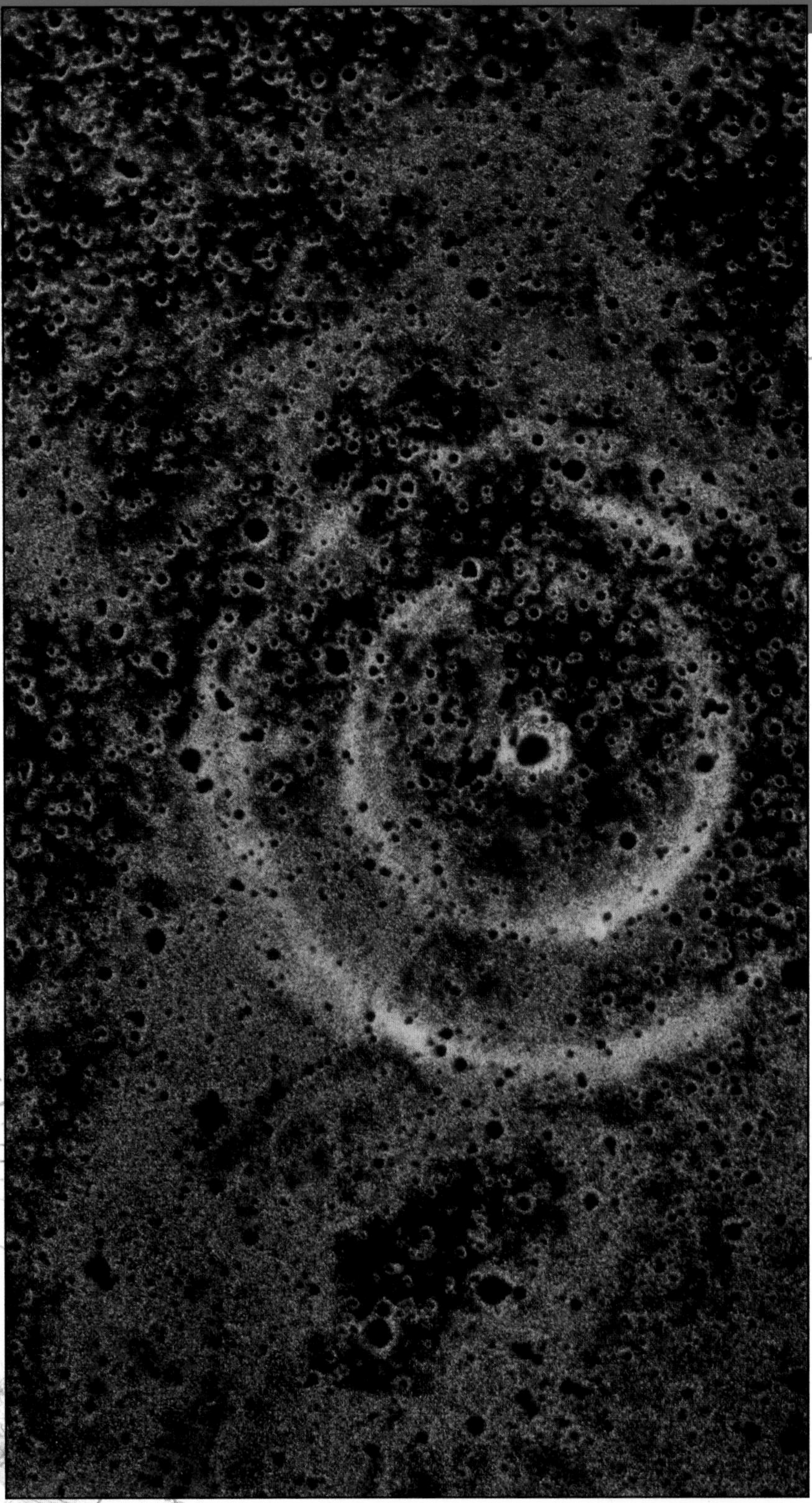

***When the core of a massive lone star*** *is depleted of hydrogen, it collapses and starts to burn helium (1). Later there are further collapses (2) and additional elements are burned. When element-building reaches iron (3), a Type 2 supernova is imminent (4).*

***Ripples of light spreading around a supernova*** *(above) are "echoes" of the event. Light from the explosion travelled for 170,000 years to reach the Earth in February 1987, while the "echoes", reflected from interstellar matter, took some months longer.*

# PULSARS

## • *Cosmic radio beacons*

In 1967 a new research programme began at the Mullard Radio Astronomy Laboratory of Cambridge University. It was humorously referred to as the only research project carried out with a sledgehammer. In fact, the sledgehammer was needed to put up posts carrying 2,048 radio antennae resembling clothes lines. These made up a radio telescope covering no less than 16,000 square metres.

When observations began Jocelyn Bell, a postgraduate student, noticed that the telescope recordings displayed some curious signals. They consisted of pulses about a twentieth of a second in length that recurred at precise intervals of about $1\frac{1}{3}$ seconds. She discussed this with her supervisor, Anthony Hewish, and they decided to investigate in more detail, using a special high-speed recording instrument.

They soon ruled out the idea that the signals could be due to interference from electrical equipment near the observatory: they were definitely celestial because they moved across the sky at the same rate as the stars.

At the time this work was being carried out there was considerable discussion of the possibility of civilizations elsewhere in the universe. Could the Mullard Observatory's pulses be signals from alien beings? The question was rightly treated seriously, even if the signals were for a while flippantly referred to as LGMs (for "little green men"). In fact this name was used as a precaution against the media getting wind of the discovery at this early stage of the investigation.

However, the signals varied in intensity in what appeared to be a completely random way, and no intelligible code could be derived from them. What is more, they did not show Doppler shifts in frequency (p. 46) of a kind that would be expected if they originated from a planet in orbit around a distant star.

Nevertheless, they proved to be at the right sort of distance to be from a star of some kind. Longer wavelengths were slightly delayed compared with shorter ones, owing to the scattering of the radio waves by electrons in interstellar space, and calculations showed that the source therefore lay at a distance of 400 light-years.

By 1970 no fewer than 50 of these pulsars had been discovered, and today hundreds are known. At first there was no optical identification. But it soon became clear that the pulses must come from bodies that were oscillating, pulsating or rotating. They must also be very compact to give such sharply defined pulses.

In 1968 the British cosmologist Thomas Gold suggested that pulsars could be rapidly spinning neutron stars, the superdense remnants of massive stars that had exploded (pp. 88–91). The central star of the Crab Nebula, the remnant of the supernova of 1054, was identified as a pulsar, and it was found to flash at optical as well as radio wavelengths. The remnant of the 1987 supernova in the Large Magellanic Cloud is also a pulsar. Thus the mystery of Hewish and Bell's LGMs has received a purely astronomical explanation.

### The nature of pulsars

Current ideas of the structure of a pulsar are deduced purely theoretically from fundamental physics. They show that neutron stars probably have a solid core of neutrons. Around this is a "superfluid" layer, consisting of neutrons, protons and electrons. (A superfluid is matter in a state in which it flows with no resistance. Since electrically charged protons and electrons are present as well, the superfluid in the neutron star is also a superconductor, which means that it possesses no electrical resistance.)

Outside this layer lies a very thin crust, probably only 600 metres thick, composed of neutrons also in a superfluid state. Finally there is a solid outer crust, only half as thick as the inner crust, containing atomic nuclei and electrons as well as neutrons.

The star from which the neutron star formed would certainly have possessed a magnetic field. During its final collapse this field would have been "frozen into" the neutron core, and would have become billions of times more intense. Charged particles – protons and electrons – are constantly given off from the surface of the neutron star. The intense magnetic field whirls these around, generating radio waves that are channelled into beams emerging from the magnetic poles.

The neutron star is spinning very fast – at a rate ranging from once every few seconds to about 1,000 times per second – and the two beams from the magnetic poles swing round. If one happens to point in the right direction, it is detected on Earth as a series of pulses.

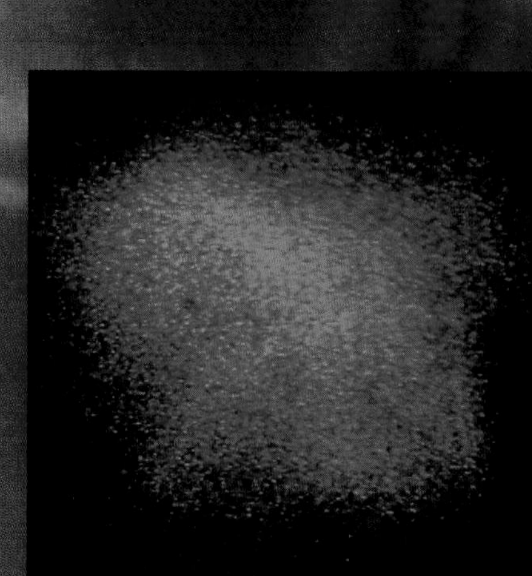

***The Crab Nebula is the wreckage of a supernova,*** *which has been spreading outward since the explosion was observed over 900 years ago. At its heart is the collapsed core of the star that exploded, now a pulsar giving out flashes of radio waves, visible light and X-rays. As it ages, it will slow down and radiate only at the lower-energy radio wavelengths.*

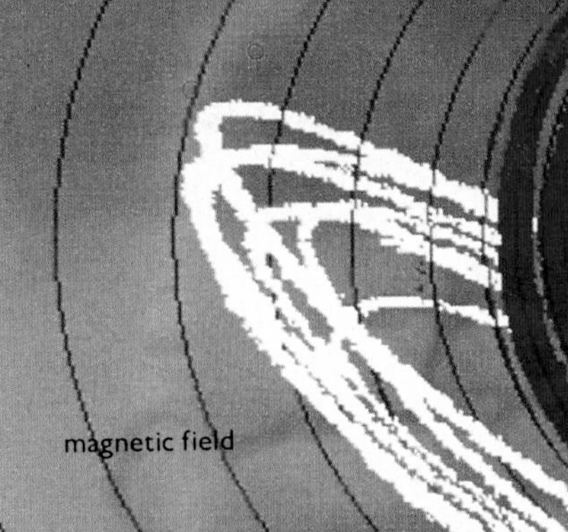

***The Crab Nebula pulsar*** *flashes on (above) and off (left) 30 times per second.*

neutron core

magnetic field

axis of rotation

beam of radio waves

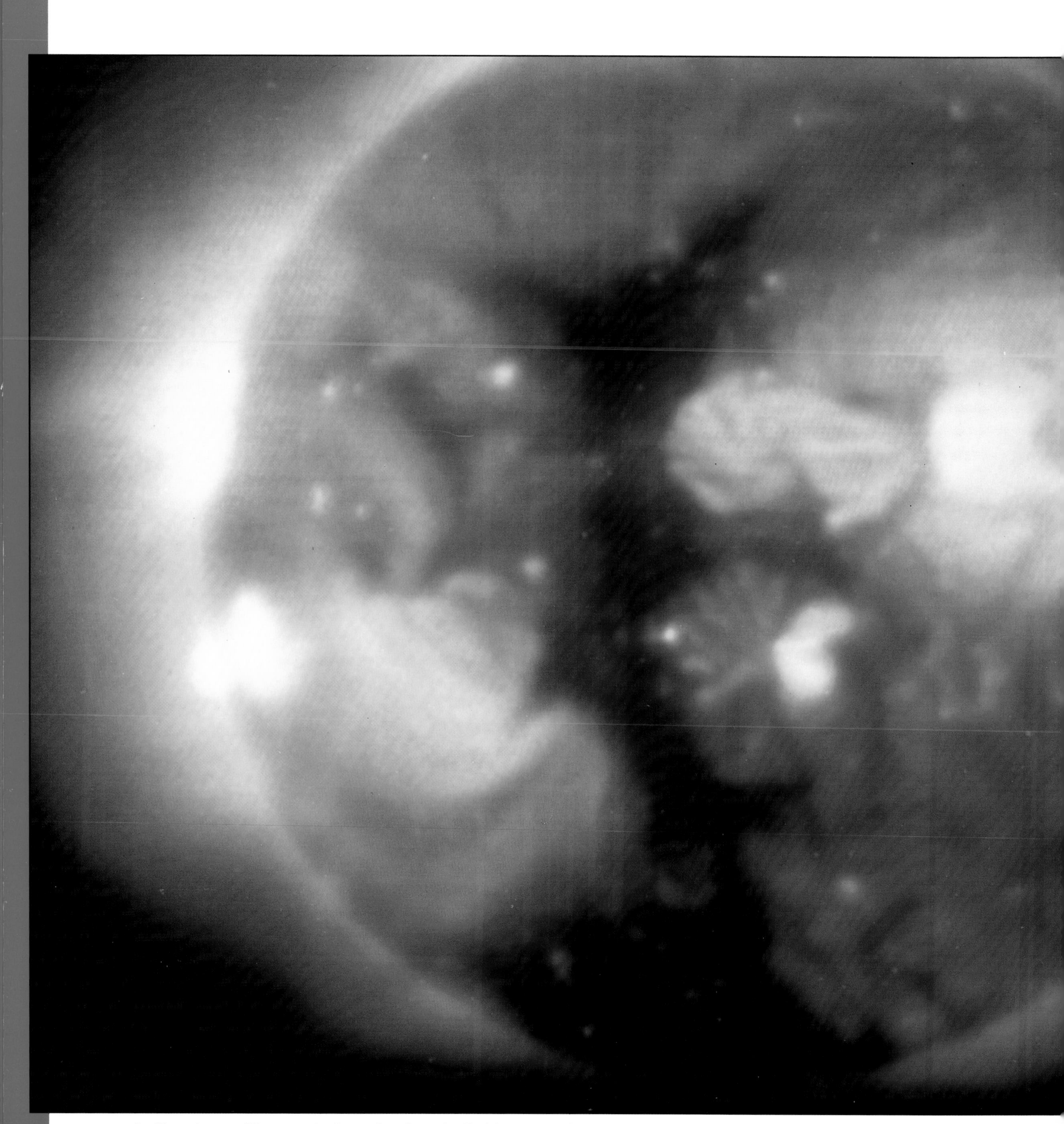

***An X-ray image of flares*** *on the Sun, taken from the Skylab space station, conveys the turmoil of its surface.*

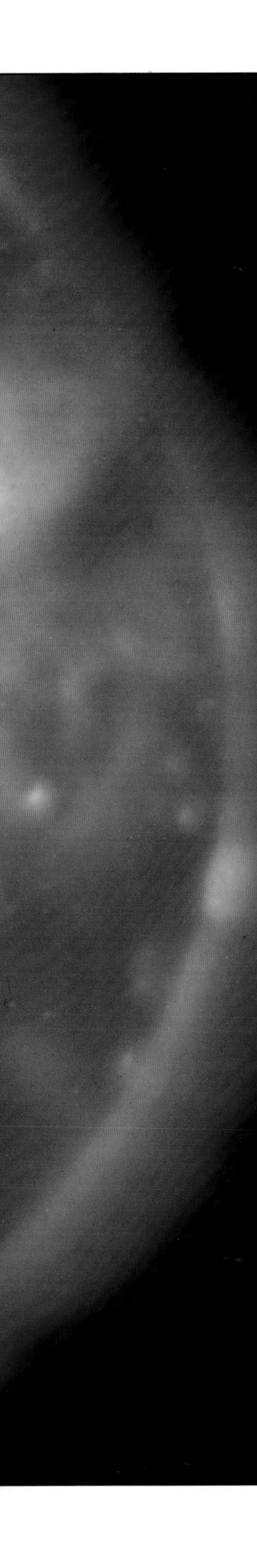

while smaller ones do so at higher levels. Viewed under very clear conditions through a telescope, the topmost cells can be seen at the surface as incessantly moving bright granules. This visible surface is called the photosphere ("sphere of light") and has a temperature of about 6,000 K.

The composition of the outer layers of the Sun can be directly determined by analysing sunlight spectroscopically. It consists of 73.5 percent hydrogen and 25 percent helium, some of which was formed in the first few minutes after the Big Bang, and the rest of which is the product of the nuclear reactions in the Sun. However, there are also traces of elements with heavier atoms, built up in earlier generations of stars that exploded, scattering the elements through the interstellar matter from which the Sun was formed. These include oxygen, carbon, iron, magnesium and so on. However, none of these elements accounts for even 1 percent by weight of the Sun.

It has been discovered that the outer skin of the Sun, 10,000 kilometres thick, oscillates, its depth increasing and decreasing by about 25 kilometres. In fact the Sun is vibrating like a ringing bell, with an oscillation period of only five minutes.

The cause of this oscillation may be a slight periodic fluctuation in solar transparency, causing a corresponding change in the pressure exerted by the radiation from the interior. Alternatively, the oscillation may be due to pressure waves triggered by turbulence in the convective region beneath. This process is similar to the one that causes the far greater pulsations of the outer layers of Cepheid variables (pp. 86–87).

There is also another oscillation, with a period of 2 hours 40 minutes. This could be due to variations in the rate of the proton-proton reactions in the core (pp. 82–83). But to account for it in this way would mean that the central temperature of the Sun must be 10 percent less than has been calculated. Yet such a lower temperature would result in a Sun that is not as bright as is actually observed. It appears that our present understanding of what is going on in the Sun's core needs reconsidering.

Another pointer to this involves the output of neutrinos from the nuclear reactions. Only one-third of the theoretically expected neutrinos are observed. This would be accounted for if the temperature were 10 percent below the figure that is generally accepted.

***Power is generated** within the innermost 30 percent of the Sun's radius. Energy is carried most of the way to the surface by radiation, but in the outer layers by convection. One oscillation pattern is shown: red areas are receding, blue are approaching.*

Lying above the photosphere, and visible only when the brilliant light of the latter is cut off in an eclipse or by special instruments, is a layer of reddish-pink gas, the chromosphere ("colour sphere"), most of which is only some 5,000 kilometres thick. Its temperature rises from about 4,000 K in its lower regions to some 50,000 K at its top. This upper surface is composed of spikes, or spicules, vertical columns of gas 1,000 kilometres broad and 10,000 kilometres high. The gas in the spicules shoots upward at velocities of about 15–30 kilometres per second and then falls back into the chromosphere.

The chromosphere also contains fibrils, horizontal strands of gas, appearing dark by contrast with their surroundings, which last for only 10–20 minutes. They are about 10,000 kilometres long and 1,000–2,000 kilometres thick. They are associated with active regions in the photosphere beneath.

The most obvious phenomena of the photosphere are sunspots, first observed and recorded by Chinese astronomers well over 2,000 years ago. They were only studied seriously in the West after 1610, when they were first observed with the telescope.

All sunspots display a relatively dark central region, the umbra, and a lighter surrounding area, the penumbra, which is nevertheless darker than the surrounding photosphere. Sunspots are not really dark; they appear so only by contrast with their surroundings. The temperature of a spot's penumbra is some

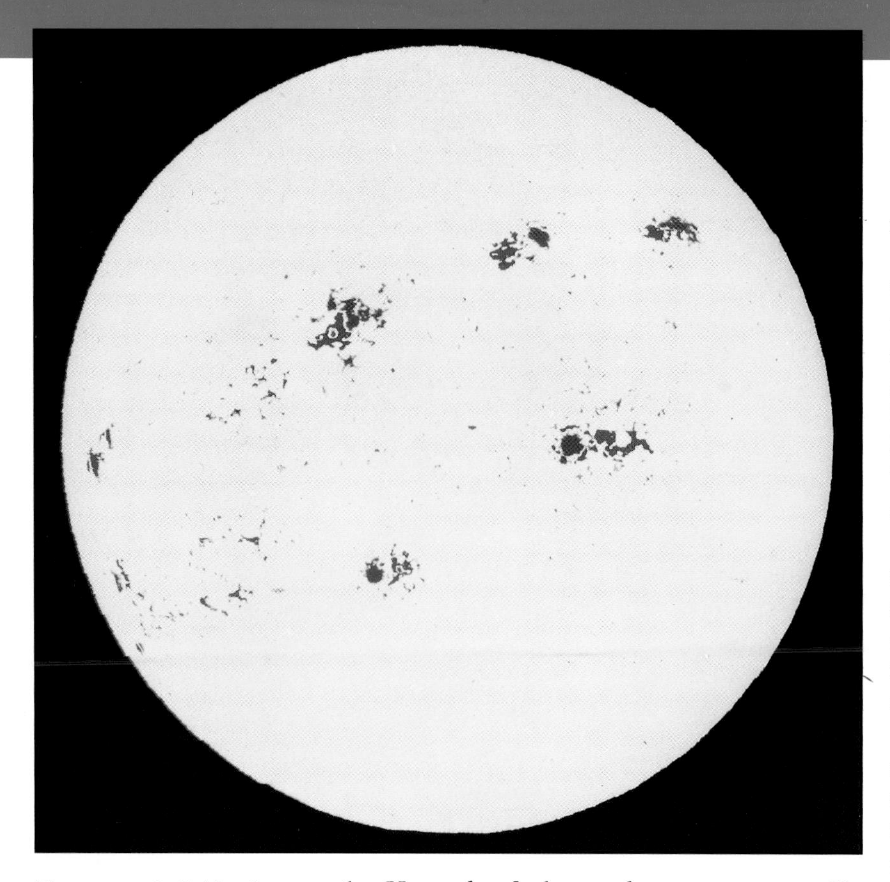

***The magnetic fields of sunspot groups*** *are displayed in this computer-coloured image. The north and south magnetic poles are shown in red and blue, respectively. In one hemisphere north magnetic polarity leads in each sunspot pair, while in the other the south pole leads. In the next sunspot cycle, the leading spots will have the opposite polarity.*

5,600 K, and of the umbra some 4,000 K, compared with the 6,000 K of the photosphere. It has been calculated that if the umbra of a sunspot could be seen in isolation, but still at the Sun's distance, it would be some 50 times brighter than the full Moon.

The movement and appearance of sunspots show that the Sun rotates at a speed that varies according to latitude. Studies of Doppler shifts show that near the solar equator it takes about 26 days to complete one revolution; at latitudes of 30 degrees the time taken is over 28 days. Near the poles the rotation period is some 37 days.

Sunspots vary enormously in size: some are no more than 1,000 kilometres across, while others far exceed the Earth in diameter. They are usually to be found in pairs, at latitudes rarely exceeding 45 degrees north or south.

Sunspots display a cycle lasting 11 years. At the beginning of a cycle the Sun is almost free of spots. A few appear at high northern and southern latitudes and then, as the cycle progresses, they appear in greater numbers at lower latitudes. They reach their maximum at latitudes of 15 degrees at the peak of the cycle. Their numbers then dwindle, although new ones continue to form closer to the equator.

This is the general rule, but the maximum number of spots varies from one cycle to another. What is more, there is sometimes a dearth of spots. During the period from 1645 to 1715 – the "Maunder minimum" – virtually none were visible.

When sunspots are observed as they approach the Sun's edge, or limb, the umbra appears to be lower than the photosphere, but this is an optical illusion. Gas can often be observed being emitted from the umbra and falling back again into the companion spot. The route of the gas always follows the lines of force of the magnetic field associated with every spot.

Although details of the sunspot phenomenon are still a matter for debate, an explanation proposed by Horace Babcock and Robert Leighton is now generally accepted. Their model involves the Sun's magnetic field, which at the beginning of a solar cycle is believed to lie in the lower part of the convective layer. Magnetic field lines stretch between the Sun's magnetic north and south poles and, being electrically highly conductive, the gas traps the field. Because different parts of the Sun rotate at different rates depending on their latitudes – faster near the solar equator and slower near the poles – the field lines become wrapped

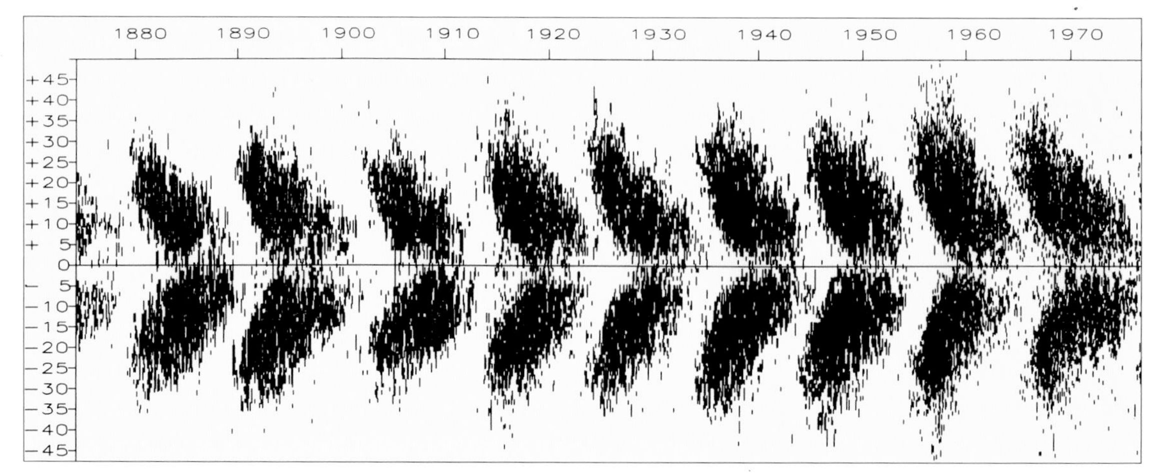


***Each 11-year sunspot cycle*** *begins with the appearance of spots at latitudes of 25 to 30 degrees. Spots disappear and new ones appear at lower latitudes, so that a chart of the latitudes of spots against time gives this butterfly diagram, or Maunder chart. Spots are very rare above 45 degrees north or south and on the equator itself.*

around the Sun. Wherever the lines come close together, the strength of the field is increased, and here "tubes" of gas form, threaded by intense magnetic fields.

Because of turbulence, the magnetic tubes also become wound up like the strands in a rope. They also tend to rise, and a time comes when the upward force on these magnetic ropes becomes so great that they float up to the photosphere, where kinks in them break through the surface. The magnetic field lowers the temperature of the electrified gas; accordingly, a pair of spots will form, one where the field lines emerge, the other where they re-enter the photosphere.

The spots are effectively magnetic poles – one a north pole where the lines emerge, the other a south pole. This has long been known from study of the effect of the magnetic field on the sunlight from the spots. In one hemisphere the leading spot in each pair is a north pole in one sunspot cycle, and a south pole in the next; the leading spots in the other hemisphere are always of the opposite polarity.

The disruption in the photosphere caused by sunspots is accompanied by other effects. Brighter patches in the upper part of the photosphere, called faculae, can be seen at places where sunspots are about to appear, and for a short time after they have arrived. The spots also give rise to areas of increased brightness called plages (the French word for beaches) and filaments of dark light-absorbing gas between the sunspot pairs. These phenomena generally disappear two to four weeks after the sunspots have gone, although some times they will last for some months.

The Sun possesses an outer atmosphere known as the corona (Latin for crown), which may have been discovered as early as the first century AD. Because it is faint, it is ordinarily invisible to the unaided eye, but becomes visible during total solar eclipses. In fact, the first knowledge of the less obvious properties of the Sun was afforded by observations made during such eclipses.

A solar eclipse occurs whenever the Moon comes between the Sun and the Earth. The apparent size of the Moon is, by coincidence, almost exactly the same as that of the Sun. If the Moon passes centrally in front of the Sun, it normally just covers the latter's disc, causing a total eclipse. If the Moon happens to be at its greatest distance, however, the rim of the Sun

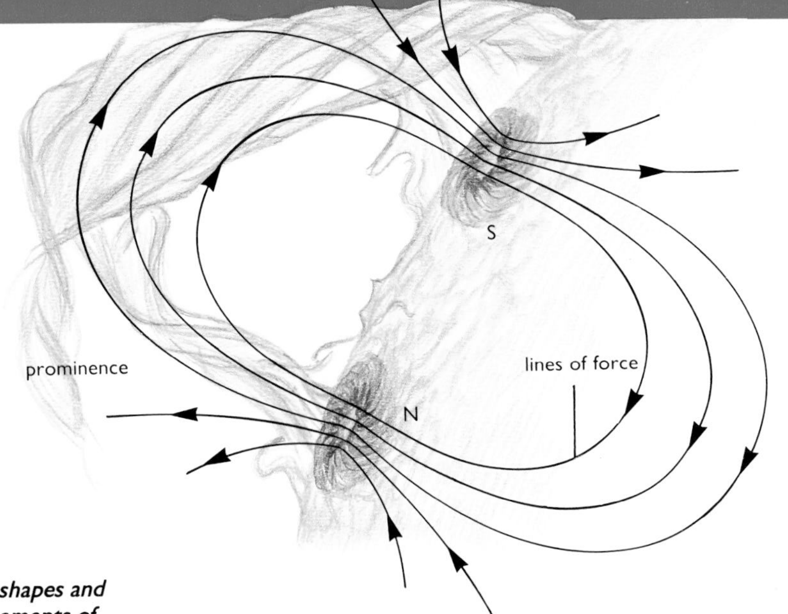


***The shapes and movements of prominences*** *are dominated by the magnetic fields of sunspots. Magnetic field lines loop between the two members of a spot pair, emerging from a spot that has north polarity and re-entering a spot with south polarity. Despite appearances, the gas in a prominence is usually moving downward. It often forms an arch following the field lines, and may show braiding and twisting effects.*

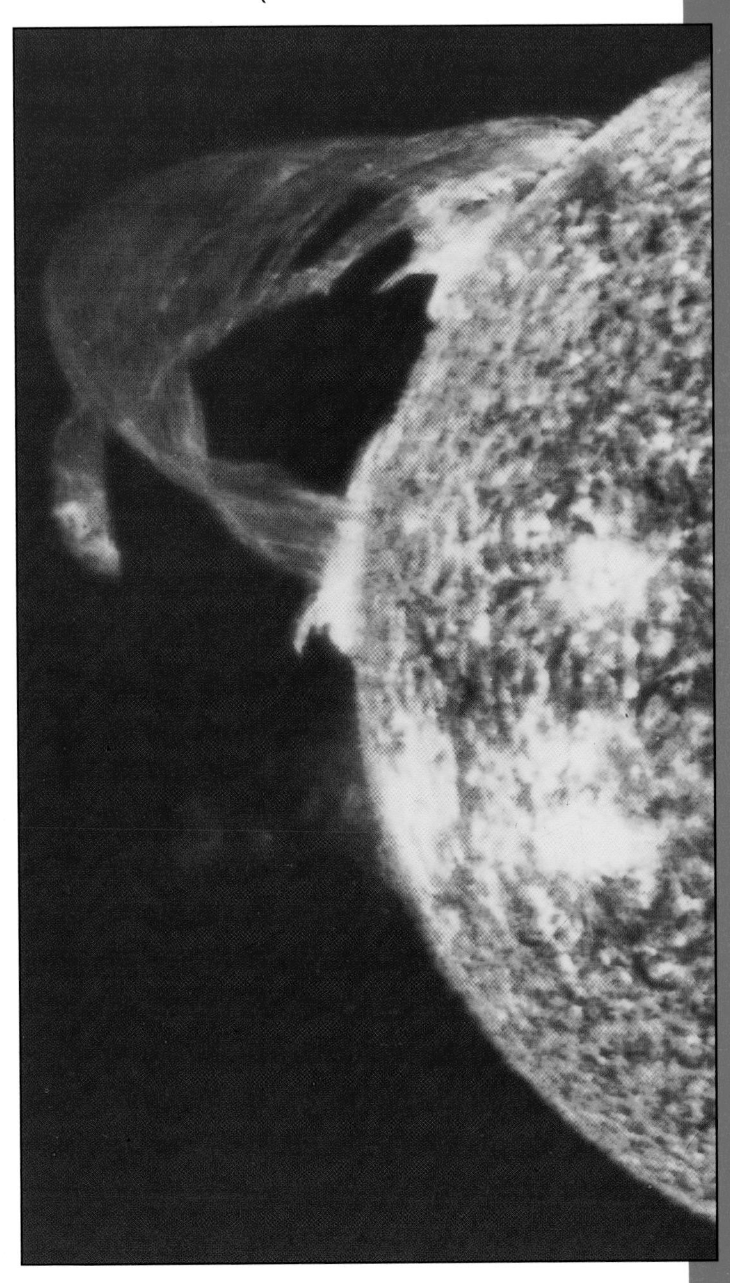

***A solar prominence is seen in ultraviolet light*** *by instruments on board Skylab on 19 December 1973. The arc of ionized helium extends 588,000 km – 45 times the diameter of the Earth – across the surface of the Sun. Before expanding to this size it had the appearance of a quiescent prominence, apparently destined to last months while showing little activity.*

*__During a total eclipse__ the pearly white corona, or atmosphere, of the Sun becomes visible. The corona is extremely tenuous but intensely hot, its outer regions reaching temperatures of millions of degrees.*

## Blotting out the Sun

The Moon as viewed from the Earth is just big enough to cover the Sun's disc when the three bodies line up. The eclipse is total only in the central part of the shadow, which is at most 268 km across on the Earth's surface. Outside this region the eclipse is partial, with part of the Sun visible around the edge of the Moon.

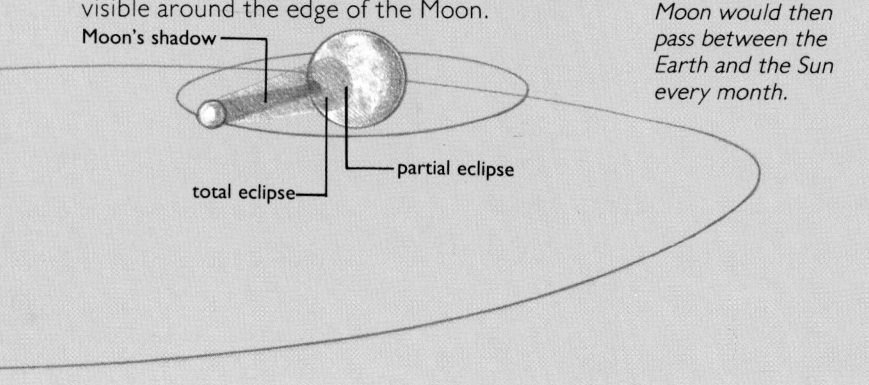


*__Solar eclipses__ occur at least twice a year, and at most five times. They would be far more common if the Moon's orbit were not tilted in relation to the Earth's orbit around the Sun; the Moon would then pass between the Earth and the Sun every month.*

remains visible around the Moon, giving an annular, or ring, eclipse.

Total eclipses occur about six times in a decade. Each eclipse is total only in a narrow band, never more than 268 kilometres wide and a few thousand kilometres long. To each side of this band, the eclipse is partial.

A total solar eclipse is most impressive: the sky darkens, stars appear, animals lie down as if it were night, and buildings look strangely flat, like a stage set. The temperature drops noticeably. Because the Moon moves quickly, the shadow it casts never stays long in a particular place: the maximum duration of a total eclipse is 7 minutes and 31 seconds. In these brief moments some of the most spectacular phenomena on the Sun can be seen.

During an eclipse, huge, flamelike clouds of hot gas called prominences may be seen. Prominences look as if they are emerging from the chromosphere. They have temperatures of around 10,000 K and radiate in the ultraviolet and X-ray regions of the spectrum as well as at visible wavelengths. Some "quiescent" prominences are vertical masses of hot gas, over 160,000 kilometres long and 5,000 to 8,000 kilometres thick, hovering more than 300,000 kilometres above the photosphere. They last a matter of months, or in some cases as much as a a year.

Occurring in the magnetically neutral areas between the two members of pairs of sunspots, quiescent prominences look as if they consist of gas moving upward from the chromosphere, especially since they display the same reddish-pink colour. This is true of some, although most are actually material from the corona, which becomes more concentrated as it descends.

Active prominences are smaller than quiescent ones, being about 40 percent as long. Their activity is measured in minutes rather than months. Some take the form of loops or arches; these are composed of material descending from the corona and following the local magnetic field associated with sunspots.

Other active prominences clearly consist of material thrown up from the chromosphere and sometimes ejected from it, at speeds of 100 to 200 kilometres per second. Such prominences are related to flares, sudden releases of energy that occur in active regions, again usually in the neutral regions between associated sunspots. The frequency of flares is closely connected with the numbers of sunspots present on the Sun at the time. The flare energy is released in the form of radiation, masses of ejected gas and fast-moving electrons and protons.

The corona is variable in form and extent. Above active sunspot regions coronal streamers may travel 140 million kilometres into space – almost reaching the Earth. The part of the corona seen at visible wavelengths is large and fairly evenly spread around the Sun at sunspot maximum, whereas at minimum it is generally smaller, but stretches in long streamers from the Sun's equatorial regions. At the poles it forms spreading, curved lines with a "brushed" appearance, revealing the form of the Sun's magnetic field.

The corona is extremely tenuous – at its densest it is 10,000 times less dense than the photosphere. Nevertheless it is very hot, its temperature ranging in different regions between 1 million and 5 million K. Yet because of its thinness it contains very little energy – it would not, for example, melt a spacecraft within it.

In pictures taken at ultraviolet and X-ray wavelengths the corona displays conspicuous dark regions, which have become known as coronal holes. These occur where the magnetic field of the Sun is weak and the field lines rising from one pole of the Sun do not return to the other but break away and trail into space. Here hot matter escapes into the solar wind. Such regions are cooler than the surrounding corona, which is why they seem dark at these energetic wavelengths.

As early as 1900 Sir Oliver Lodge suggested that electrified gas might be ejected from the Sun, but it was not until 1958 that it was realized that the matter in the corona is continually expanding, becoming the solar wind as it moves outward. The electrified material consists of protons, electrons and ionized heavier atoms, ejected at average speeds of 450 kilometres per second.

The solar wind carries with it a magnetic field, which affects the magnetic fields of the planets, including the Earth; when charged particles enter the atmosphere they give rise to the beautiful aurorae, or polar lights. What is true of the Sun is true also of other stars: spots have been detected on some of the nearer ones, and others show evidence of stellar winds, consisting of electrified particles.

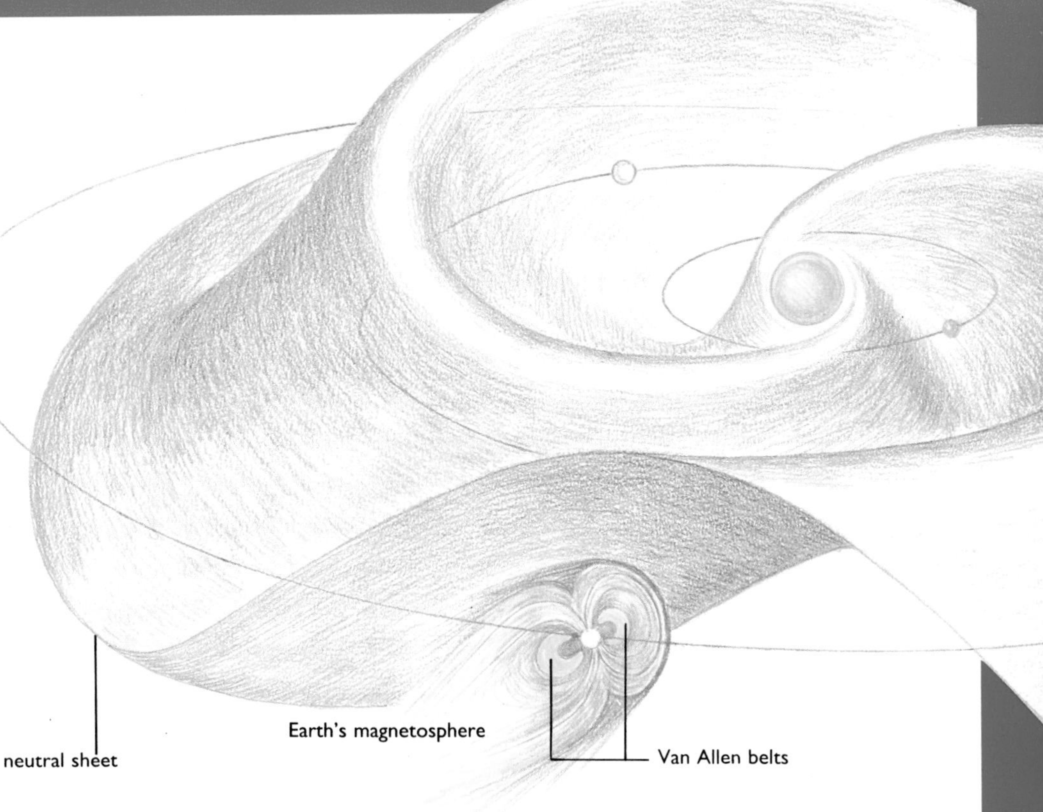


***The solar wind** distorts the magnetic fields of the Sun and the Earth. Fluctuations in the wind – a stream of electrically charged particles from the Sun – make the Sun's field wave up and down, as revealed by the buckling of the central, neutral sheet. (Above this sheet the field points away from the Sun, below it in the opposite direction.) The solar wind blows the Earth's magnetosphere into a long tail, and some of its particles are trapped in the Van Allen belts.*

### Lights in the sky

Aurorae are luminous displays that appear in the night sky at high latitudes, taking the form of coloured arcs, streamers or draperies. They are caused by charged particles from the Sun, such as electrons and protons, interacting with atoms and molecules in the Earth's atmosphere, usually more than 100 km up. The particles arrive in the solar wind and succeed in penetrating the Earth's magnetosphere, the region occupied by our planet's magnetic field. They are trapped for a time in the Van Allen belts, but they gradually escape and follow the Earth's field down toward the magnetic poles.

***The influence of the solar wind** on the Earth shows up clearly in this colourful aurora. The best chance of observing aurorae comes when the Sun is active, for then the solar wind is most intense.*

# BIRTH OF A SOLAR SYSTEM

## • *From planetoids to planets*

When the Sun formed from gaseous material some 4.56 billion years ago something like 99 percent of the gas cloud formed the protosun before condensing down into the Sun itself. The remaining material formed a disc – often referred to as the solar nebula – around the condensing star. This solar nebula was not composed only of the lighter gases such as hydrogen and helium; there were other, heavier elements present as well.

It seems that the Sun formed in company with a number of other stars, some of which were probably very large. The comparatively slow contraction of the Sun continued while these large stars quickly passed through their life cycles to reach the supernova stage.

Moreover, some of an earlier generation of stars within our galaxy had reached that stage too. When stars become supernovae they manufacture and then spew out heavier elements. Thus there were two sources to provide heavy elements in the solar environment. Some of these elements were swept up into the Sun, but 1 percent, at least, remained in the solar nebula.

To begin with the solar nebula was hot and opaque. Gravitation made it contract, causing its temperature to rise, and its opaqueness helped retain infrared radiation in its central regions. Yet although the central regions, at least, became very hot, the heat did radiate away slowly, and the nebula began to cool.

As the solar nebula cooled various chemical substances condensed out. Once the temperature was less than 2,000 K, compounds of aluminium, calcium, magnesium and titanium formed; below 1,000 K, compounds of silicon and oxides of metals appeared. At 180 K ice formed from water vapour, and in the outer parts of the nebula, where the temperature would have dropped to 20 K, methane solidified. Such chemical condensation resulted in the formation of tiny grains.

Mathematical studies of what happens in thin discs of condensing material show that the material soon ceases to be evenly distributed throughout and pockets of denser material form. In the solar nebula the grains of the various compounds began to collect together to form lumps of material that measured a few kilometres across. By cosmic standards the whole cooling, condensing and coalescing process described so far did not take long – only about 1,000 years.

Thus at the end of that period, the solar nebula had condensed and accumulated into a rotating disc composed mainly of small lumps or planetoids. Two processes then took place.

In the first, collisions occurred between planetoids. In some cases, when the collisions were fast enough to be disastrous, the planetoids broke up. In others, where the collisions amounted to more gentle encounters, fragments coalesced and fused, owing to mutual gravitational attraction, to form larger units, perhaps as much as 1,000 kilometres across. This left the solar nebula containing a considerable number of protoplanets. The second process was the coalescence and fusion, again due to gravitational attraction between them, of protoplanets into true planets.

Both these processes swept up a considerable amount of the planetoid material, and the resulting orbits of the planets so formed were almost circular. Thus the newly formed planets became isolated from each other, each pursuing its own orbit. These two processes took of the order of 100 million years.

The gravitational attraction of the Sun on the solar nebula caused the orbits of the planets to lie in the same plane, although this was not the case for the remaining planetoid material. Some of it coalesced and fused to form satellites, which were then attracted into orbits around the larger planets. Our Moon, once thought to have been a part of the Earth that broke away owing to centrifugal effects, is now thought almost certainly to have formed as a result of the fusion of planetoid or, considering its size, protoplanetary material.

Planetoid material that did not coalesce into larger units formed the asteroids or minor planets (pp. 142–43). Moreover, calculations show that some of the smaller primary planetoid material did not coalesce at all. Instead it formed icy conglomerates, sometimes called planetesimals, which became comets (pp. 144–45) and meteors (pp. 147–48). These objects are particularly interesting from the point of view of the history of the solar system because they provide astronomers with evidence as to what chemical elements made up the material in the very early solar nebula.

***A planetary system is born*** *from the gas and dust remaining after the formation of a protostar (top). The first stage is the building up of countless rocky lumps called planetoids by the accretion of grains of dust (centre). In further collisions the planetoids break down and re-form, finally forming a small number of protoplanets (bottom). Nuclear reactions begin in the star, while its radiation blows away remnants of the cloud. Light gases, such as hydrogen, helium, methane and ammonia, cannot be held by the small protoplanets in the hotter inner regions of the system, but are retained as deep atmospheres around the larger, colder bodies farther out.*

# OTHER SOLAR SYSTEMS

## • *Do they exist?*

***A system of planets in the making*** *may be revealed in this picture taken at infrared wavelengths by the IRAS satellite. A disc of gas and dust is seen almost edge-on, reaching 100 billion km from the star Beta Pictoris, the light of which is blocked out by a dark mask at the centre.*

The prospect of other stars possessing planetary systems in orbit around them has long intrigued astronomers. The debate about whether they have or not has, at any given time, been much influenced by the prevailing opinion about how our own solar system was formed.

When it was thought that the planets were the result of a rare set of circumstances, it was considered extremely doubtful that many stars possessed a set of orbiting planets. In the 1940s and even earlier, the generally accepted scenario was that a chance close encounter by the Sun with another star was the cause.

In such a case, according to theory, mutual gravitational attraction would have drawn out material from both stars into an elongated filament, which was fatter in the centre than at the ends. This would have broken up in due course, leaving one part in orbit around the Sun. It was claimed that gradually this would condense into planets, the terrestrial ones being those which formed at the thin end near the Sun, with the gas giants forming from the more distant bulge.

However, later mathematical analysis

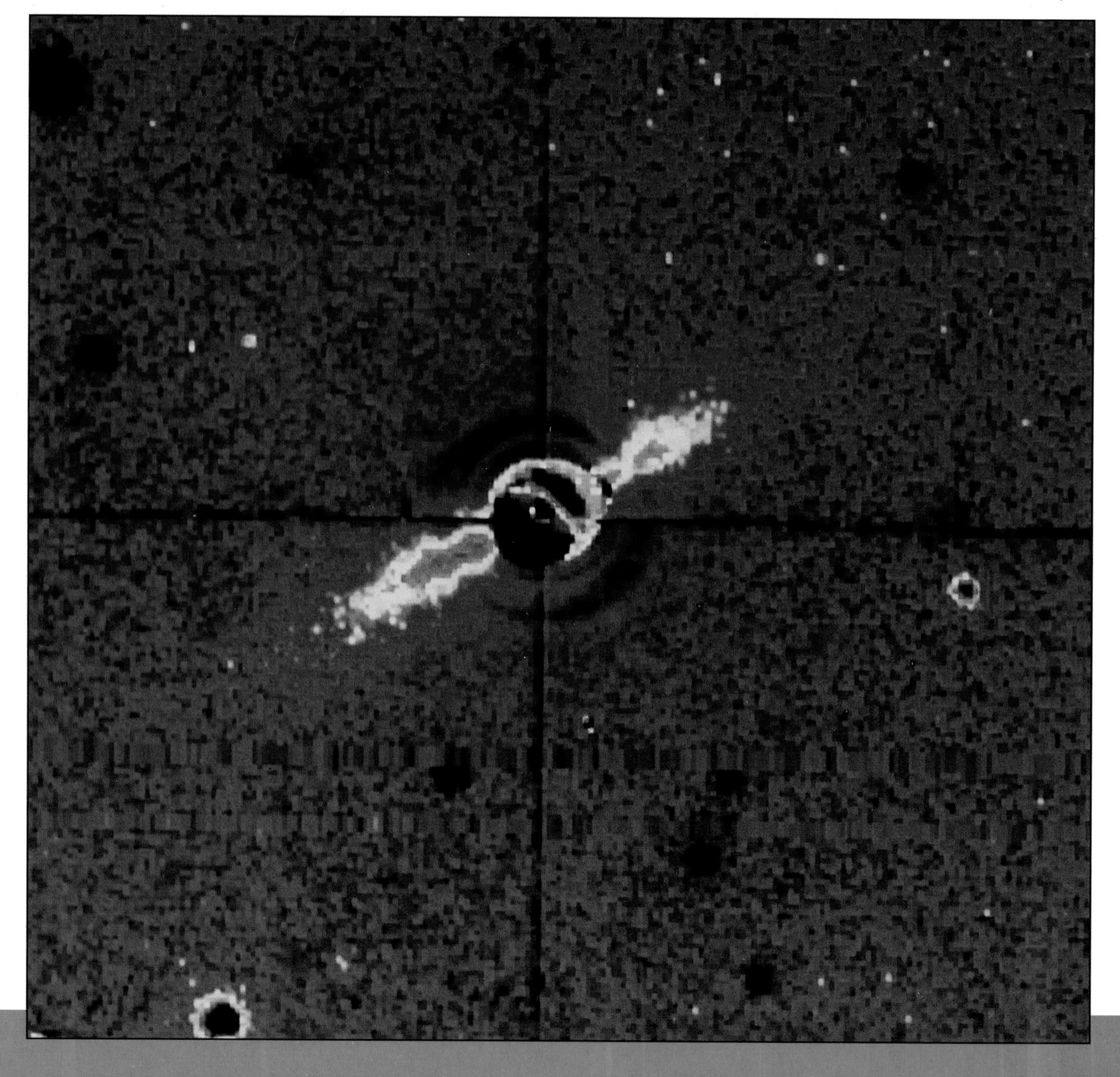

has shown that such a scenario would not give rise to a solar system such as ours. Furthermore, increased knowledge about how stars are born makes the accretion theory (pp. 102–3) more likely. Mathematical studies confirm its validity but, above all, observational evidence now makes it almost certain.

Evidence indicating the existence of planets around other stars has come from work done at Sproul Observatory in Pennsylvania. Observations of nearby Barnard's star show that as it moves across the sky against the background of more distant stars it "wobbles". This wobble is just what one might expect if the star is orbited by two planets, each about the size of Jupiter.

Barnard's star is not an isolated case. Another nearby star, called Epsilon Eridani, also displays a similar wobble. But not all astronomers are convinced that the deviations from a straight path in space are real or, if they are, that orbiting planets are the cause.

Accepting accretion in a solar nebula as the explanation of how the solar system was formed opens the way to seeking proof by direct observation. Put simply, the task is to see whether a solar nebula can be detected around some not too distant stars.

During 1983 the Infrared Astronomical Satellite (IRAS) did, in fact, detect a bright radiating dust structure around the star Vega (Alpha Lyrae) – a white, class A0 star about 60 times as bright as the Sun. This structure extends out almost half a light-day, or a distance 2.8 times that of Neptune from the Sun, and its mass is about the same as that of our own solar system.

Another star with a nebula detected by IRAS is Fomalhaut (Alpha Piscis Austrini), a first-magnitude A-type star. But observationally the most exciting is Beta Pictoris, only 78 light-years away. Examination by IRAS showed it had a solar nebula and astronomers at the University of Arizona managed to obtain a picture. Tilted some 7.5 degrees to our line of sight, the disc-shaped nebula has been traced out to a distance of some 100 billion kilometres from the parent star, or approximately 3.7 light-days.

There is yet another group of celestial bodies which appears to possess solar systems in the making. These are the T Tauri stars, named after a typical example in the constellation Taurus. They are extremely young stars, no more than a few million years old, settling themselves into the main-sequence stage of their life cycles and varying irregularly in brightness as they do so. During 1989 and 1990, astronomers analysed radiation from T Tauri stars at various wavelengths and concluded that these stars are surrounded by flattish discs of material, with masses similar to those of our solar system.

There does therefore seem to be a great deal of evidence for the existence of other solar systems. Indeed, Gibor Basri of the University of California at Berkeley has stated that "something like one-third of newly forming stars of one solar mass or less appear to satisfy most of the conditions we believe necessary for the formation of planets."

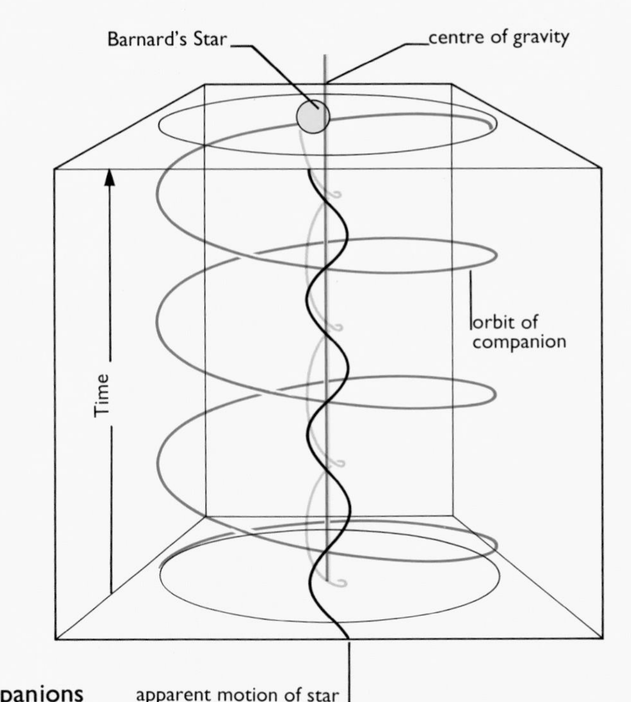


**Unseen companions**

A "wobble" in the movement of Barnard's star, one of the Sun's closest neighbours, suggests the presence of invisible companions. Barnard's star is a red dwarf, about six light-years from the Earth. It has the fastest apparent motion of any star, and also seems to move in an undulating path. It is believed that the star belongs to a multiple system, the centre of gravity of which moves in a straight line. The star's deviation from this line is tiny – no more than a few hundredths of an arc second.

Measurements made by Peter van de Kamp over many years suggested that the star probably had a companion that was about the size of Jupiter, and therefore could well be a planet. He concluded from further research that there could be two planetary companions, one orbiting the star in about 11.5 years, the other farther out and taking some 20 to 25 years.

# OUR SOLAR SYSTEM

## *The Sun's family*

The orbits of all the major planets display characteristics which are to be expected if the solar system was formed by accretion from a disc-type solar nebula (pp. 102–3). For example, the orbits all lie in the same plane, or very nearly so. There are exceptions, however. Venus, for instance, has an inclination of 3.4 degrees and that of Mercury, the closest planet to the Sun, is 7 degrees. All the others have inclinations less than this, except for one maverick, Pluto, the farthest planet from the Sun, with an inclination of just over 17 degrees.

This fact, along with other characteristics, leads astronomers to believe that Pluto is not actually a major planet (p. 133). Pluto aside, the orbits of the major planets of the solar system really do lie virtually in the same plane.

This is not so true, however, of the asteroids, the comets with their associated streams of meteors, and other small items of debris. Cometary orbits, in particular, vary widely in inclination. Some are inclined so much that they move around the Sun in a retrograde fashion; in other words, they move backward compared with the major planets and the asteroids. Halley's Comet is a good example as its inclination is 162 degrees.

The second significant characteristic of the orbits of the major planets is that they are stable and approximately circular. Again there are exceptions. Mercury, for example, orbits in a pronounced ellipse with an eccentricity of 0.206; this is five times greater than the average elliptical eccentricity of the other major planets. But even this is not large.

A third characteristic is the observed fact that all the major planets orbit the Sun in the same direction: counterclockwise when viewed from above. This is another indication that they were formed in a rotating solar nebula.

The major planets can conveniently be divided into two groups. Closest to the Sun lie the terrestrial planets, so-called because they have some similarities with Earth. These are Mercury, Venus, Earth and Mars. The second group, comprising the so-called gas giants, is farther away, beyond Mars. This group is made up of Jupiter, Saturn, Uranus and Neptune.

The gas giants all have solid cores surrounded by vast, cold atmospheres where methane, ammonia, helium and hydrogen are found in quantity. Such light gases were present originally on the terrestrial planets, but were lost.

Because the terrestrial planets are much closer to the Sun than the gas giants, they received far more heat and the light gas molecules began to move very fast. As the terrestrial planets are less massive than the gas giants, each had an insufficient gravitational pull to retain such quickly moving molecules.

The most distant giant planet, Neptune, lies at an average distance of 4,497 million kilometres or just over 4 light-hours. As the nearest independent star is Proxima Centauri, at 4.3 light-years, the scale of the solar system seems minute by interstellar standards. However, the solar system does not consist only of the major planets; there are also many small bodies such as asteroids and comets. Although there is no firm evidence that asteroid orbits extend out to great distances (most orbit between Mars and Jupiter), comets are different.

It is now generally accepted that there are clouds of debris in orbit around the Sun which, when disturbed, disgorge material that we recognize as comets. There seems to be more than one such cloud. The first is one which gives rise to the short-period comets with orbital periods of no more than 150 years. This is the Kuiper belt (named in memory of the astronomer Gerard Kuiper), which extends from 6 to 24 light-hours from the Sun. Next, there may well be another narrower one lying some 20 times farther out.

But most significant is the Oort belt (named after the astronomer Jan Oort), which is thought to be the major source of cometary material. This lies over a wide band of space, stretching from about 4,500 billion to 15,000 billion kilometres out – between almost 6 and 18 light-months. The Oort cloud therefore takes us out to greater distances: just over one-third of the distance to Proxima Centauri.

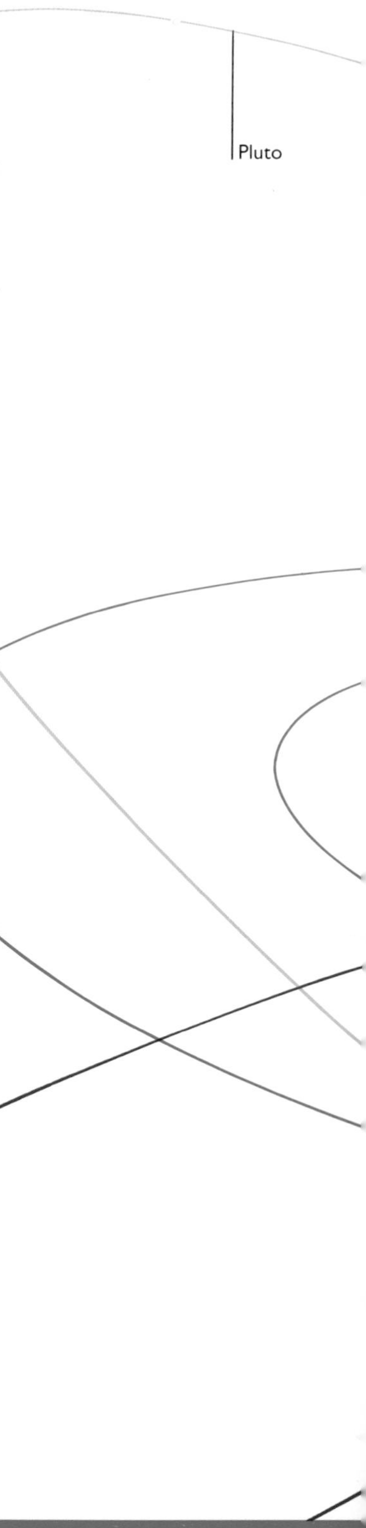

Pluto

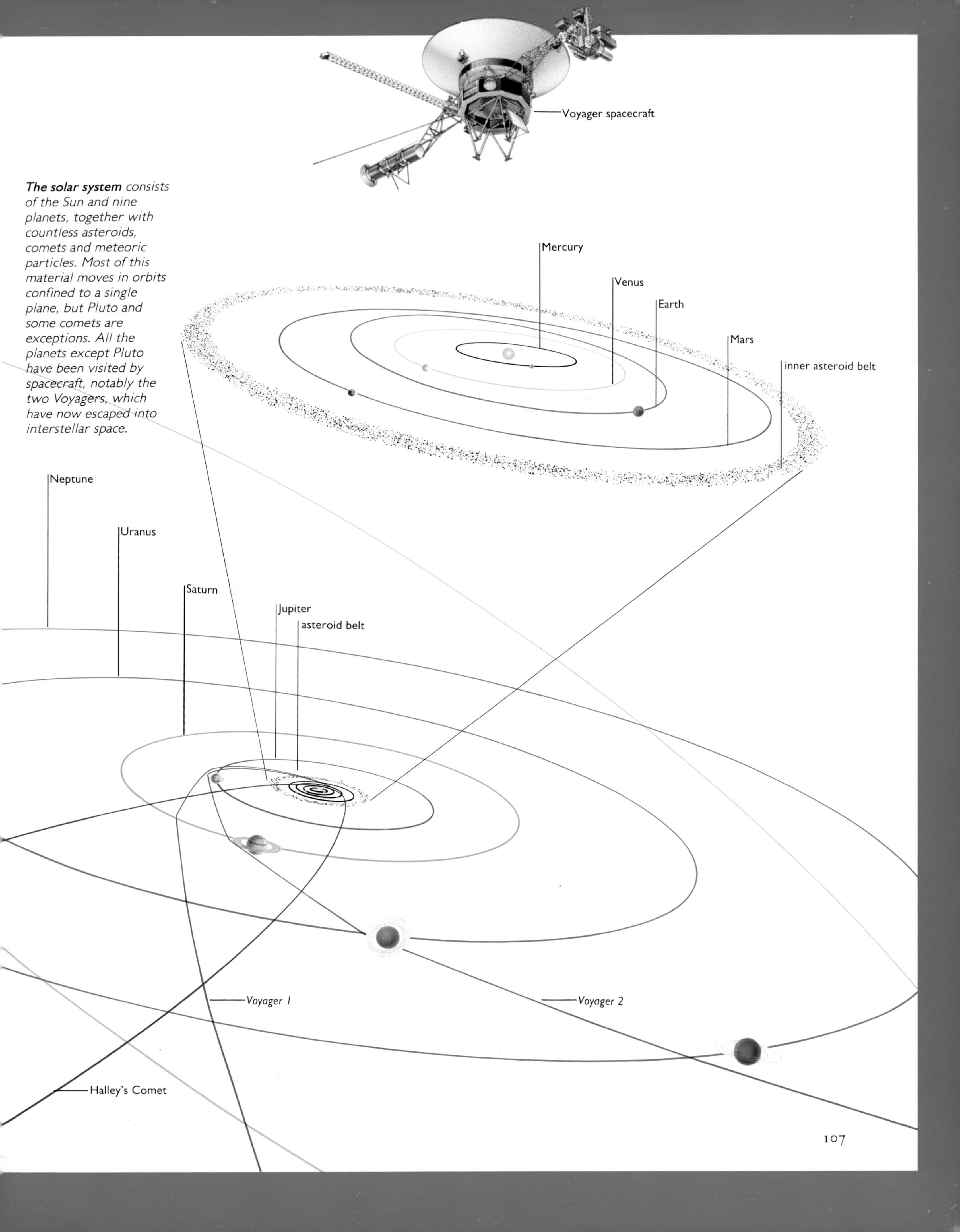


***The solar system*** *consists of the Sun and nine planets, together with countless asteroids, comets and meteoric particles. Most of this material moves in orbits confined to a single plane, but Pluto and some comets are exceptions. All the planets except Pluto have been visited by spacecraft, notably the two Voyagers, which have now escaped into interstellar space.*

# MERCURY

## • *Attendant of the Sun*

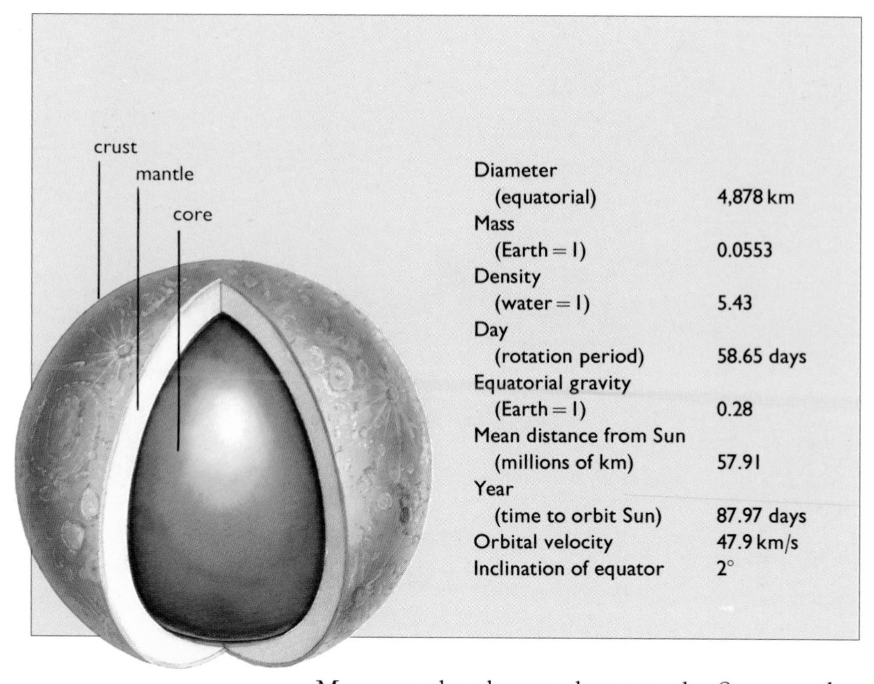


| | |
|---|---|
| Diameter (equatorial) | 4,878 km |
| Mass (Earth = 1) | 0.0553 |
| Density (water = 1) | 5.43 |
| Day (rotation period) | 58.65 days |
| Equatorial gravity (Earth = 1) | 0.28 |
| Mean distance from Sun (millions of km) | 57.91 |
| Year (time to orbit Sun) | 87.97 days |
| Orbital velocity | 47.9 km/s |
| Inclination of equator | 2° |

Mercury, the closest planet to the Sun, can be observed either just before dawn or after sunset as a bright, silvery, starlike object. Planetary scientists have discovered that it is a parched, airless, cratered wilderness, baking in the fierce heat of the Sun.

Since its orbit lies within the Earth's, Mercury presents phases similar to our Moon. So when at its nearest to Earth Mercury is seen merely as a relatively featureless thin crescent, even through the largest telescopes. The whole disc is seen only when it is farthest away on the opposite side of the Sun. This has made it very hard for ground-based observers to examine the planet.

The problems of observing Mercury made it difficult for astronomers to make accurate measurements and observations of the surface. Only in 1965, when radar pulses were sent out and successfully received back on Earth after bouncing off the planet's surface, did it first become possible to get an accurate fix on Mercury's rotation period. But it was not until a decade later that the value could be confirmed at 58.6461 days by the *Mariner 10* spacecraft, which passed close to Mercury in March and September 1974 and March 1975.

The surface of Mercury is crater-pitted and looks just like our Moon's, but Mercury does not have the large lava plains, or "seas", which dominate parts of the lunar landscape. Of the multitude of Mercurian craters, well over 230 have been named. The largest – called Beethoven – is 625 kilometres across. Planetary astronomers believe that craters are still being formed on Mercury, as they are on the Moon. But, because of the general appearance of Mercury's surface, the rate of impact there is reckoned to be about 20 percent greater.

Mercury is never farther from the Sun than 69.7 million kilometres and, with its elliptical orbit, it gets as close as 45.9 million kilometres. Thus the Sun dominates the planet. Because it is so close, it gets at least 4.7 times more heat, light and other radiation per unit area than the Earth. Because of this its surface temperature can reach as high as 467°C.

The heat, combined with Mercury's low gravitational field, caused the planet's original atmospheric gases to boil off into space aeons ago. Today, such atmosphere as there is con-

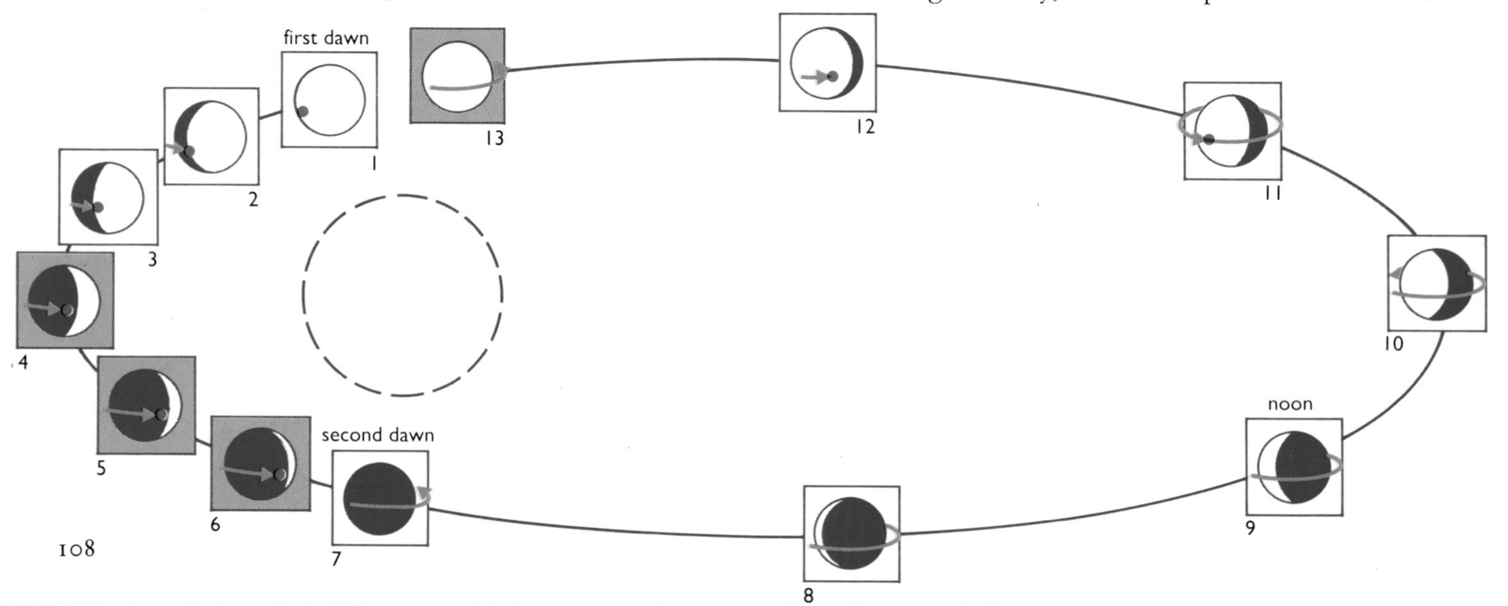


***A double dawn is seen*** *at some places on Mercury. Such a location (red dot) experiences its first dawn (1) when Mercury is accelerating as it nears the Sun. The planet turns through only a small angle (2–3) before it swings past the Sun, so night overtakes the given spot (4). The second dawn comes a little later (7). Daylight (8–13) lasts almost a Mercurian year.*

sists of hydrogen and helium, gases from the solar wind briefly held back as it streams past the planet. But at only one million-billionth the density of Earth's atmosphere, it is far too thin to have any measurable effect on surface conditions.

Mercury's low mass contributes to its inability to retain an atmosphere. It is only 5.5 percent as heavy as Earth, so its escape velocity is 2.6 times less. Although only 1.4 times larger than our Moon in diameter, and apparently quite like it, with its cratered surface, Mercury is 1.7 times as dense. This is 5.4 times the density of water, almost the same as that of the Earth. Planetary scientists have concluded that Mercury must resemble the Earth in its internal structure, with a central iron-nickel core.

But to fit the theory, it is calculated that its core must contain about twice as much iron as Earth's and have a diameter of 3,600 kilometres. So it seems that Mercury's very dense solid core is actually larger than our entire Moon. Outside this huge core there is thought to be a relatively thin rocky mantle some 600 kilometres deep. Over the mantle is a light crust that at its deepest is no more than 66 kilometres thick.

Despite its bleak surface, Mercury is not totally inert. It has "hot" regions, caused by heat from volcanic activity below the surface, which are usually known as "hot poles" because they lie on opposite sides of the planet. *Mariner 10* also showed that Mercury has a weak magnetic field, about one-hundredth the strength of Earth's. However, it is stronger than those of the Moon, of Venus and of Mars. As with the Earth, the magnetic poles of Mercury do not coincide with its poles of rotation; they are about 11 degrees away. But the existence of the field does not tie in with the idea that the core is solid iron, since magnetic fields are thought to be generated by molten iron cores.

A magnetic field does offer some protection against the radiation from the Sun but the weak field of Mercury, and the virtual absence of atmosphere, mean that the surface is continually bombarded by dangerous ultraviolet radiation and X-rays. This makes Mercury one of the most inhospitable members of the Sun's family of planets.

***Mercury is covered with craters*** *(above), caused by a heavy meteoric bombardment very similar to that which the Moon has experienced. The spacecraft* Mariner 10 *obtained this mosaic of views of Mercury during March 1974.*

***A relatively new crater*** *(left), about 12 km across, lies within an older crater basin. This picture was taken from a distance of about 20,700 km.*

physical features of Venus were the subject of much speculation. Most theories were based on similarities in size between Earth and Venus, on the fact that, since it orbits much nearer to the Sun, Venus receives twice as much solar radiation as the Earth does, and on the fact that the albedo (reflectivity) of the planet is relatively high. Scientists deduced that vast oceans covered much, if not all, of the Venusian surface. This view remained popular until the spacecraft showed otherwise.

So far there have been two types of spacecraft investigation of Venus: landers and orbiters. Thirteen probes – some more successful than others – have landed on the planet. Some of these took measurements of the atmosphere on the way down before crashing into the surface, while others made soft landings and sent back video images of the surface and analysed the soil. Most did not transmit for long – a couple of hours at the most.

Both *Venera 13* and *Venera 14* – which arrived at Venus in 1982 – made soft landings. Each craft sent back video images of a complete panorama (180 degrees) of a different part of the surface. The first, *Venera 13*, showed a relatively smooth terrain, with a surface which seemed to have a sandy covering of small grains, probably eroded – some scientists suspect originally by water – and then cemented together by atmospheric droplets and gases. The surface was also strewn with debris of small rocky pieces of various sizes.

*Venera 14*, which landed almost 950 kilometres from *Venera 13*, showed a slightly different scene. Again, some small rocks and pebbles appeared scattered on the surface, but otherwise the terrain was covered by flat pieces or plates of material, which showed signs of having suffered erosion followed by cementation. The plates had a sharp, angular appearance, but without any sandy or granular covering. Planetary geologists suggest that this is an area that was formed comparatively recently, probably about 10 million years ago. Analysis of the soil itself showed it to be a kind of basalt, similar to that found on the seabed on Earth, though with a stronger concentration of potassium.

The second method of investigating the surface by spacecraft has been by radar measurements made during close approaches to the planet. Short-wavelength radio pulses penetrated the cloudy atmosphere and reached the surface of Venus, from which they were reflected back and received by the same radar equipment. Computer analysis of the results has made it possible to chart the surface; the Magellan spacecraft in particular has provided astonishingly detailed pictures.

Most of the surface – some 70 percent – is composed of flat plains. Of the remainder, 30 percent is land depressed below the level of the plains. The balance consists of highlands, though these are mainly concentrated in two

***The surface of Venus*** *(below), shown in a composite image from the Russian soft-lander spacecraft* Venera 13, *part of which is visible in the foreground of the picture. Precisely what causes the weathering of the rock is not known, as there is no water and little wind at the surface to cause erosion. A radar picture taken late in 1990 by the Magellan spacecraft (left) shows three craters on one of the northern plains of Venus. The bright colours indicate rough ground, the darker colours smoother terrain. The diameters of the craters range between 35 km and 65 km.*

areas, one in the north and the other almost in line with the Venusian equator. The name given to the northern region is Ishtar Terra, with its Maxwell and Akna mountain ranges in the east and toward the west of the high region respectively. Ishtar Terra is not all highland and covers a large area, larger than the continent of North America. Most of it lies a few kilometres above the general flat surface but the Maxwell mountains tower 12 kilometres above that already raised surface, while a large flat plateau to the west covers an area over 2,500 kilometres across.

The highland area on the equator is known as Aphrodite Terra and it covers an area half the size of Africa. It seems rougher and also more complex than Ishtar, with some deep straight canyons in the eastern central region. The latter are hundreds of kilometres wide and over 1,000 kilometres long; some are almost 3 kilometres deep.

The Venusian lowlands are either round or broad and long. Here and there are some slight circular depressions, often with a central "mountain". These depressions range from 40 to 1,700 kilometres in diameter. They may well have been caused by impacts. If so, this indicates that part at least of the Venusian surface is several billion years old.

Taking all these radar-mapped features into consideration, planetary scientists appear now to be convinced that the surface of Venus has suffered impacts, volcanic eruptions and the folding and faulting of strata. The actual surface is surprisingly varied, and this suggests that the geological changes which gave rise to the surface occurred before the Venusian atmosphere evolved into its present state.

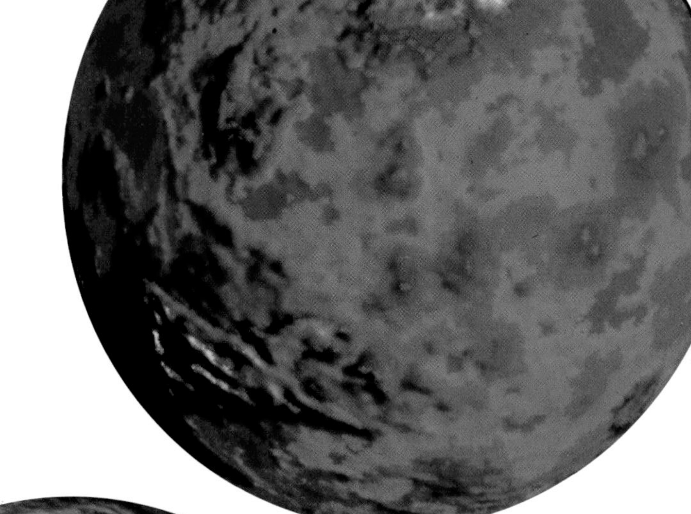

***Venus mapped by radar*** *(right and below), shown in computer-generated images. The upper globe shows the highland regions of Ishtar Terra in yellow, at the top near the unmapped north pole. The large equatorial highland zone of Aphrodite Terra appears at the bottom left in green. Most of the planet is lowland, shown in blue. The lower globe gives a good view of the Aphrodite Terra region.*

# EARTH

## • *The living planet*

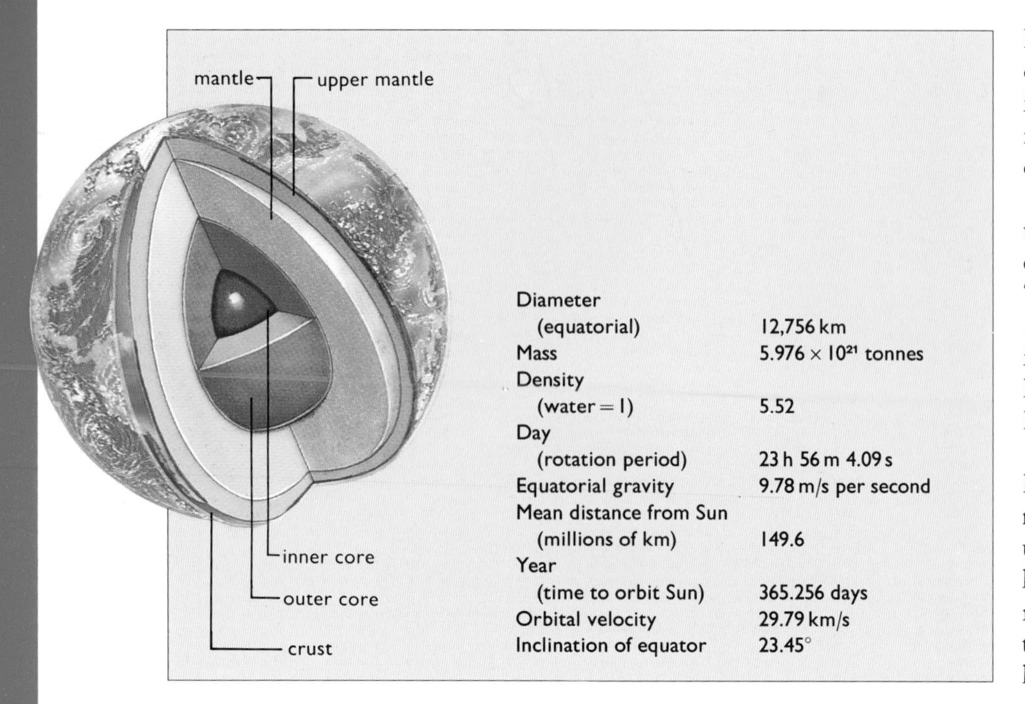


| | |
|---|---|
| Diameter (equatorial) | 12,756 km |
| Mass | $5.976 \times 10^{21}$ tonnes |
| Density (water = 1) | 5.52 |
| Day (rotation period) | 23 h 56 m 4.09 s |
| Equatorial gravity | 9.78 m/s per second |
| Mean distance from Sun (millions of km) | 149.6 |
| Year (time to orbit Sun) | 365.256 days |
| Orbital velocity | 29.79 km/s |
| Inclination of equator | 23.45° |

Naturally enough, scientists know more about our beautiful, temperate world than any other in the universe. They are on hand to experiment directly, to interpret the evidence and to draw conclusions.

From the available evidence, planetary scientists are now reasonably certain that at the centre of the Earth there is a nickel-iron core. This is divided into two parts: a metallic centre 2,500 kilometres in diameter and, surrounding it, a layer where the iron and nickel are in a liquid state. This liquid section is about 2,200 kilometres thick.

Above the core, the Earth has a thick, rocky layer about 2,900 kilometres deep, called the mantle. The lower mantle is fairly rigid; the upper mantle consists of a rather more plastic layer known as the asthenosphere and a thin, rigid outermost layer, some 100 kilometres thick, the lithosphere. The topmost part of the lithosphere is the Earth's crust – the ground on

### The Earth's crust

Geologists now think that the Earth's crust is divided up into six major, and a number of minor, rigid plates, floating on a "plastic" layer of the mantle. These plates are moved about by rising convection currents, which bring heat from the Earth's molten core to the surface.

Wherever this happens, usually in an ocean bed, new ocean floor is created, pushing the crust each side outward. The ocean bed thus created is made of the comparatively dense rock basalt. At present the rate of drift of Europe and North America means they are moving apart by about 2 centimetres a year. Where new material is being created there is an underwater ridge at which red-hot new crust material comes into contact with icy cold sea water.

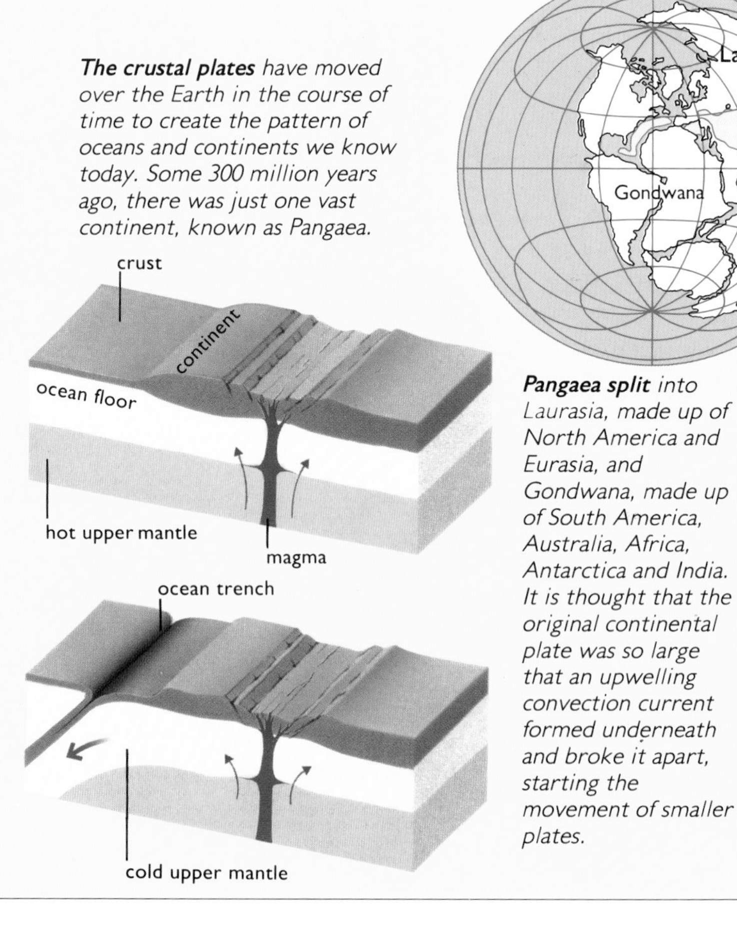


***The crustal plates** have moved over the Earth in the course of time to create the pattern of oceans and continents we know today. Some 300 million years ago, there was just one vast continent, known as Pangaea.*

***When magma wells up** beneath a continent from the hot regions of the mantle (right, above), it can force the crust to separate. The thinner crust at this point may form new ocean floor. Where two plates collide (right, below), one may dive beneath the other, creating a trench.*

***Pangaea split** into Laurasia, made up of North America and Eurasia, and Gondwana, made up of South America, Australia, Africa, Antarctica and India. It is thought that the original continental plate was so large that an upwelling convection current formed underneath and broke it apart, starting the movement of smaller plates.*

which we live. This varies in thickness; the continental crusts are between 30 and 40 kilometres thick, while the crust of the seabed is only 5 kilometres thick.

The lithosphere layer is constructed of many plates, which move across the mantle. Evidence for large-scale movement of continental land masses and of the sea floor comes from examination of rock formations, fossils, the magnetic polarity of rocks and the shapes of the continents. This evidence also shows that at an earlier stage of the Earth's history – probably some 160 million years ago – the world's continents formed two vast land masses. These gradually split apart to make the continents with which we are familiar today. The drifting of the continents still continues.

Above the Earth's surface lies the atmosphere, which is absolutely vital for the existence and maintenance of life on Earth. It is made up of a number of gases: nitrogen (77 percent); oxygen (21 percent); water vapour (1 percent); and the inert gas argon (0.93 percent). There are also some traces of carbon dioxide, neon, helium and sulphur.

One most important feature of the atmosphere is that it acts like a blanket around the Earth, preventing heat from the Sun's warming from being radiated away into space. This is the so-called greenhouse effect, whereby certain gases in the atmosphere reflect back infrared radiation instead of allowing it to escape. These gases include nitrous oxide, methane and, by far the most important, carbon dioxide.

Since the start of the Industrial Revolution, it has been estimated that the amount of carbon dioxide in the atmosphere has nearly doubled as a result of burning "fossil" fuels, such as coal, oil and gas, and of large-scale deforestation. Scientists currently forecast that the average temperature of the Earth's atmosphere might rise significantly because of the increase in carbon dioxide. The effects of this are unpredictable and, in the worst scenario, could be disastrous.

The weather we experience is caused by the complex circulation of the atmosphere driven by heat from the Sun. Warm, moist air rises over the tropics and moves in the direction of the poles; at around 30 degrees latitude in both hemispheres, it cools and sinks, then returns to the equator. There are also eddies, generated by temperature differences between oceans and continental land masses and, to complicate things further, the comparatively rapid rotation of the Earth whirls these currents and eddies around into an involved and ever-changing pattern.

In principle, however, the circulation of the Earth's atmosphere can be described fairly simply. East winds, coming from the poles, meet west winds from the temperate latitudes; while in the tropics there are trade winds, from the northeast in the northern hemisphere and the southeast in the southern. The trade winds are separated by low-pressure "depressions" in the equatorial regions.

***To an observer from space*** *the Earth's constantly changing weather patterns contrast strongly with the fixed surface features of the continents and the oceans. A storm occurs when a heated mass of air rises. As it does so it cools and water vapour condenses, causing cloud and rain. New air is sucked in to replace that which is rising. Because of the Earth's rotation it spirals in, creating the characteristic cloud patterns of a storm.*

# MARS

*The red planet*

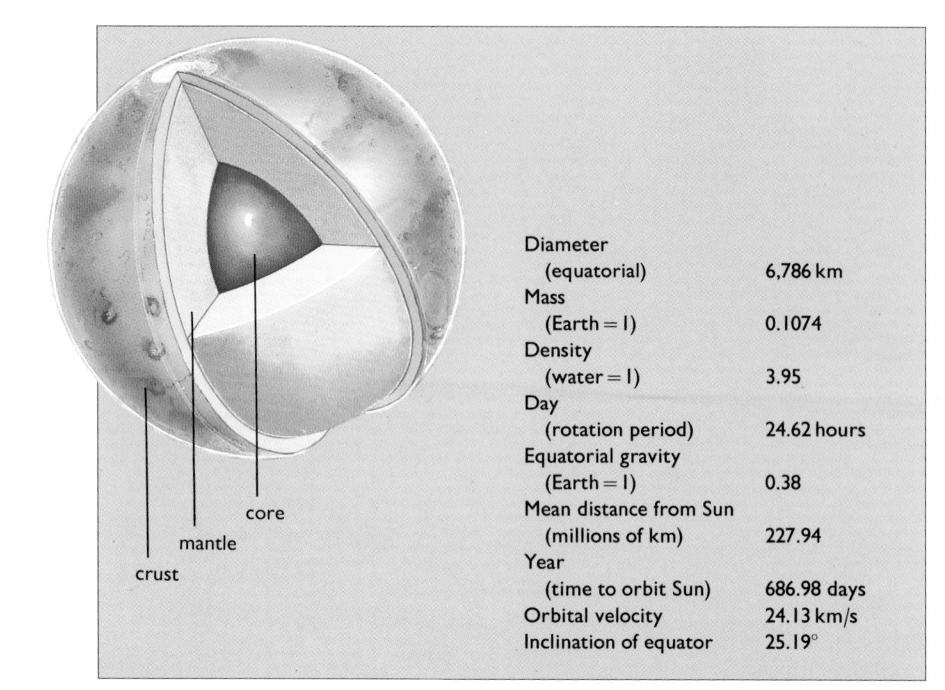


| | |
|---|---|
| Diameter (equatorial) | 6,786 km |
| Mass (Earth = 1) | 0.1074 |
| Density (water = 1) | 3.95 |
| Day (rotation period) | 24.62 hours |
| Equatorial gravity (Earth = 1) | 0.38 |
| Mean distance from Sun (millions of km) | 227.94 |
| Year (time to orbit Sun) | 686.98 days |
| Orbital velocity | 24.13 km/s |
| Inclination of equator | 25.19° |

Mars, the brilliant red planet with white poles, has long intrigued humankind. This, the last of the terrestrial planets, has exerted its fascination most especially through comparisons with features of our own world – including the possibility of life. But Mars has become an object for serious scientific investigation only in the last two centuries. Of modern studies, most significant are the revelations of space missions, which contradict the conclusions of earlier, Earth-based observations.

The diameter of Mars is only just over half that of the Earth, and its average density some 30 percent less. Since most of the mass of a terrestrial planet is concentrated in the core, that of Mars cannot, therefore, be large. Astronomers now favour a model of the interior similar to that of the other terrestrial planets, namely a core of iron and iron compounds with a diameter of some 3,000 kilometres. Outside the core is a mantle of silicate materials some 1,800 kilometres thick, overlain by a crust a little more than 100 kilometres thick.

Taking 686.98 days – close on 1.9 Earth years – to complete an orbit around the Sun, Mars rotates on its axis once every 24.623 hours, giving a day length just a little more than the terrestrial one. But because its orbit is five and a half times more eccentric than the Earth's, Mars undergoes more marked changes in the length of its seasons.

Seasons on Mars are similar to those on Earth. This is due to the inclination of its axis

***Valles Marineris,*** *the giant Martian canyon that seems to show evidence of past erosion by flowing water, is captured above by the cameras on board* Viking 1, *which visited Mars in 1976. A mosaic of images from* Mariner 9 *was used to create the close-up picture (right) of the north polar ice cap in 1971. The striking spiral pattern was probably caused by wind erosion.*

of rotation of just over 25 degrees (Earth's is 23.4 degrees). The eccentricity of its orbit means, however, that the seasons on Mars are more unequal than on Earth: in the northern hemisphere of Mars spring lasts 194 days and is 51 days longer than autumn.

The seasonal changes on Mars have engaged astronomers' attention since the 19th century, for two very noticeable associated effects can readily be observed from Earth. The first is a discoloration of the reddish surface: areas that are less red – and sometimes even grey or green – begin to spread during the Martian spring. The second effect is that, during the spring in each hemisphere, the white polar cap starts to shrink. Early astronomers linked the two phenomena, and it became widely accepted that the discoloration was due to areas of vegetation spreading as they were irrigated by water released by the partial melting of a polar cap.

Like Venus, Mars has an atmosphere composed mainly of carbon dioxide (95 percent), with 2.7 percent nitrogen and 1.6 percent argon. Oxygen accounts for 1.3 percent and water vapour for no more than 0.3 percent. But the Martian atmosphere is much thinner than that of Earth. At ground level, it exerts a pressure only 0.7 percent of Earth's atmosphere. And because carbon dioxide is a very efficient radiator of heat (infrared) radiation, temperatures near ground level can drop at night to well below −53°C, and to as low as −133°C at the winter pole.

Because Mars has no oceans, its entire surface responds quickly to changes in temperature. Temperature differences lead to the setting up of strong winds, which follow the hotter air that is warmed by the Sun. These Sun-following, or tidal, winds can reach velocities of 45 to 90 metres per second at the Martian surface.

At such speeds these winds can make the sand grains skip along the ground, and send dust up into the atmosphere. The dust grains are so small – about one-hundredth of a millimetre across – that they stay suspended in the atmosphere for months. Such dust storms are mainly local, but twice in each Martian year the disturbances are so widespread that great tracts, or even the whole, of the Martian surface become invisible from Earth.

***The characteristic white caps*** *at the north and south poles of Mars are made up of a mixture of solid carbon dioxide and water ice.*

Under good observing conditions it is possible to see from Earth long straight streaks on the Martian surface. The astronomer Giovanni Schiaparelli observed and, indeed, plotted a network of these. He called them *canali* – Italian for channels – but some astronomers linked the word to artificial canals, partly because such networks of intersecting straight lines gave every appearance of an artificial construction.

The notion of an "unnatural" canal system led to the implication that intelligent beings must exist on Mars. These creatures would be partially supported, presumably, by cultivation of plants, of which the grey-green areas seemed to be evidence. The greatest protagonist of this widely held view was the astronomer Percival Lowell, who at his personal foundation, the Lowell Observatory in Flagstaff, Arizona, plotted the canals with diligence during the 1890s.

Until the space probes of the 1960s and '70s, most astronomers accepted the existence of long straight streaks on Mars and were happy to call them "canals", although almost no one thought that they were certain evidence for the existence of Martians. Attitudes remained unchanged until 1965, when the *Mariner 4* spacecraft took some close-up pictures of the Martian surface. Although not detailed, they revealed conclusively that there were no canals; instead Mars was a barren land, with craters scattered over it. It looked more like the Moon than the Earth.

Other Mariner probes provided similar visual evidence, but in 1971 when *Mariner 9* took pictures showing details as small as 100 metres across, it became clear that the Martian surface displayed not only impact craters but also vast canyons, volcanoes, and what appeared to be dried-up river beds. Five years later, two Viking spacecraft landed on the surface. There they took close-up pictures of Martian rocks and made a chemical analysis of the ground itself.

*Viking 1* landed on a fairly cratered area in the Chryse Planitia (Chryse Plain). Spacecraft pictures showed that rocks were strewn about nearby, while on the horizon a number of craters could be seen; these reached up to 600 metres in diameter. The surface was "sandy".

The second spacecraft, *Viking 2*, landed 8,846 kilometres to the northeast on a flat surface with fractured features. Some 200 kilometres to the south was a large crater now named Mie, with a diameter of some 100 kilometres. It seems that *Viking 2* came to rest on the crater wall. The rocks near the Viking landings were pitted with holes, possibly evidence of the escape of gas once contained in them, or perhaps of ongoing erosion by windswept dust.

The surface of Mars, as revealed in close-up, is certainly red. Two-thirds of the sand particles consist of silicon and iron, and there is a strong concentration of sulphur – more than 100 times that found in terrestrial material. The red coloration is due, quite simply, to rust (iron oxide with other iron impurities, notably iron sulphide). Other evidence shows the surface to be an "iron-rich clay". The discoloration once thought to be vegetation is now known to be merely a chemical reaction on the rock-strewn surface. And both Vikings showed that the Martian sky is pink, not blue, a result of the suspension of fine dust particles of iron oxide.

The orbiter sections of the Viking spacecraft circled the planet and mapped the Martian surface down to details as small as 150 metres across, with some selected places at a resolution of 8 metres. The Vikings revealed some notable features, including the giant volcano Olympus Mons, and the Valles Marineris, a system of vast canyons in the equatorial regions which is well over 5,000 kilometres

***A Martian dune field**, pictured in the early morning light by* Viking 1 *in 1976, shows remarkable similarities to an Earthly desert. The sharp dune crests and small deposits of rock some distance from the dunes have been created by wind. The large rock at the extreme left, "Big Joe", is about 2 m long. The strip down the centre of the picture is the spacecraft's meteorology boom, an extending arm carrying weather instruments.*

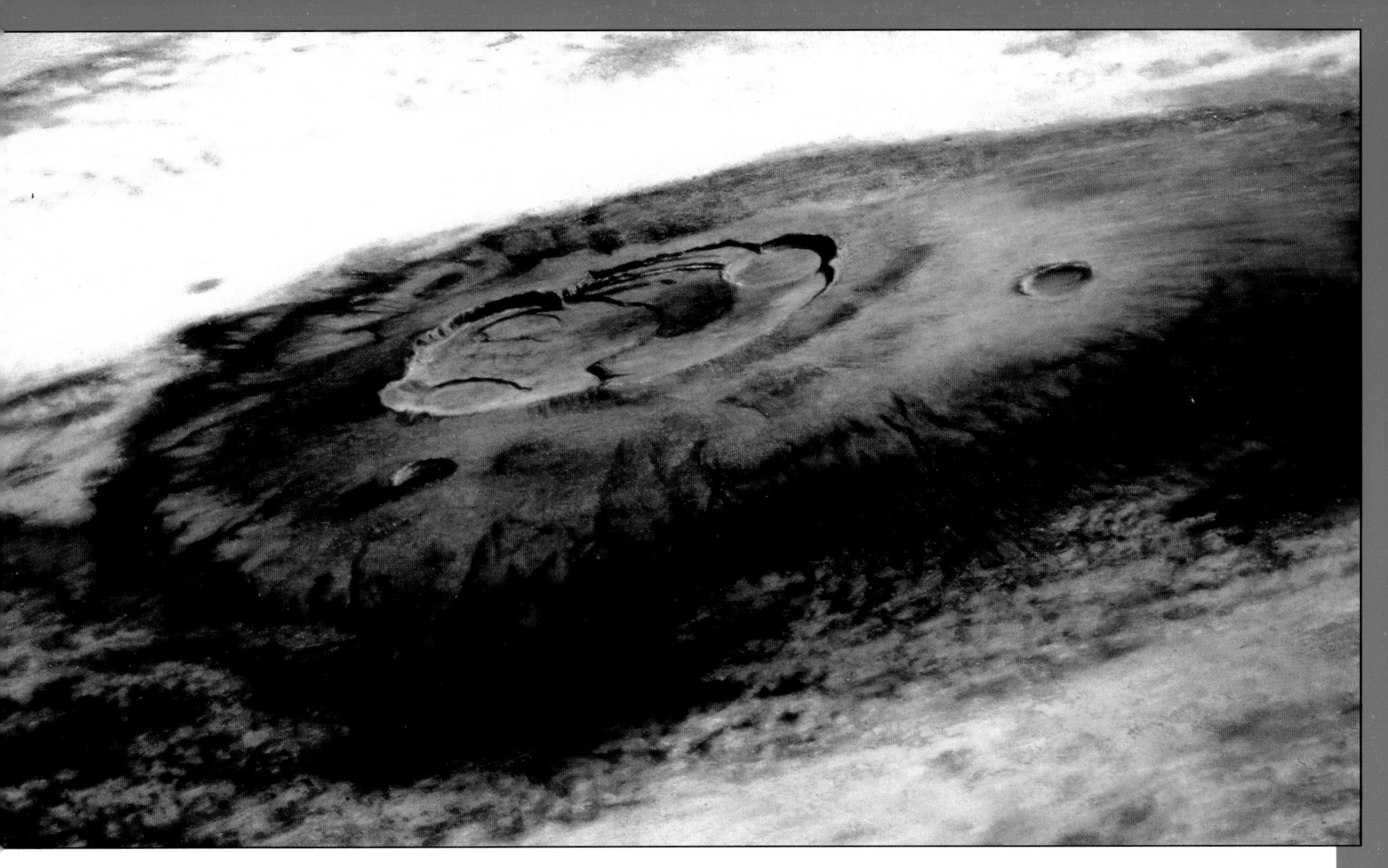

long and some 7 kilometres deep. Many other canyons have also been mapped, two of them over 1,000 kilometres long.

Arguably the most intriguing fact brought to light by the Vikings is a confirmation that the surface of Mars does contain features which seem to be dried up water channels. (These are *not* the "canals" previously studied; those have been shown to be an optical illusion.) There is no liquid water on the Martian surface now, but the channels seem to support the idea that, early on in the history of the solar system, during the time when the Sun was developing, it underwent a brief period of intense radiation. If so, this could have removed the original planetary atmosphere, which could well have been rich in water vapour and carbon dioxide, so allowing the presence of liquid water on the planet.

Another theory is that the channels may have been formed by the melting of ice during previous periods of intense volcanic activity. What seems certain, however, is that there is no evidence of life forms on Mars.

***Olympus Mons** (above), a vast shield volcano built up from successive eruptions of lava, is 600 km across and more than 24 km high. The largest equivalent structure on Earth is Mauna Kea in Hawaii, which stands only 9.7 km above the seabed.*

# JUPITER

## *Lord of the planets*

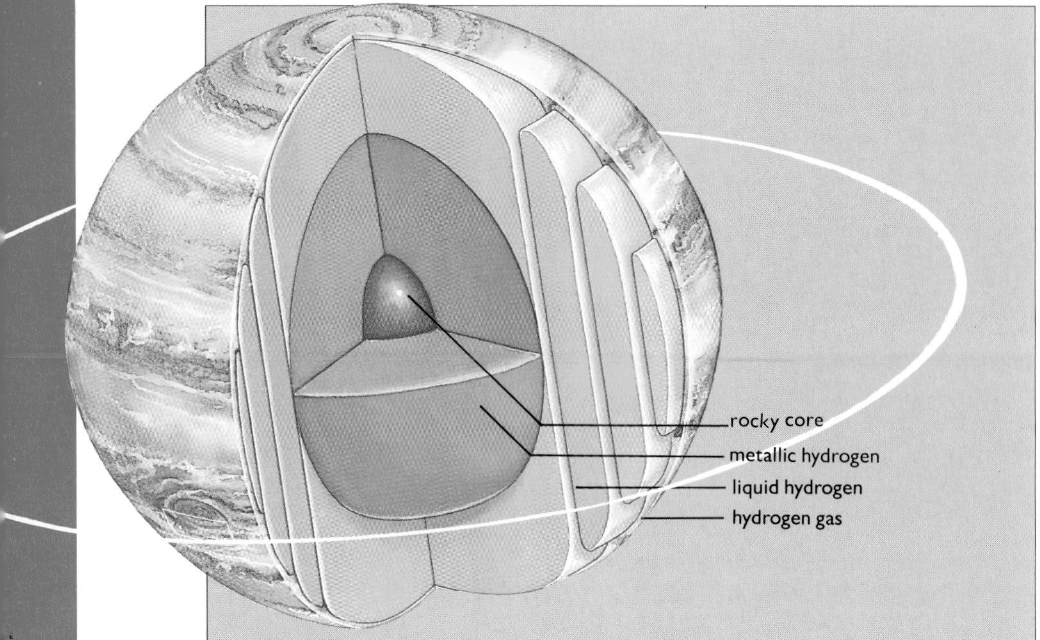

| | | | |
|---|---|---|---|
| Diameter (equatorial) | 142,984 km | Equatorial gravity (Earth = 1) | 2.34 |
| Mass (Earth = 1) | 317.94 | Mean distance from Sun (millions of km) | 778.33 |
| Density (water = 1) | 1.33 | Year (time to orbit Sun) | 11.86 years |
| Day (rotation period) | 9.841 hours | Orbital velocity | 13.06 km/s |
| | | Inclination of equator | 3.12° |

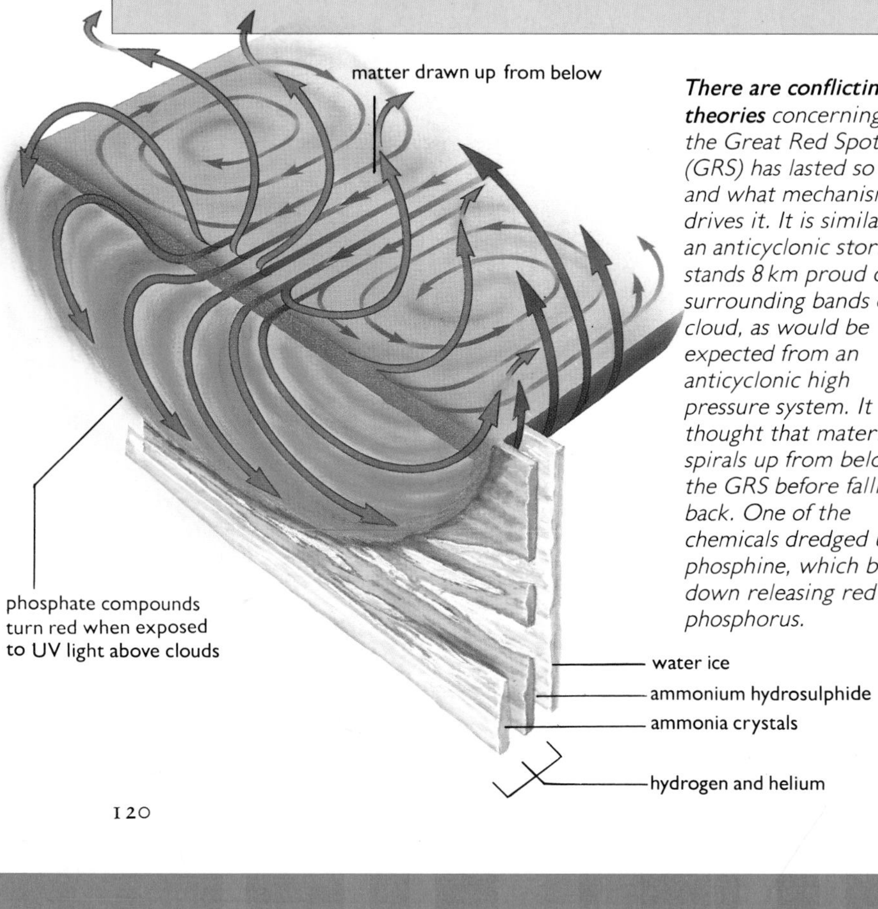

***There are conflicting theories*** *concerning why the Great Red Spot (GRS) has lasted so long and what mechanism drives it. It is similar to an anticyclonic storm; it stands 8 km proud of the surrounding bands of cloud, as would be expected from an anticyclonic high pressure system. It is thought that material spirals up from below in the GRS before falling back. One of the chemicals dredged up is phosphine, which breaks down releasing red phosphorus.*

Jupiter is the closest to the Sun of the solar system's gas giants – huge planets possessing vast, dense atmospheres extending thousands of kilometres into the space around them. They contain the major proportion of the system's orbiting planetary mass and Jupiter alone accounts for more than 71 percent of this.

In essence, Jupiter is a vast ball of gas with a dense central core, with a diameter of some 30,000 kilometres, and a temperature between 20,000 to 30,000 K at the centre. According to calculations, the core appears to be mainly a mixture of iron and silicates, with some ices of water, ammonia and methane converted into metallic form by the immense pressure of the overlying material. The pressure at the surface of the core is 45 million times the atmospheric pressure on Earth, or 450 million kilograms per square centimetre.

Outside the central core is a zone of hydrogen, also under great pressure, in this instance 2 million times our atmospheric pressure – over 2 million kilograms per square centimetre. Under such pressure hydrogen becomes metallic, and its density is four times what it would be if it were in a gaseous state. This outer core extends for some 30,000 kilometres above the central core.

Outside the metallic hydrogen lies another zone of hydrogen which, while still under enormous pressure, is not metallic but takes the form of liquid molecular hydrogen. This extends out a further 25,000 kilometres. Above this is a gaseous, hydrogen-rich atmosphere 1,000 kilometres thick. It is this outer layer that can be seen from Earth and has been seen at close range by the Voyager spacecraft.

Even in a relatively small telescope Jupiter is a fine sight with its banded atmosphere and four bright satellites (pp. 138–39). What can be seen of Jupiter is a squashed, or oblate, disc, wider at the equator than at the poles. This form of the planet is due mainly to its largely fluid nature and to its fast axial rotation; a Jovian day lasts not 24 but only 9.8 hours.

The observable Jovian atmosphere is not only banded but displays many features which show that there is a strong circulation pattern in the outer gaseous regions. The circulation is most rapid in the equatorial regions and slower toward the poles. Because of the different rotation speeds, the atmosphere presents an ever-changing series of features. Sometimes it is enlivened by the appearance of the Great Red Spot (GRS), which has been observed through ground-based telescopes ever since the 1650s. In addition, the four larger satellites can be observed as they orbit the planet, becoming eclipsed as they move behind it and then crossing the disc, casting shadows on the cloudy surface.

Yet if Jupiter is spectacular using Earth-based telescopes, it proved to be even more striking when observed close up by the *Voyager 1* and *Voyager 2* spacecraft in March and July 1979. From the evidence of these observations, we now know that the surface clouds are 90 percent hydrogen. Helium accounts for almost all the remaining 10 percent, but there are traces, in descending order, of ammonia, methane and water vapour.

The continual circulation of the bands of clouds effectively distributes the heat from the interior of the planet. The wind velocities, measured with reference to the eastward rotating mass of the atmosphere, range from easterlies travelling at up to 120 metres per second to westerlies (where the winds are slower than the general rotating mass) travelling at over 50 metres per second.

Using cameras and infrared detectors, the Voyager spacecraft have probed a little way below the outer cloud surface and also up into the atmosphere above this layer. Just above the clouds, where the pressure is five times that of Earth at sea level, are further brown clouds thought to be made up of water vapour and some as yet unknown compounds, probably of sulphur. The temperature here is 7°C. About 30 kilometres higher, there are reddish-brown clouds of ammonium hydrogen sulphide along with other unidentified compounds. Here the temperature has dropped to −73°C.

At about 65 kilometres above the brown clouds, there are wispy cirrus clouds of ammonia and the temperature is −133°C. However, at 90 kilometres – at Jupiter's tropopause – the

***The Great Red Spot** is the dominant feature of the southern hemisphere of Jupiter. This colour-enhanced* Voyager 1 *picture clearly shows the circulating nature of the GRS and shows small puffy features within the spot itself. It rotates counterclockwise with a period of approximately 6 days and is currently about 26,200 km long by 13,800 km wide.*

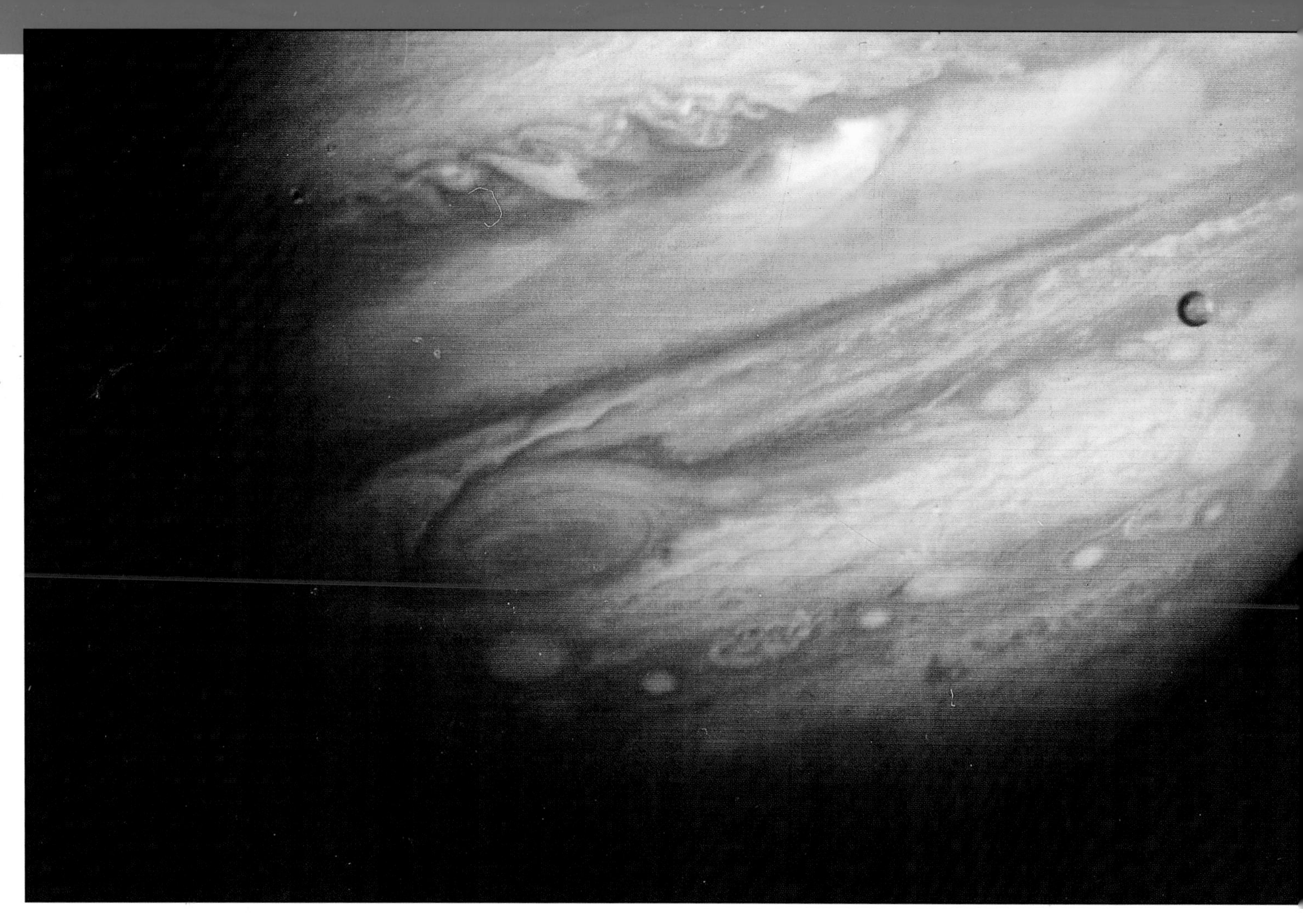

*Surrounding Jupiter* in a huge belt is a plasma ring, or torus (below). It consists of electrically charged particles left behind in a trail by volcanic activity on the moon Io. The particles are trapped by the lines of force of Jupiter's magnetic field. The torus is approximately the same shape and size as Io's orbit but is inclined so that it lies in the plane of Jupiter's magnetic equator.

temperature begins to rise again; but even at 150 kilometres it never reaches more than −113°C as Jupiter is 780 million kilometres away from the warming radiation of the Sun.

Below the clouds, within the molecular hydrogen layer, planetary scientists think there may be a series of cylindrically shaped envelopes lying one outside the other. Each has its own period of rotation giving rise to the wind currents observed in the upper cloud layer at the differing latitudes.

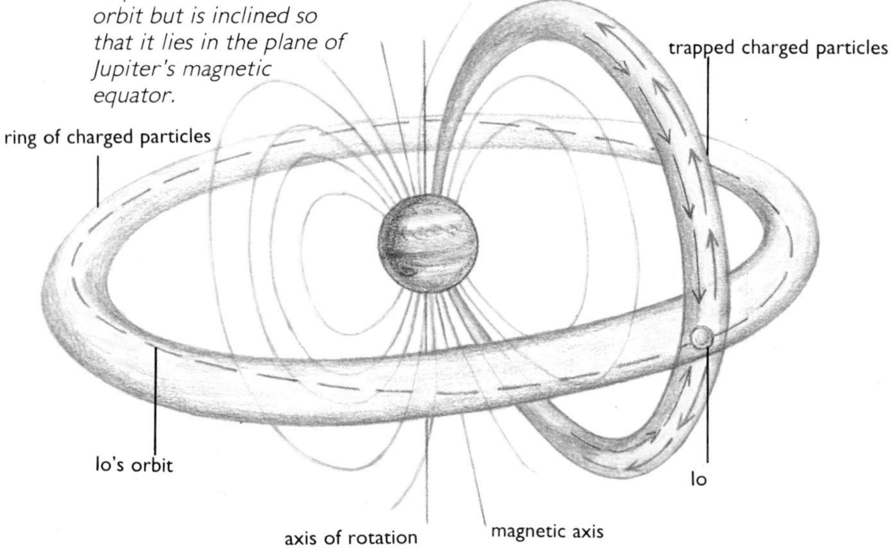


The Great Red Spot (GRS) is only one of a number of long-lasting oval patches in the cloudy Jovian atmosphere which can persist for months or even years. But since the GRS has been visible for more than 340 years, its longevity seems exceptional. Moreover, it is unusual in that it rises some 8 kilometres above the surrounding cloud mass. Exceptional also is the size of the GRS, which covers 10 degrees of Jovian latitude, making it about as wide as the entire Earth.

The ovals, including the GRS, keep to the same latitude but drift in longitude, seeming to roll between different layers of the high-velocity clouds. Voyager close-up pictures of the GRS, showing details down to 30 kilometres across, reveal that it looks like a form of cyclone, with a complex pattern of swirling atmospheric motions. Currently, planetary meteorologists are uncertain as to the causes of such atmospheric features in general and of the GRS in particular.

Jupiter emits radio waves and has a large and strong magnetic field. Like that of the Earth, this field is a dipole, similar to a bar magnet. The planet's magnetic poles are inclined at 11 degrees to Jupiter's axis of rotation and, in

***The complicated nature of the cloud belts*** *on Jupiter is shown by a* Voyager 1 *photograph (left) taken in February 1979 when the craft was 28.4 million km from the planet. Also visible are the inner moon Io and the moon Ganymede.*

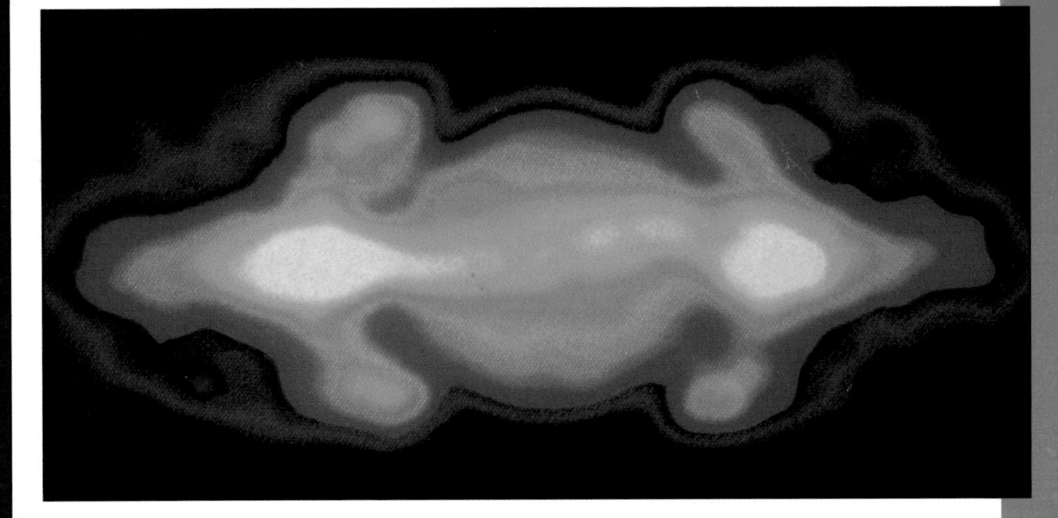

***Jupiter's radiation belts*** *shown in a false-colour image taken by the Very Large Array radio telescope in New Mexico. Jupiter's strong magnetic field traps electrons in belts equivalent to the Earth's Van Allen belts. These electrons radiate synchrotron radiation at radio frequencies. This image was taken at a wavelength of 21 cm.*

addition, the magnetic axis is offset by one-tenth of Jupiter's radius – that is, some 71,400 kilometres – from the planet's centre. As a result, the magnetic field at Jupiter's cloud surface is not equal at every point; it differs most notably between the northern and southern hemispheres.

Because of its magnetic field, Jupiter possesses a magnetosphere which extends into space around the planet. Although it has a dipolar field, Jupiter's magnetosphere is very different from that of the Earth. Two main factors contribute to this. First, Jupiter's magnetic field is about 100 times larger than the Earth's and, second, the effect on it of the solar wind is some 25 times less because Jupiter is so much farther from the Sun.

On the sunward side of Jupiter the magnetosphere extends some 2 million kilometres above the planet, but on the side away from the Sun it is more extensive. Its intense central region has a diameter of 4 million kilometres. The magnetosphere extends farther than this and is elongated away from the Sun, "blown" in that direction by the solar wind. This elongation is immense, and it is possible that the tail of Jupiter's magnetosphere extends as far as the orbit of Saturn, more than 600 million kilometres away. The magnetosphere on the sunward side is also drawn out a little toward the Sun, but nowhere near as far as that on the opposite side.

The solar wind and electrified particles in the Jovian magnetosphere together give rise to auroral displays, as they do on Earth, and an auroral arc in the Jovian sky was indeed observed by *Voyager 1* on its close approach to the planet. Such displays occur over the whole surface of Jupiter, not just in polar regions as on Earth, possibly due to low-speed electrons which have leaked through to react with the Jovian ionosphere.

Jupiter emits more radiation than it receives from the Sun. This is due to the heating caused by shrinkage of the planet's core, a process which may still be continuing, and to radioactive heating. Because of thermal radiation from inside the core, Jupiter transmits at radio wavelengths. Some of these are centimetre wavelengths, due to the motions of electrified particles in the magnetosphere; others are longer: between 10 metres and 3 kilometres. These transmissions are disturbed by one of Jupiter's nearest satellites, the eruptive Io (pp. 138–39), which passes well within 350,000 kilometres of the upper cloud layers of the planet and thus close enough to disturb the Jovian magnetosphere.

Jupiter also has a system of thin rings. The outer edge of the main ring is 50,000 kilometres above the cloud surface, with a very tenous ring extending farther out.

# SATURN

## • *The ringed world*

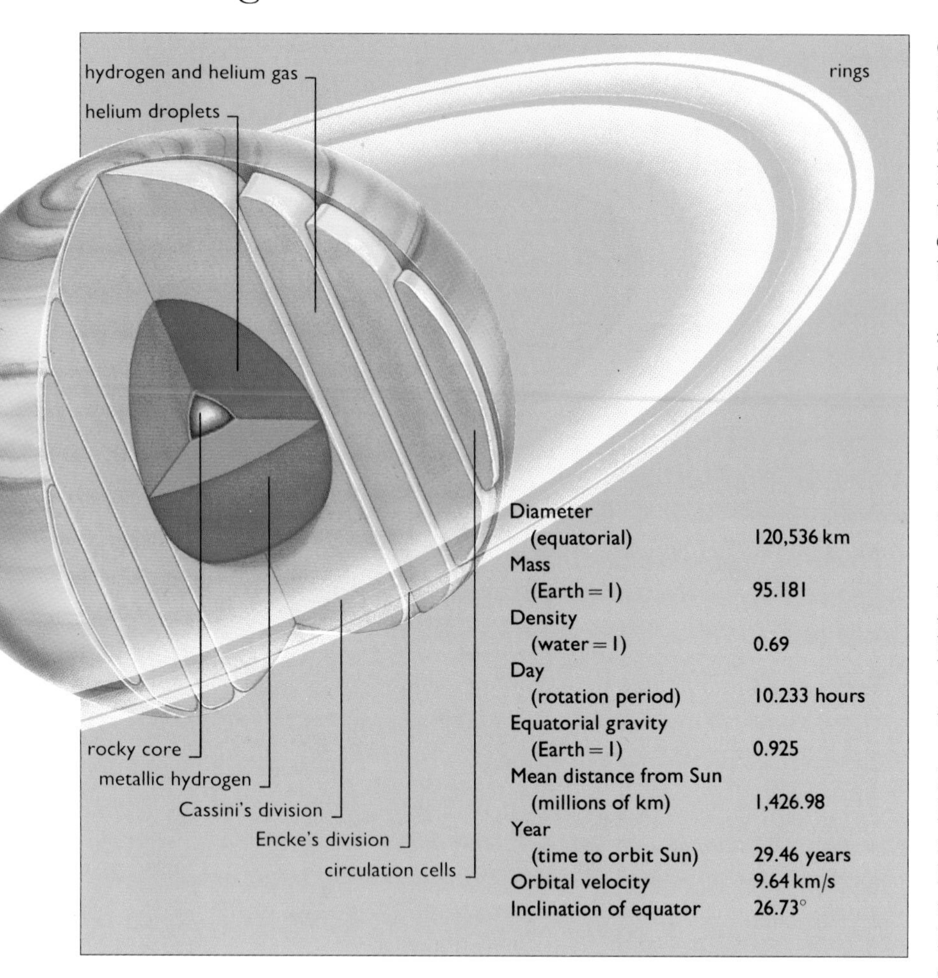


| | |
|---|---|
| Diameter (equatorial) | 120,536 km |
| Mass (Earth = 1) | 95.181 |
| Density (water = 1) | 0.69 |
| Day (rotation period) | 10.233 hours |
| Equatorial gravity (Earth = 1) | 0.925 |
| Mean distance from Sun (millions of km) | 1,426.98 |
| Year (time to orbit Sun) | 29.46 years |
| Orbital velocity | 9.64 km/s |
| Inclination of equator | 26.73° |

Of all the planets in orbit around the Sun, Saturn, with its ring system, is the most spectacular. Yet Saturn is not alone in being surrounded by rings: Jupiter, Uranus and Neptune also possess ring systems. But none of these rivals the Saturnian system either in extent or variety, and only Saturn's rings can be directly observed by telescope from Earth.

Though not as large as Jupiter, Saturn is another gas giant and it accounts for 21 percent of the total mass of all the planets. It is, however, less dense than Jupiter; while the average density of Jupiter is 1.33 times that of water, Saturn's average density is only 0.69. Thus, if Saturn could be put into a big enough bath of water, it would float.

Calculations indicate that Saturn has a central core about 25,000 kilometres across, composed of silicates, minerals and various types of ice. The temperature of the core is some 14,000 K. The overlying pressure is 10 million times the atmospheric pressure on Earth.

Above the core, stretching up some 11,460 kilometres, lies a shell of metallic hydrogen, less than a tenth of the thickness of Jupiter's metallic hydrogen region. Beyond the metallic hydrogen is a layer, some 4,200 kilometres thick, of helium droplets. Beyond this again lies the remainder of the Saturnian material, extending outward for around 29,000 kilometres. Being composed of 93 percent molecular hydrogen and 7 percent helium, it is light in weight. This forms the outer layer of Saturn, which is visible from Earth and which was photographed by the Voyager spacecraft.

Like Jupiter, Saturn is oblate (flattened at the poles), though because of its lightness and a 10-hour rotation period, its degree of flattening is more than 50 percent greater than Jupiter's. Saturn is, in fact, the most oblate of all the planets. The surface we see is crossed by bands of clouds, although these appear more curved than those on Jupiter.

The cloud patterns are, nevertheless, similar, with jet streams of fast-moving clouds giving rise, as on Jupiter, to eddies and even, at a latitude of 55 degrees south, a permanent, oval reddish spot. This is similar to the Great Red Spot (GRS) of Jupiter, but it is much smaller, with a length of only 6,000 kilometres in an east-west direction, compared with 26,200

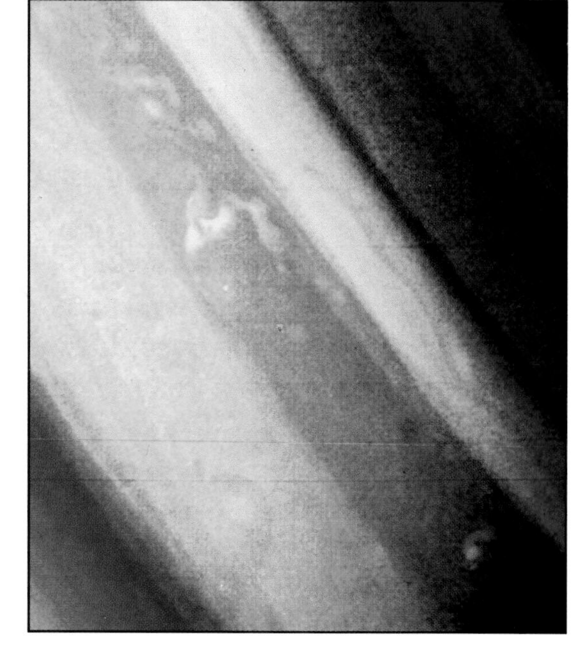

***The visible features on Saturn*** *consist of bands and other cloud patterns – ovals, eddies and interacting spots. This false-colour image of Saturn's northern hemisphere, taken by* Voyager 1 *in 1980 at a range of 9 million km, shows an isolated convective cloud in the light brown region and a thin longitudinal wave in the blue belt. The smallest features discernible in the picture are 175 km across.*

*Saturn and its magnificent ring system* *are here seen in open aspect by Voyager 2's cameras. The picture was taken four and a half days after the spacecraft made its closest approach to the planet. Saturn can never be seen in half-phase like this from Earth, as from our standpoint the Sun illuminates virtually the whole of the disc of the planet.*

kilometres for the GRS. Saturn also has other ovals at different latitudes, mainly in the northern hemisphere. On Saturn, when spots and ovals meet, they move around each other and do not merge as such features do on Jupiter.

Saturn also possesses a permanent, distinctive dark wavy line in its cloud cover; this lies at 45 degrees latitude north and extends some 5,000 kilometres in an east-west direction.

Much has yet to be learned about the behaviour of Saturn's cloud layer and the atmosphere above it, although it is now known that, as on Jupiter, the bands of cloud circulate at high velocities. Just above and below Saturn's equator these reach speeds of about 480 metres per second; yet near its poles the speeds reduce to zero, which means that they merely keep pace with the planet's axial rotation.

Above the cloud surface, the atmosphere of Saturn extends outward and, at a height of 30 kilometres or so, there are ammonia clouds. The Voyager spacecraft have shown that the upper atmosphere of Saturn has some coloured clouds as well, but the heights of these have still to be determined.

Saturn, like Jupiter, emits more radiation than it receives from the Sun. In Saturn's case this is 1.76 times more than it receives, and the internal heat, or energy, is thought to be caused by contraction of Saturn's core or by helium droplets condensing and falling deep into the planet's interior.

Saturn has a magnetic field generated by electric currents in the metallic hydrogen of the core. The field's magnetosphere – much like Jupiter's magnetosphere, although less extensive – stretches outward into space, trailing away in a direction opposite to that of the Sun but with a bulge on the sunward side. This bulge is elongated, extending 1 to 1.5 million kilometres. But Saturn's magnetic poles are very close to its poles of axial rotation, making

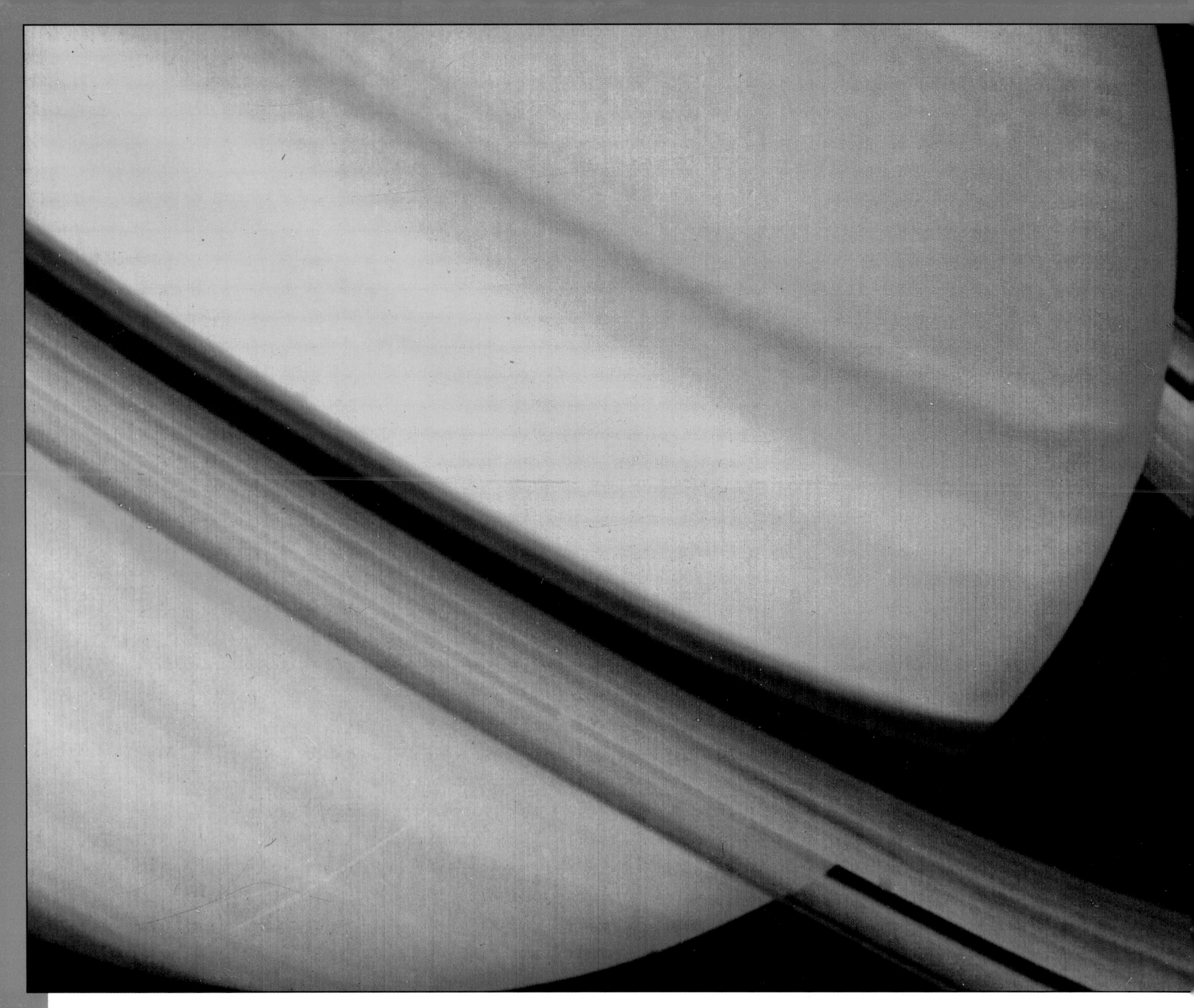

***Saturn's rings*** *are no more than 1,000 m thick. How the rings were formed is something of a mystery. There are two competing theories. The first is that they are the debris left behind after a satellite was torn apart by the gravitational forces of Saturn. The second is that they consist of material that failed to coalesce into a satellite at the time the planet was forming. The second theory is now thought to be the more likely.*

its magnetosphere almost symmetrical.

The magnetic field and magnetosphere point to Saturn's having an ionosphere, which proves to be the case. It lies thousands of kilometres above the visible cloud surface, and, since it is composed mostly of ionized atoms of hydrogen, it reflects radio waves, though it is less concentrated than the Earth's ionosphere.

Saturn also emits electromagnetic radiation at radio wavelengths of between 15 and 300 metres. The emissions are intense and come from two sources, the stronger only 10 degrees from the north pole on the sunlit side of the planet, the other from a similar spot in the southern hemisphere. The strength of both changes as Saturn rotates on its axis, and they reach a maximum every 10.65 hours. It seems likely that the sources are linked to, and indicate, the period of rotation of the magnetic field deep within the planet.

The cause of these transmissions is not fully explained, but it seems likely that they are associated with particles from the solar wind accelerated inside the magnetosphere, for they coincide with the appearance of aurorae at Saturn's poles. These auroral displays, which are widespread over the planet's surface, are caused by low-speed electrons interacting with the Saturnian ionosphere. Six of Saturn's satellites orbit within the magnetospheric regions, and so does the famous ring system.

When Saturn's ring system is observed from

Earth, its precise appearance depends on how the planet is situated in the sky with reference to the Earth's orbit. Every 14 or 15 years the rings seem to disappear; this is because we are viewing them edge-on and, since they are extremely thin, they merge with the background of the sky. Between the occasions when they are visible in an edge-on aspect, they are seen from so much above or below their plane that they blot out large areas of the planet.

Saturn's ring system was first seen in July 1610 by the famous Italian physicist and astronomer, Galileo, with his newly developed telescope. His instrument was not good enough to show the rings clearly, even though they were at their most open aspect at the time. All Galileo could report was that Saturn appeared to be a triple planet. When he was observing it some seven years later, it was near its edge-on aspect and the rings were invisible to him. Saturn, he said, seemed to have swallowed its own children. Because of the poor optical quality of the telescopes then, astronomers could not explain the phenomenon.

The mystery was finally cleared up in 1655 by the Dutch astronomer Christiaan Huygens, who managed to observe that there was, indeed, a ring around Saturn. But the true nature of Saturn's rings was not discovered until two centuries later, in 1856, when James Clerk Maxwell analysed the evidence and showed that the gravitational field of Saturn would tear any solid ring to pieces. Maxwell concluded, therefore, that the rings could be composed of tiny particles in orbit around the planet. Subsequent studies, including results obtained from the Voyager probes, confirm Maxwell's conclusion.

Despite the fact that Earth-based observers had detected a number of separate rings around the planet, no one expected there to be as many as the astounding photographs from the Voyagers revealed. These photographs also confirmed astronomers' suspicions that the ring system was extremely thin. However, the Voyager measurements have made it clear that the rings are no more than 1 kilometre thick, much thinner than previously believed. For their size, they are far and away the thinnest rings of any planet in the solar system.

The extreme thinness of the rings is thought to be caused by the gravitational effects of some small satellites, called "shepherd satellites", which orbit Saturn close to the plane of the rings. These satellites prevent the rings from spreading out above and below.

The rings are a magnificent spectacle and extend outward from 7,000 kilometres above the cloudy surface of Saturn to more than 74,000 kilometres. The most accurate counts to date show that there are at least 10,000 rings.

Voyager pictures have also revealed dark patches, like the spokes of a wheel, in the ring system. These patches, which are more than 10,000 kilometres long and about 2,000 kilometres wide, rotate around Saturn. They are continually breaking up and re-forming, and were a complete mystery for some time. It is now known that their rotation period coincides with that of Saturn's magnetic field, and it is almost certain that these dark spokes consist of particles affected by that field.

***Dark radial markings*** *resembling spokes constantly dissolve and re-appear, revolving with Saturn's rings. They seem to consist of clouds of electrically charged particles, about a thousandth of a millimetre in size, suspended above the plane of the rings and moving under the influence of Saturn's magnetic field. They have been seen to form in a matter of minutes where the rings emerge from Saturn's shadow, and they survive for only one or two revolutions of the planet before vanishing.*

# URANUS

## • *The tilted planet*

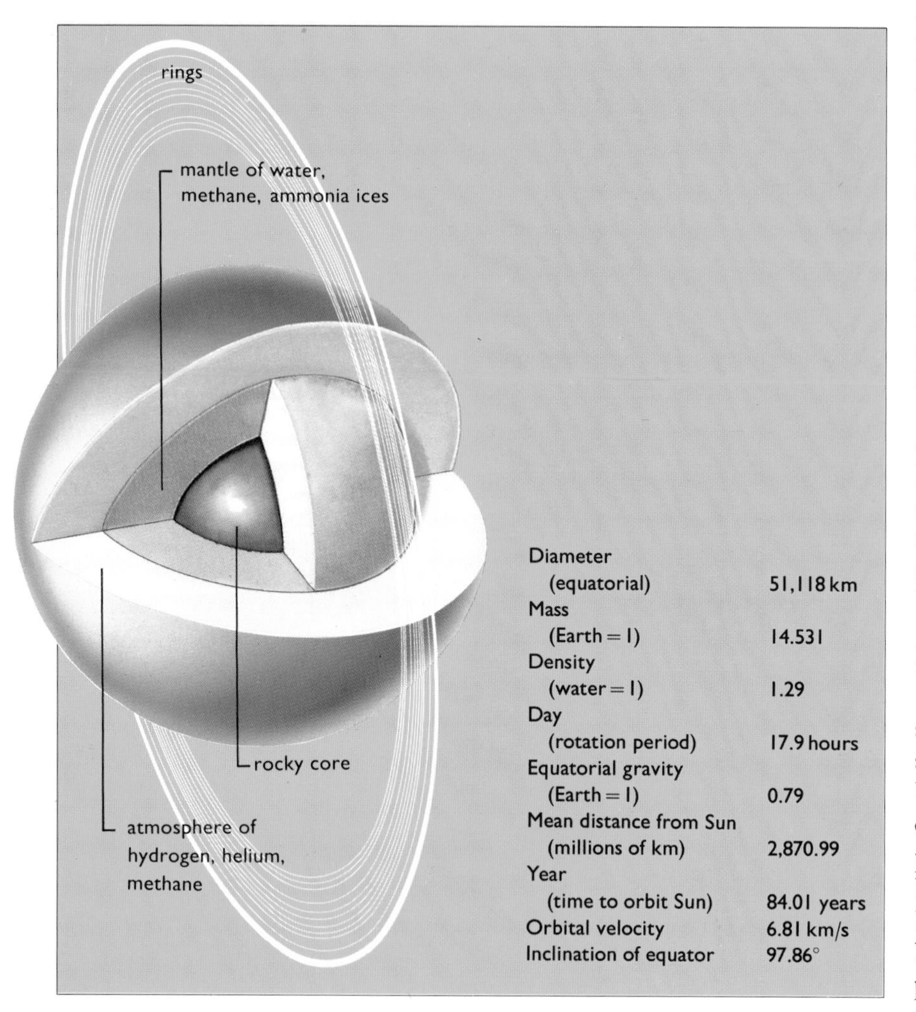


| | |
|---|---|
| Diameter (equatorial) | 51,118 km |
| Mass (Earth = 1) | 14.531 |
| Density (water = 1) | 1.29 |
| Day (rotation period) | 17.9 hours |
| Equatorial gravity (Earth = 1) | 0.79 |
| Mean distance from Sun (millions of km) | 2,870.99 |
| Year (time to orbit Sun) | 84.01 years |
| Orbital velocity | 6.81 km/s |
| Inclination of equator | 97.86° |

Uranus, another of the gas giants, was the first major planet to be discovered in relatively modern historical times; the others – Mercury to Saturn – have been known since earliest antiquity. Uranus was discovered in 1781 by William Herschel, probably the best visual observer in the history of astronomy.

Since then, knowledge about the planet has increased by leaps and bounds, especially with the evidence from *Voyager 2*. Calculations show that Uranus probably has an iron silicate core about 14,500 kilometres in diameter; its core is, therefore, a little larger than the entire Earth. Outside the core is a mantle, just over 10,000 kilometres thick, which is thought to be composed of water ice along with ammonia and methane, also in an icy, or possibly liquid, state.

Beyond this is a 9,000-kilometre-deep layer of molecules of hydrogen, helium and, perhaps, some methane. Methane certainly forms the planet's thick atmosphere, which is what we observe from Earth, and which was photographed by *Voyager 2* when it approached the south pole of the planet in 1986. Clearly, the structure of Uranus differs from that of Jupiter and Saturn, as does its internal temperature of about 7,000 K. Despite the fact that it emits slightly more heat – 0.1 percent – than it receives from the Sun, the core of Uranus appears to have few elements to give it an internal source of heat.

The pressure at the centre of the planet's core, though equivalent to 20 million times the Earth's atmospheric pressure, is not high enough to liquefy hydrogen (which would make it electrically conductive). Although its mass is just under 5 percent of Jupiter's, Uranus has about the same average density as Jupiter, some 1.3 times that of water, and is, therefore, denser than Saturn.

The visible surface of Uranus rotates in slightly less than eight hours, which is fairly slow compared with Jupiter and Saturn. But Uranus' axis of rotation is unique, for it is tilted over so that it is almost in the plane of the planet's orbit. The north pole is in fact just below the plane of the orbit, which means that Uranus rotates on its axis in a retrograde, or backward, direction compared with the other planets. This is thought to have been caused by a collision Uranus suffered with another substantial body early in its history.

In spite of having no electrically conducting liquid hydrogen in its interior, Uranus does possess a magnetic field, most probably generated in the mantle. The field is dipolar and inclined at 60 degrees to the planet's axis of rotation, the most inclined magnetic field of any major planet in the solar system.

The upper cloud layer of Uranus is banded and rotates in the same retrograde (east to west) direction as the solid body of the planet, but faster. In fact, the nearer the clouds are to the poles the faster they rotate; this is opposite to what happens on Jupiter and Saturn.

The clouds in the upper atmosphere of Uranus are similar to those on the two larger gas giants. Unfortunately, details of the planet's weather system have so far been hidden, even from *Voyager 2*, for much of the upper Uranian atmosphere is shrouded by haze caused by sunlight acting on acetylene and ethane present there.

Uranus has a ring system, discovered in March 1977, even before the arrival of *Voyager 2*. When Uranus passed in front of a ninth-magnitude star, Earth-based observers noticed that it "winked", or suffered very brief eclipses, five times before being blotted out by the disc of Uranus. From this evidence it was deduced that Uranus is surrounded by a ring system, and this was confirmed by *Voyager 2*.

Uranus' rings show some similarities to Saturn's, although they are nowhere near as spectacular. We know now that there are at least nine rings – not five as was assumed in 1977. They lie between 42,000 and 52,000 kilometres from the centre of Uranus, which means they are 16,000 to 26,000 kilometres above its cloudy surface. The rings are extremely narrow, with eight of the nine having an average width of less than 10 kilometres.

Some parts of them vary quite noticeably from this figure: the outermost, or Epsilon, ring is anything between 20 and 100 kilometres thick. This ring also has an elliptical shape, bringing it 800 kilometres closer to the planet at one point. Five of the remaining eight rings are also elliptical; only three are circular. Only the elliptical rings vary in width, but all of them, elliptical and circular, are very thin.

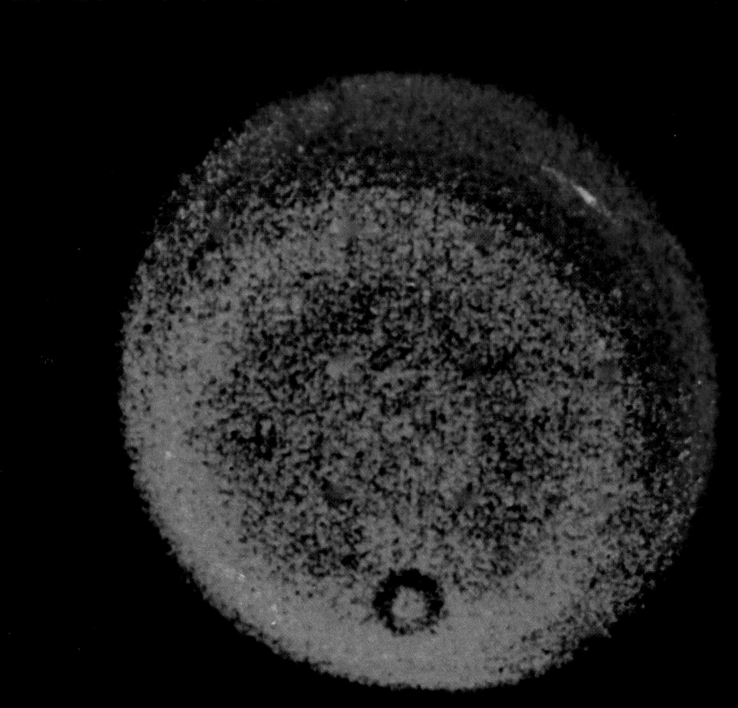

***A methane haze shrouds the planet*** *(left), obscuring the banded cloud systems. But the false colours of this computerized image enhance details such as the high-altitude cloud at top right.*

***The rings of Uranus*** *(above) are made up of particles in two distinct size ranges. Many are between centimetres and metres across, but most are no more than a few thousandths of a millimetre across.*

# NEPTUNE AND PLUTO

## • *Frontiers of the solar system*

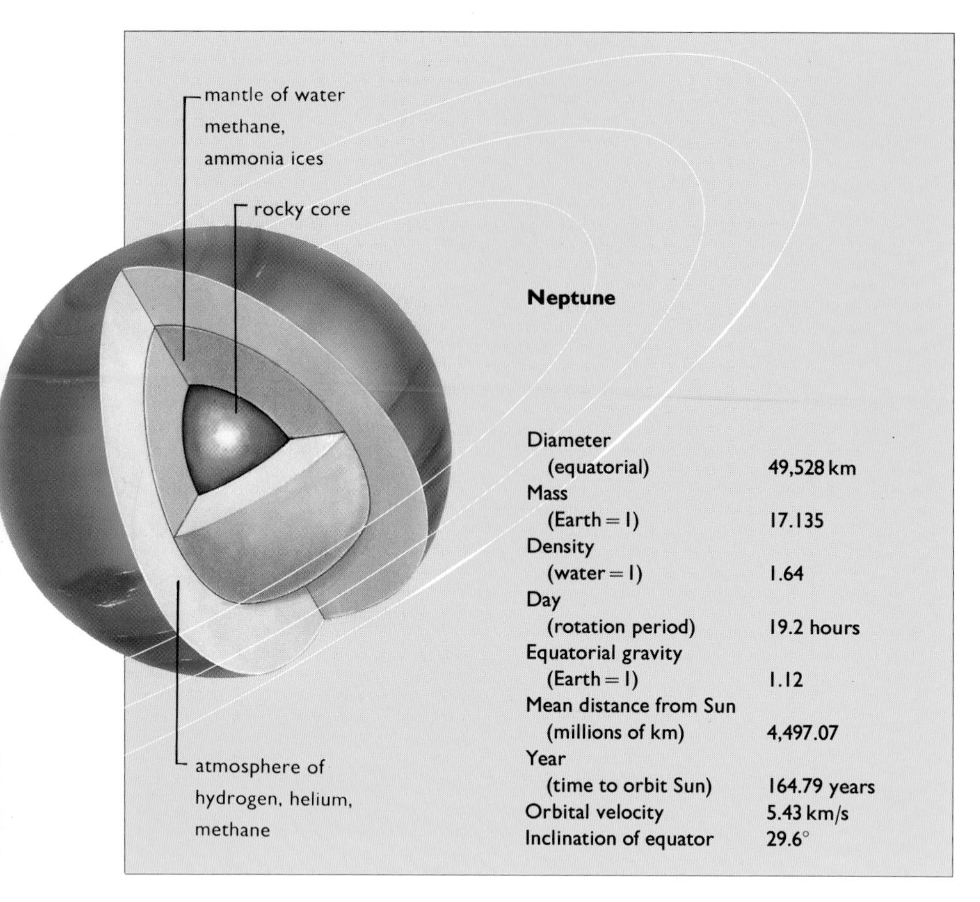


| Neptune | |
|---|---|
| Diameter (equatorial) | 49,528 km |
| Mass (Earth = 1) | 17.135 |
| Density (water = 1) | 1.64 |
| Day (rotation period) | 19.2 hours |
| Equatorial gravity (Earth = 1) | 1.12 |
| Mean distance from Sun (millions of km) | 4,497.07 |
| Year (time to orbit Sun) | 164.79 years |
| Orbital velocity | 5.43 km/s |
| Inclination of equator | 29.6° |

The existence of Neptune, the last known gas giant out from the Sun, was predicted in the 19th century by two astronomers: John Couch Adams in England and Urbain Jean Joseph Leverrier in France. Both had been stimulated to search for an eighth planet by their attempts to account for apparent irregularities in the orbit of Uranus.

Like all astronomers of the time, these scientists assumed that Newton's law of universal gravitation was precisely correct and, using Newton's formulae, both independently concluded that the cause of Uranus' orbital irregularities must be the gravitational attraction of another planet beyond its orbit. Proving their case was no easy task: to discover where such a planet might lie in the sky they had to assume that it had a certain mass and orbit and then work out the consequences. A thorough knowledge of solar system dynamics helped reduce the possible choices, but to most astronomers the challenge was daunting.

In 1843, however, Adams cracked the problem, in his first research project after obtaining his first degree. At the age of 24, he was considered too young, as well as too inexperienced, and so failed to have his result taken seriously. And at that time, no observatory in Britain possessed an up-to-date chart of the region of the sky in which Adams' calculations showed that the planet might be. Without such a chart, checking would have been a long and tedious task.

Leverrier, who was eight years older than Adams, met with less resistance from colleagues when he announced similar findings in 1846. Fortunately, the Berlin Observatory had recently completed a new chart of the correct region of the heavens, and the director, Johann Galle, set a search in motion. Neptune was first observed on 23 September 1846, less than one degree from its predicted position.

Orbiting at an average distance of 4,497 million kilometres from the Sun, Neptune is not only the most distant known gas giant of the solar system but also very difficult to observe from Earth. Only recently has our knowledge of the planet been revolutionized by information sent back by *Voyager 2* in 1989.

Before *Voyager 2*'s mission, astronomers had based their calculations concerning the structure of Neptune's inner regions on the planet's known mass, size and position. From these they deduced correctly that, like other gas giants, it probably has a central rocky core made of iron, probably with a mixture of silicates. This core is similar to that of Uranus.

Outside the core, Neptune's mantle is thought to differ from that of Uranus, being made up of ionized molecules of water and ammonia and probably hydroxyl. Indeed, it is sometimes described as an "ion ocean". Above the mantle is an envelope of helium, hydrogen and methane similar to that of Uranus.

Like other gas giants, Neptune emits more energy than it receives from the Sun, by a factor of 2.8. This is much greater than the energy emission of Uranus, which is only 0.1 percent greater than the energy received.

Earth-bound observations also led astronomers to think that Neptune possessed a magnetic field – a theory confirmed by measurements taken by *Voyager 2*. However, the field proved to be weaker than those of the

**Blue, cold and remote,** *Neptune is the outermost of the known gas giants. This Voyager image shows the blue cloud surface and the white wispy clouds of methane above it. The high clouds rotate with the planet, rather than with the faster-moving, deeper-lying clouds.*

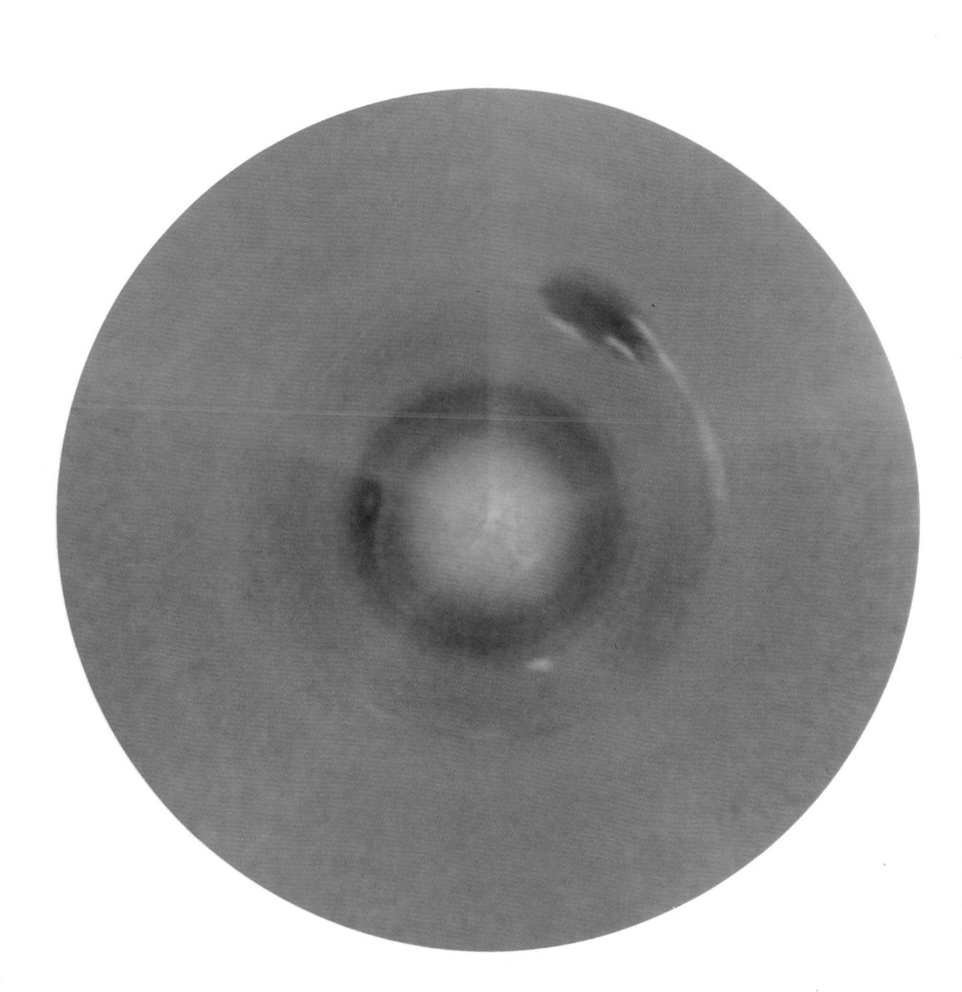

***White clouds of methane crystals*** *are striking features of Neptune, visible in this* Voyager 2 *image. They lie high above the blue cloud layers and move independently of them. One, called the scooter, moves so fast that it catches up and overtakes the Great Dark Spot every few Neptunian days.*

other three gas giants, and is inclined at an angle of 50 degrees to Neptune's axis of rotation. Surprisingly, the field does not pass through the planet's centre, but is set some 10,000 kilometres to one side. It is thought, therefore, that the electric currents within the planet, which give rise to the field and its associated magnetosphere, must exist fairly close to the surface.

Because of the nature of Neptune's magnetic field, the point at which the magnetosphere meets the solar wind is more than 800,000 kilometres from the planet – about twice as far as that of Uranus. Neptune also transmits at some radio wavelengths, and experiences auroral displays over its equatorial regions.

Like the other gas giants, Neptune has its own ring system. It has three rings, one of which is rather dim, and all are considerably less substantial than those of Jupiter, Saturn and Uranus. Before the *Voyager 2* flypast, it was thought that the rings might be in the form of fractured arcs of matter, but they are, in fact, complete. They do, however, have arcs which are brighter than other parts. These are probably due to uneven distribution and clumping of the particles of which they are composed. Some ring particles measure a few kilometres across, and are termed "moonlets".

The dimmer sections of Neptune's rings are of such low brightness that *Voyager 2* was only just able to detect them. A single "shepherd" satellite appears to be associated with each ring, the two brighter rings being linked with two of Neptune's newly discovered satellites. Because the material in the rings is widely spread out, it occupies only about a tenth of the ring space. However, a disc of tiny particles accompanies the rings; this is spread throughout the entire space the rings occupy.

Of all the information transmitted back by *Voyager 2*, that concerning Neptune's cloud system was the most dramatic. The planet's cloud surface, coloured a beautiful blue, is swept by strong winds, blowing in some latitudes at speeds up to 30 metres a second. Neptune also has a long-lived dark blue spot, named the Great Dark Spot (GDS), which moves more slowly than the clouds surrounding it in the southern hemisphere.

Neptune's GDS is 14,000 kilometres long and measures almost 6,667 kilometres in the north-south direction. The entire Earth would fit inside its east-west extension. There is another similar, but smaller, feature in the southern hemisphere, 33 degrees nearer the planet's south pole.

Suspended about 50 kilometres above the blue surface clouds are white, wispy, cirruslike clouds of methane. These do not take part in the fast relative motion of the cloud layer below. Instead, like clouds above some high mountains on Earth, they seem to hang still, moving in step with the planet's rotation.

Neptune's atmosphere extends above the methane clouds. It is composed mainly of hydrogen, which makes up 85 percent of the total, with 13 percent helium and 2 percent methane. Under the influence of ultraviolet radiation from the Sun, the methane undergoes an unusual cycle of chemical change. The ultraviolet splits the methane into carbon,

methane frost
mantle of ices
rocky core

**Pluto**

| | |
|---|---|
| Diameter (equatorial) | 2,300 km |
| Mass (Earth = 1) | 0.0022 |
| Density (water = 1) | 2.03 |
| Day (rotation period) | 6.387 days |
| Equatorial gravity (Earth = 1) | 0.04 |
| Mean distance from Sun (millions of km) | 5,913.52 |
| Year (time to orbit Sun) | 248.54 years |
| Orbital velocity | 4.74 km/s |
| Inclination of equator | 122.46° |

hydrogen and a mixture of hydrocarbons including acetylene.

These hydrocarbons then sink down to cooler levels of the atmosphere where they condense into hydrocarbon ices. In turn, these ices sink still lower, to where the atmosphere is warmer. Here they evaporate and recombine to form methane, which rises again.

Even after Neptune had been discovered, astronomers continued to observe irregularities in the motions of Uranus and even of Neptune itself. So, following the example of Adams and Leverrier, Percival Lowell and William Pickering calculated the position of another planet, which they believed to be the cause of the disturbances. In 1905, a search for the planet began, but it was not until 1930, at the Lowell Observatory in Arizona, that a young astronomer, Clyde Tombaugh, photographed a tiny object some 5 degrees from the predicted position. The object possessed a planetary orbit and was named Pluto.

The planet Pluto is extremely difficult to observe from Earth because it is so distant and so small – it is only some 2,300 kilometres across, that is, about one and a half times the size of Triton, Neptune's largest satellite. Pluto's mass is equal to 0.22 percent of Earth's.

Pluto's orbit has been revealed as quite unlike that of any other planet in the solar system. It is extremely eccentric – considerably more so than Mercury – and although it sometimes orbits more than 2,800 million kilometres beyond Neptune for part of its circuit around the Sun, at other times it comes closer to the Sun than Neptune ever does. Its orbit is inclined at 17 degrees to the plane of the solar system, more than any major planet.

Its size, mass and orbit all support the view that Pluto is not a major planet at all but a large asteroid, or even perhaps an escaped planetary satellite. Its gravitational attraction is far too small to disturb the orbits of Neptune and Uranus on its own, which means that these disturbances must have some different cause, possibly the influence of some other, currently undiscovered, planet.

Only a thin, hazy atmosphere has been detected around Pluto, but infrared observations with a spectroscope have shown that it is covered by ice. Much of this is frozen methane, turned a reddish colour by sunlight, but there are also water and ammonia ices present.

If Pluto is a giant asteroid, it is unique in that it has a satellite. This was discovered in 1978 by James Christy at the Lowell Observatory. Named Charon, this satellite orbits Pluto at a distance of about 20,000 kilometres. The axial rotations of Pluto and Charon are synchronized, owing to tidal forces between these closely orbiting neighbours. Charon is too small to have retained any methane it may once have possessed, and is covered with water ice.

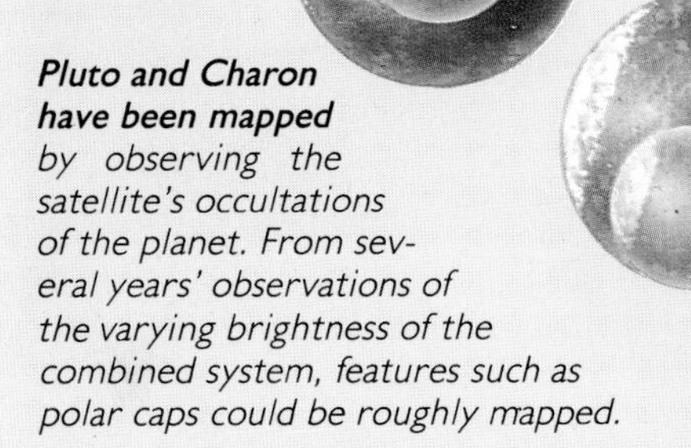

***Pluto and Charon have been mapped*** *by observing the satellite's occultations of the planet. From several years' observations of the varying brightness of the combined system, features such as polar caps could be roughly mapped.*

# SATELLITES

## • *Companions of the planets*

Earth is not alone in possessing an orbiting satellite. This fact remained unknown, however, until 1608 when Galileo, looking through the recently invented telescope at the heavens, observed the four largest satellites of Jupiter. Since then it has been discovered that satellites orbit every other major planet, except for Mercury and Venus.

It is no surprise that there are such satellites in a solar system that condensed out of the solar nebula (pp. 102–3). During the process much debris, in the form of planetesimals, was left over, and some of this was captured by the gravitational fields of the major planets. The evidence for this comes from examination of the orbits of the various satellite systems.

As far as Earth is concerned, the Moon's orbit is inclined at 5 degrees to the ecliptic which marks out the plane of the Earth's orbit around the Sun. But this is exceptional and the orbits of most of the larger members of other satellite systems lie closer to the plane of the equator of their parent planet.

As for the major planets, Mars has two tiny satellites that orbit in the planet's equatorial plane. The four largest satellites of Jupiter also orbit in its equatorial plane, as do its four nearest but tiny ones, Metis, Adrastea, Amalthea and Thebe. Jupiter's outer satellites, however – which are also small and were, presumably, captured some time after the solar system formed – orbit at greater inclinations.

A group of four – Leda, Himalia, Lysithea and Elara – orbit Jupiter at an inclination of some 27 degrees. A second group comprising the four outermost satellites – Ananke, Carme, Pasiphaë and Sinope – orbit at an inclination of around 150 degrees, which means that they are in retrograde motion.

All Saturn's satellites, with the exception of the outermost, Phoebe, orbit in the equatorial plane. Phoebe is thus probably a late addition as it moves in an eccentric orbit, and is some 13 million kilometres beyond the next nearest satellite, Iapetus – four times as far out. Uranus is similar in that every one of its 15 satellites orbits in the planet's equatorial plane.

Neptune's six inner satellites have equatorial orbits, but the two outermost do not. Of the latter, Triton, the larger with a diameter of 2,272 kilometres, has an orbit inclined at no less than 160 degrees, giving it a retrograde orbit. The other, Nereid, which orbits 16 times farther out, is inclined at some 27 degrees.

The number of satellites possessed by the major planets varies widely. The inner, terrestrial planets have few, with one for the Earth, two for Mars, but none for either Mercury or Venus. The gas giants are in a different league. Jupiter possesses no fewer than 16 satellites, Saturn 18, Uranus 15 and Neptune 8. This is not surprising since all four have much greater mass and, therefore, stronger gravitational fields and must be expected to have captured more planetesimal material.

The mass of the Earth is only 81 times that of the Moon, whereas the mass of Jupiter is more than 12,000 times greater than the most massive of its satellites, and Saturn is more than 4,000 times greater than its giant satellite Titan.

In the case of Uranus, measurements show that it is 4,000 times greater in mass than its densest satellite, Oberon, and Neptune is 800 times more massive than its comparatively immense satellite Triton. Clearly, the Earth-Moon system is significantly unusual.

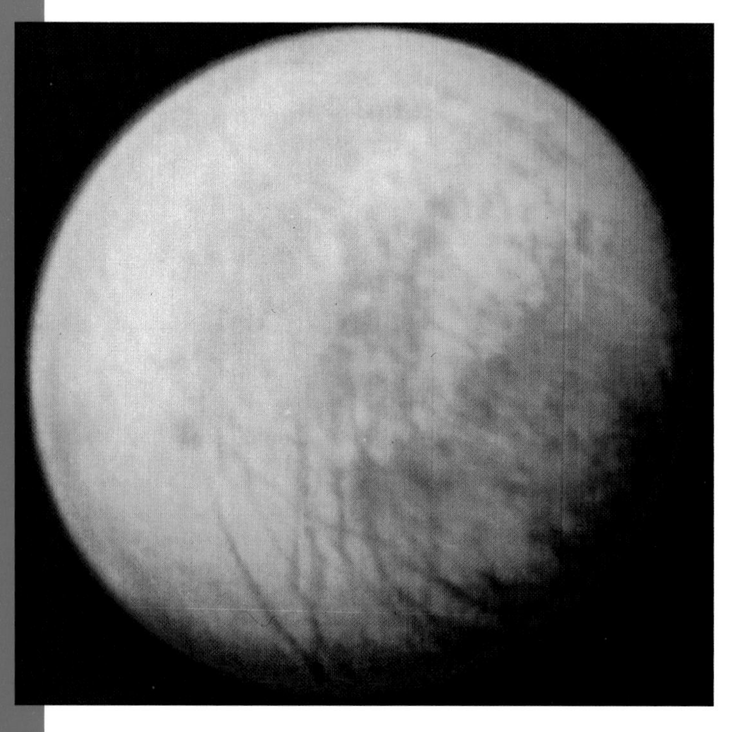

***Jupiter's large satellite Europa*** *orbits in the planet's equatorial plane and, at about 3,100 km across, is only a little smaller than the Earth's Moon.*

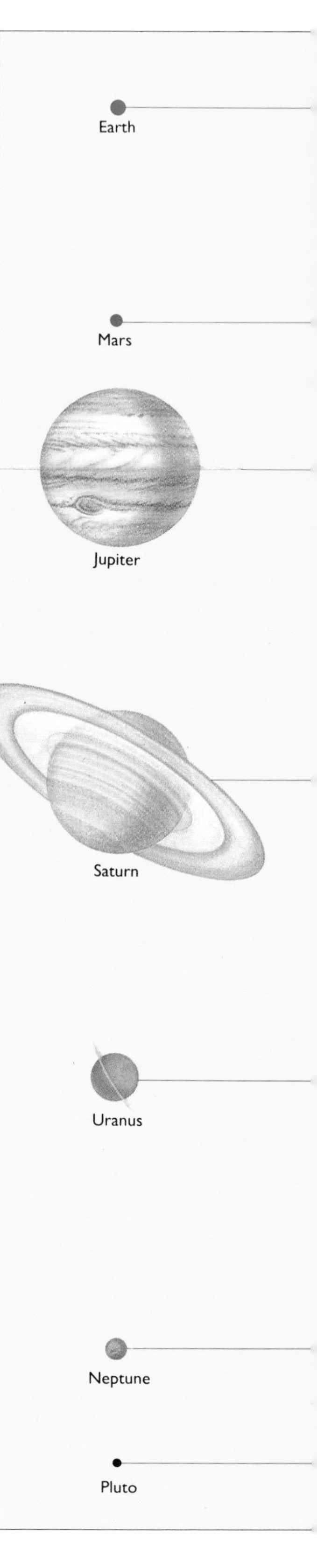

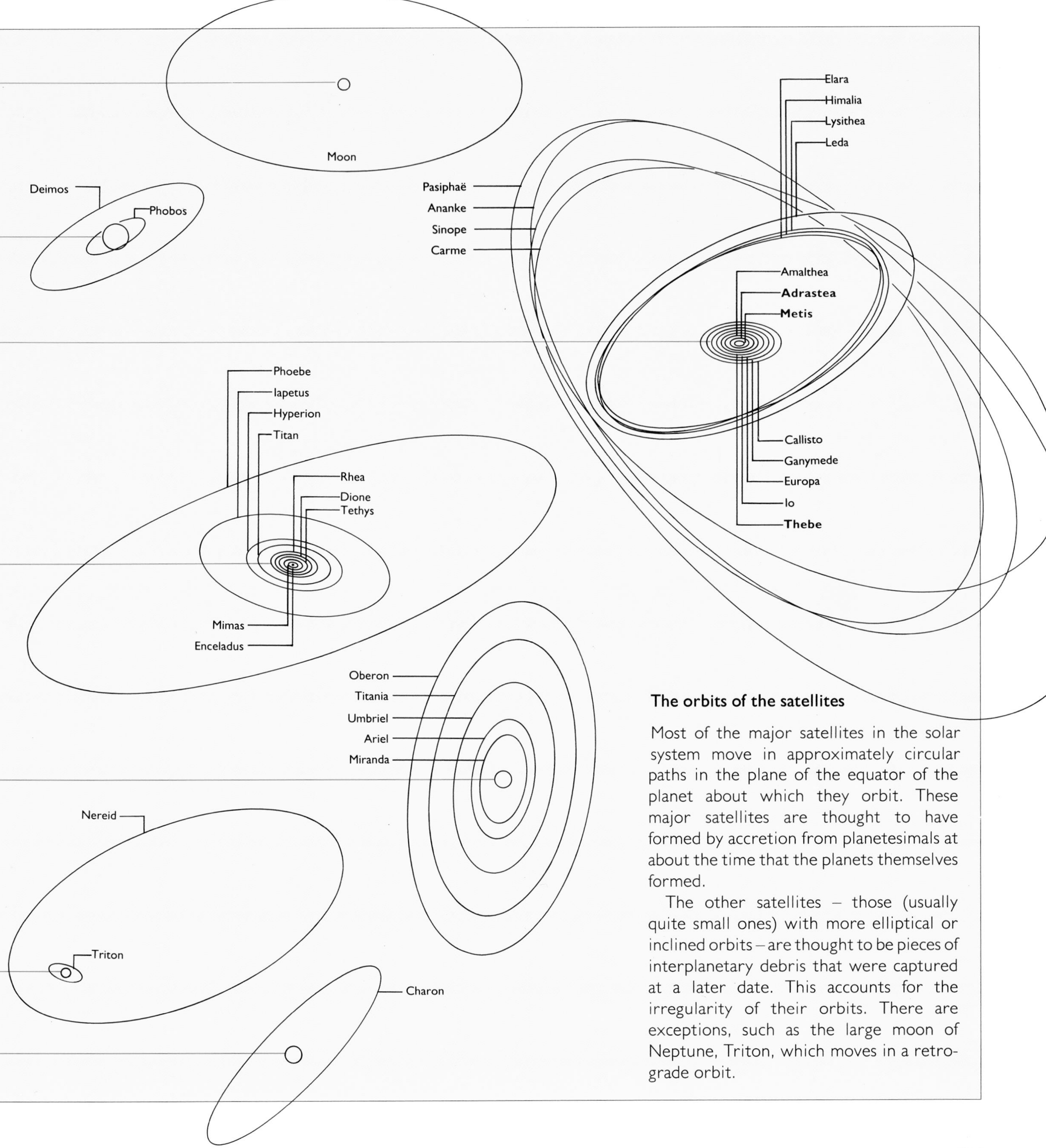


## The orbits of the satellites

Most of the major satellites in the solar system move in approximately circular paths in the plane of the equator of the planet about which they orbit. These major satellites are thought to have formed by accretion from planetesimals at about the time that the planets themselves formed.

The other satellites – those (usually quite small ones) with more elliptical or inclined orbits – are thought to be pieces of interplanetary debris that were captured at a later date. This accounts for the irregularity of their orbits. There are exceptions, such as the large moon of Neptune, Triton, which moves in a retrograde orbit.

*The Earth's Moon* shines by light reflected from the Sun. It displays phases, ranging from a thin crescent to a full Moon, because different proportions of its illuminated side are presented to us at different times. Compared with other satellites, the Moon is large in relation to the Earth. Thus the Moon and the Earth may be regarded as a binary planetary system.

You do not need a powerful telescope to observe the Moon, it is there for all to see. It has fascinated astronomers over the years and there have been various theories about the Moon's formation and its internal structure.

Earlier this century it was generally believed that the Moon broke off from the Earth when the Earth was rotating at high speed while still in a "plastic" state. However, opinion has now changed and today it is becoming accepted that the Moon was formed by accretion independently from the Earth. It is therefore a planetesimal, but a large one, which has allied itself to the Earth to make a binary planetary system in which both components are in orbit around each other.

Shining in the sky by reflecting sunlight – a characteristic of all the solar system's planets and satellites – the Moon thus shows phases to Earth-based observers as it orbits our planet. The length of time for a complete cycle of phases, called a synodic month, is 29.5 days. But if we measure a month by the time taken for the Moon to complete a circuit across the stars in the night sky, it is only 27.3 days, because the motion of the Earth around the Sun has also to be taken into account.

Revealed even to the unaided eye during the changing phases is the Moon's mottled appearance; it displays large dark areas, once thought to be oceans and seas. These are still referred to by their Latin names of *oceanus* and *mare*. Yet we now know, from optical observations, data from orbiting spacecraft, and manned landings on the Moon, that these are not seas at all, but vast flat plains formed by outflows of molten lava from beneath the crust. These plains, which are often bounded by mountain ranges and pitted here and there with craters, also show geological faulting.

Although there are craters on the plains they are more numerous elsewhere on the lunar surface. These craters have an enormous range of sizes. Many display a central peak, and the majority, at least, are generally agreed to be the result of bombardment of the surface by cometary material and planetesimals.

The plains and the faulting, though, give evidence of geological change from within the Moon. More evidence comes from seismic equipment set up by astronauts who have landed on the Moon, and this has provided independent confirmation that the Moon is still geologically active. Together with results of analysis of samples of lunar rock brought back from the manned landings, it is now possible to obtain an insight into the Moon's internal structure, as well as its likely early history.

It appears that the Moon probably has a central core, perhaps 600 kilometres in diameter, with a partially molten rocky outer core stretching some 350 kilometres above this. On top of this outer core lie a mantle and a crust – the lithosphere – together stretching upward some 1,070 kilometres. Deep moonquakes have been recorded in the molten (outer) core, and there have been others nearer the surface.

Samples of lunar rock give an age of 4.5

billion years; this is more than the oldest known terrestrial rocks, which have an age of 3.8 billion years. The discrepancy may well be due to still older terrestrial rocks having been destroyed during erosion, mountain building and volcanic eruptions. The lunar age does fit in well with ages of meteorites (pp. 146–47), and goes back to around the time when the solar system was forming. Probably, then, the Earth and Moon were both born at about the same epoch.

Certainly, there is evidence that during the first few hundred million years of the Moon's history the outer layers of the lunar surface were completely melted to a depth of several hundred kilometres. This was either because of violent bombardment by other planetesimals or because of heating due to the quick decay of the radioactive metal aluminium 26.

While the Moon was cooling, continued bombardment by planetesimals, perhaps as large as 250 kilometres in diameter, created huge basins like the Mare Imbrium (Sea of Showers) and Mare Orientale (Eastern Sea). About four billion years ago this prodigious bombardment slowed down, leaving highlands peppered with craters and a deep covering of broken rock. Heat produced below the crust by radioactive materials such as uranium then led to melting some 200 kilometres below the surface. This gave rise to the upwelling of great lava flows over the surface for perhaps as long as 500 million years, and resulted in those vast dark plains now so clearly visible.

After that active age, some 3.1. billion years ago, things quietened down and there has only been sporadic bombardment by much smaller material. Although the Moon is far more quiescent today, it is clear that seismic events and probably a little bombardment continue. Our satellite is not as dead as was once thought.

***The lunar mountains*** *in the Hadley-Apennine area were explored by the* Apollo 15 *mission. Covering about 23 km, the Lunar Roving Vehicle was used to help collect rock samples. These showed a maximum age of about 4.5 billion years.*

**The Moon's changing face**

When the Moon had formed by accretion it was initially molten, and no permanent surface features had formed. But when it cooled it began to retain the scars of its bombardment by debris, consisting of planetesimals up to 250 km across. This left it, about 4 billion years ago, heavily cratered. After the era of bombardment, layers 200 km below the crust melted, probably owing to radioactive heating. The maria were formed when lava welled up and flowed over the lower areas of the Moon. This happened about 3.3 to 3.1 billion years ago, and obliterated many of the craters. Little has happened between the formation of the maria and the present, except for the occasional impact of interplanetary debris to form new craters, such as Copernicus, on the relatively smooth surface of the maria.

**4 billion years ago** **3.1 billion years ago** **The present**

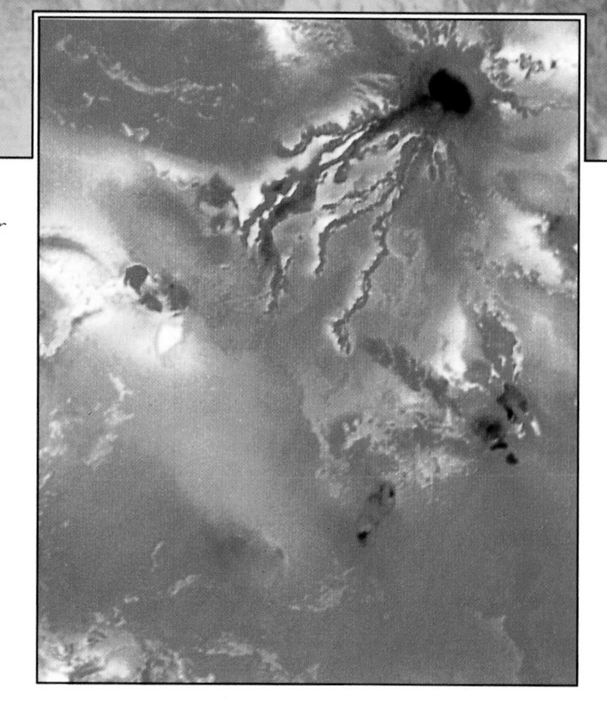

***Erupting volcanoes on Io*** *(above) spew out sulphur and other chemicals. Some of this material is left behind in a trail along the satellite's orbital path around Jupiter. The eruptions are caused by volcanic activity brought on by Jupiter's powerful gravity, which squeezes Io out of shape and generates internal heat. The red and orange colour of Io's surface (right) is caused by deposits of sulphur compounds.*

The discovery of the satellites of the other planets in our solar system had to wait until the invention of the telescope. More recently much new information has come from the revelations provided by the Voyager spacecraft. The available evidence shows that the satellites of Jupiter, Saturn, Uranus, Neptune and Mars vary widely in terms of shape, size, structure and surface features.

Beginning with Jupiter, the four large satellites that were originally discovered by Galileo (Io, Europa, Ganymede and Callisto) display some astounding features. Of these Io, orbiting at an average distance of 421,600 kilometres from the centre of the planet, is perhaps the most fascinating of all.

With a diameter of 3,642 kilometres – a little larger than our Moon – Io orbits in Jupiter's equatorial plane. This orbit carries it within Jupiter's magnetosphere, and this causes Io to generate a billion watts of electrical power from one side to the other. This is a greater output of electrical energy than can be generated by all the power stations in the whole of the United States of America. It is the reason why Earth-based radio observations of signals from Jupiter were found to be linked with Io's orbital motion.

At 1.77 days the orbital period of Io is half that of the next satellite out, Europa, which orbits Jupiter every 3.55 days. This causes Europa to modulate periodically the tidal effects of Jupiter on Io, so that Io undergoes a regular series of strong gravitational effects, which squeeze it out of shape and, in doing so, heat up the satellite's interior. The outcome is that Io experiences great volcanic activity, as Voyager observed.

Io's surface is composed of sulphur, which at the very low temperature there – some 120 K (–153°C) – should be white, not yellow as we are accustomed to see at Earth surface temperatures. However, volcanic activity and hot spots on Io's surface melt the sulphur, which then becomes orange or red as it flows over the exterior. When it cools and solidifies the orange and red colours remain. This is why Io's surface is such a spectacular reddish colour, and so smooth. It does, however, display some dark spots where sulphur has been heated in eruptions to 300°C, or more.

As for Europa, it is a little smaller than our Moon and is unusual in displaying no impact craters. Instead it is covered by ice crisscrossed with veins. This veined appearance is caused by muddy ice that welled up and flooded the surface after a period of bombardment by meteorites (pp. 146–47) some four billion years ago. Liquid water may still lurk under this frozen surface.

Farther out, at an average distance more than 2.5 times that of Io, is Jupiter's giant satellite Ganymede. Just a little larger than the planet Mercury, although less massive, it is probably composed of ice and silicates in approximately equal proportions. Its icy surface is pitted with craters and seems to be divided into two types of terrain.

The first is made up of dark patches with craters and large furrows; between these furrows, the second type of terrain, made up of light-coloured streaks, predominates. The streaks contain depressions several kilometres wide and hundreds of kilometres long. The evidence suggests that Ganymede is an ancient satellite, with a surface once heavily bombarded, but which later iced over.

The other large Jovian satellite is Callisto. Almost as large as Ganymede, and of similar density, it has a highly cratered surface. Of note is Callisto's so-called Valhalla basin, a circular area about 600 kilometres across. This feature displays some 15 concentric ridges and it appears to have been caused by the impact of a large meteorite or asteroid. Callisto's surface may well be even older than Ganymede's.

The satellites of Saturn also show many interesting features. For instance, the "shepherd satellites" have the effect of retaining ring material in place, rather than letting it spread out into space. But more remarkable are two other smaller satellites., Janus and Epimetheus, which orbit just outside the rings.

They travel in almost identical orbits, with only 50 kilometres separating their two paths. In consequence, their orbital speeds are similar, although one catches up with the other every

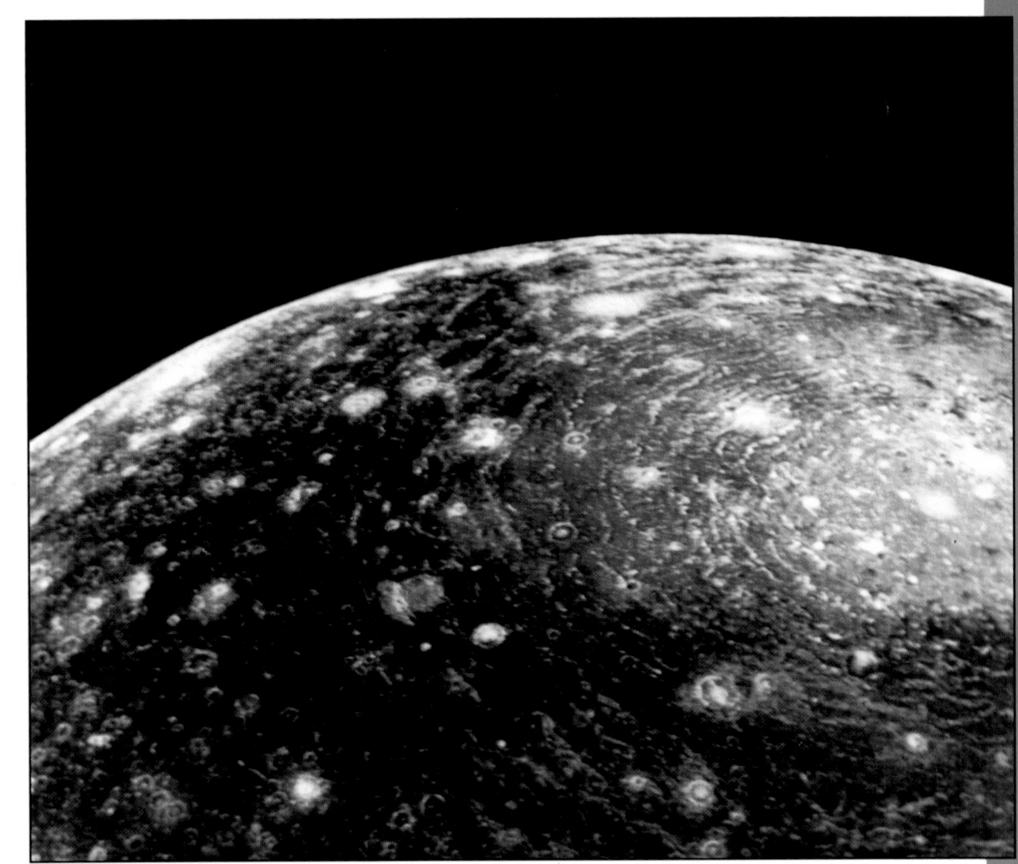

***Jupiter's second largest satellite, Callisto,*** *has a prominent "bull's-eye" feature, believed to be a large impact basin. This image, taken by* Voyager 1 *from a distance of 350 km, shows features down to 7 km across. Callisto, composed of rock and ice, with a dirty ice surface on which the temperature is –153°C, has more craters than any of the other large Jovian moons, leading astronomers to believe that it has the oldest surface.*

four years. Small and oblong, they could have collided long ago, but are prevented from doing so because, when they approach each other, mutual gravitational effects cause them to exchange orbits. The faster-moving becomes the slower-moving, and vice versa.

The shapes of these two satellites lead to the supposition that both once formed a single satellite, from which Epimetheus broke away. Perhaps they were originally like the next more distant satellite Mimas, which is a shade more substantial, having a body with a diameter of 398 kilometres. A satellite composed of ices, its cratered surface is dominated by one enormous crater, indicating that something extremely large once struck Mimas, and probably very nearly tore it apart.

Enceladus, Saturn's next most distant satellite, is still something of a mystery. Part of it is cratered, the other part smooth, and it appears also to have suffered severe bombardment. This seems to have ceased at least four billion years ago, after which something occurred to make material well up from inside the satellite, covering part of the surface which now looks like one large flat plain.

Farther out still are some larger satellites – Tethys, twice the diameter of Enceladus; Dione, which is a little larger at 1,120 kilometres; and Rhea, with a diameter of 1,528 kilometres. All are cratered, with Dione showing some strange wispy markings.

Probably the most intriguing of Saturn's satellite family is the giant Titan, which has a diameter of 5,150 kilometres, making it almost 1.5 times the size of our Moon. Titan is massive enough to retain an atmosphere, but this is an orange-coloured smog thick enough to obscure the surface completely.

Infrared observations from *Voyager 2* show clearly that although the atmosphere is 90 percent nitrogen, methane is also present – as expected from earlier Earth-based observations. But there are also complex organic molecules such as ethane and acetylene. *Voyager 2* also discovered the presence of hydrogen cyanide, a molecule which can combine with others to form adenine, one of the components of the DNA helix, and a substance found in animal and vegetable tissue.

Although there is no water present on Titan to allow substances such as amino acids to form (pp. 156–61), it may well be that Titan is in the kind of pre-life state that once occurred on Earth. If so, future study of it may well lead to a deeper understanding of how life began on our own planet.

Beyond Titan lies Iapetus, with a diameter of 1,436 kilometres; it is also heavily cratered. The outermost satellite is Phoebe, whose odd orbit suggests an origin separate from that of the other Saturnian satellites (p. 134).

Most of the 15 satellites of Uranus are small, not more than 160 kilometres in diameter, but the largest five – Miranda (472 kilometres), Ariel (1,158 kilometres), Umbriel (1,169 kilometres), Titania (1,578 kilometres) and Oberon (1,523 kilometres) – are all of interest. Pictures of all five taken by *Voyager 2* show that they are made up of ice and rock; the rock, being heavier, has sunk to the interior. All display a series of darkish cratered surfaces.

Detailed pictures of the smallest of them, Miranda, revealed an intriguing surface. Mostly cratered, it possesses two almost rectangular areas which look as if they have been smoothed by the flat side of a giant knife. One of these is covered with lines, while the other displays a chevron of lighter material.

Planetary geologists suggest that Miranda, the closest of the larger satellites to Uranus, was once broken up so that for a time it turned into a number of boulders in orbit around Uranus; later these clumped together to reform into a satellite. If this interpretation is correct, it can provide a clue to the reason why the axis of rotation of Uranus is so unusual (pp. 128–29). It could well have been that large lumps of material once fell toward Uranus, material which not only tilted Uranus over, but

***Uranus' satellite Miranda**, seen here in a composite image from Voyager, is 500 km across, and is the smallest and closest of the planet's major moons. Its unusual surface has areas that are predominantly cratered and others with fractures, grooves and fewer craters. The grooves and troughs are a few kilometres deep.*

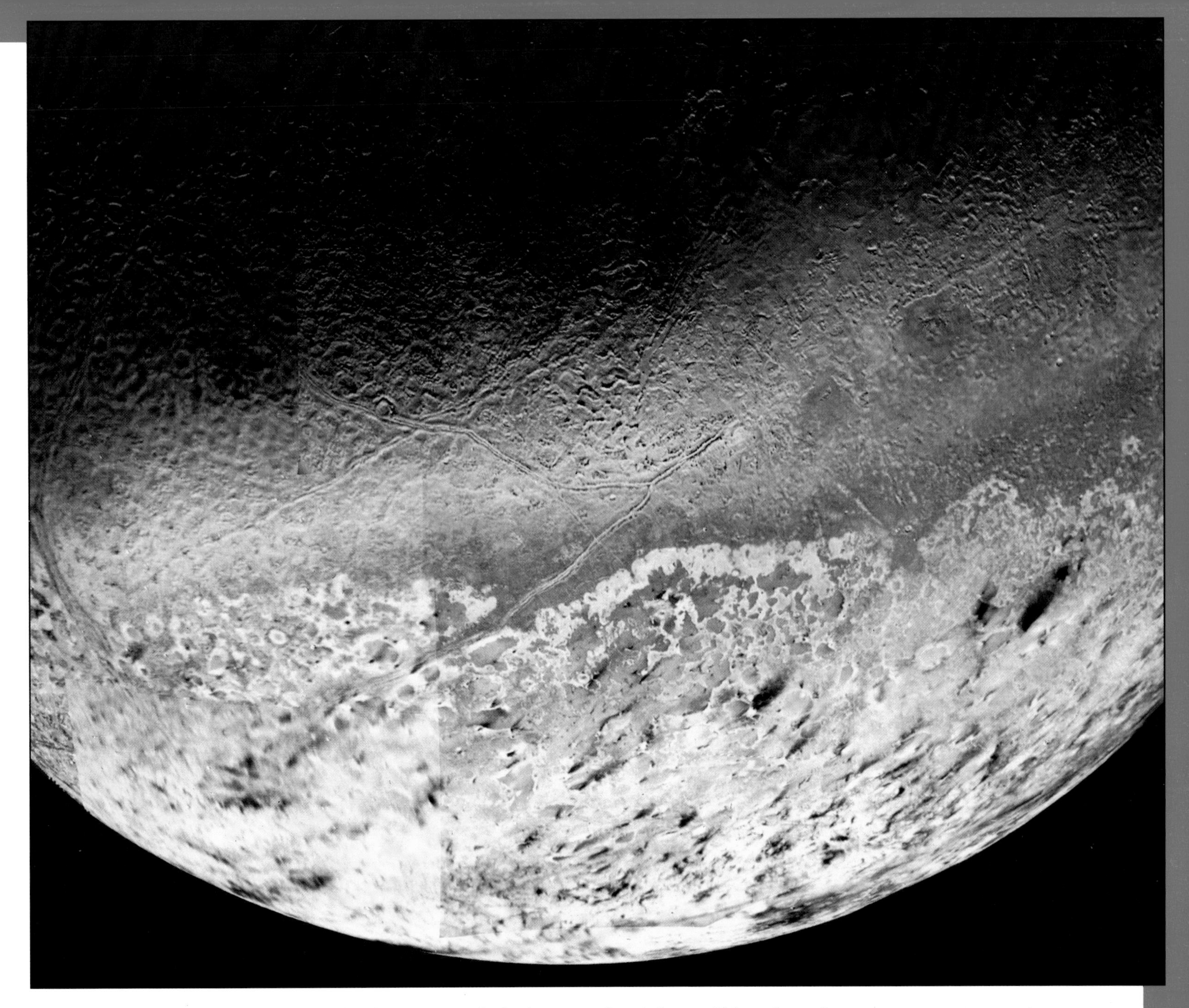

was large and fast-moving enough also to disrupt Miranda.

Neptune has only one large satellite, Triton, with a diameter of 2,720 kilometres. This is an extremely cold place, with a temperature of no more than 37 K. Even liquid nitrogen is warmer than Triton's surface. Study reveals that the surface is covered with water ice over which lie methane and nitrogen ices. The nitrogen ice forms a transparent coating about 1 metre thick.

The poles of Triton are inclined at almost 160 degrees to Neptune's equator, and as it orbits once every 5.9 days, one of Triton's poles will point toward the Sun for half of every Neptunian year, that is, for 82.4 terrestrial years, followed by the other for a similar period. As a result, night on Triton lasts for about a human lifetime, and during this time the ice builds up, especially around the satellite's poles, to about 1.5 metres.

Coming back closer to the Sun, the two tiny potato-shaped moons orbiting Mars are both very small, with Deimos only 15 kilometres and Phobos 27 kilometres long. Deimos takes 1.26 days to orbit; Phobos does so in just 7 hours 39 minutes, and thus rises and sets three times a day. Much pitted from bombardment by meteors, these are not so much moons as small captured asteroids.

Clearly the range and diversity of the many satellites in the solar system point to several different processes of formation and of capture by the parent planets.

***The south polar cap*** *of Neptune's large satellite Triton is highly reflective and pinkish in colour. Astronomers think it is made up of nitrogen ice deposited during the very long winter. Away from the cap the surface becomes redder and darker. This colouring may be due to the action of ultraviolet radiation and charged cosmic-ray particles, affecting methane in the atmosphere and on the surface.*

# ASTEROIDS

## The minor planets

Asteroids, sometimes referred to as "minor planets", are small pieces of planetesimal material orbiting the Sun. The first was observed in 1801, but astronomers had begun hunting for them 16 years before this, following the discovery by Johann Titius of a numerical relationship between the distances of the planets from the Sun.

Titius found that if he wrote down the numbers 0, 3, 6, 12, 24, 48, 96 and 192 (each one double the last), and then added 4 to each, the number series he ended up with was 4, 7, 10, 16, 28, 52, 100 and 196. He realized that if he then took the distance from Sun to Earth to be 10 units, the number 4 represented the actual distance from the Sun to Mercury, 7 that from the Sun to Venus and so on. The distance from the Sun to Mars was 16 units on this scale, 52 represented the Sun to Jupiter, 100 the Sun to Saturn. The numbers 28 and 196 were spare.

Titius published this relationship only as a footnote to a German translation of a French book on science, but it was rescued from there by Johann Bode, a young astronomer who drew wide attention to it. It became known as Bode's law, or the Titius-Bode law.

When William Herschel discovered Uranus in 1781, it was found that its distance fitted the number 196 in Bode's law. Before this Bode had suggested there should be a planet between Mars and Jupiter, with a distance expressed by the number 28; after Herschel's discovery his suggestion was taken seriously.

In 1785, therefore, the Hungarian Baron Xavier von Zach began to search for the missing planet. Despite dogged persistence he had no success for the next 15 years, and even when he organized a cooperative effort in 1800, the "missing" planet could not be found. It was, in fact, a chance observation in 1801 by Giuseppe Piazzi, who was busy making a star catalogue, that detected a planet at the correct distance. It was named Ceres, after the goddess of Piazzi's native Sicily.

Once Ceres had been observed, other dim planets were also found orbiting between Mars and Jupiter. Each appeared as a starlike point, being too small to show a detectable disc, and in 1802 the name asteroid was coined to describe them. Thousands are now known, but only 33 of them are larger than 200 kilometres in diameter and not all orbit between Mars and Jupiter.

Those that do have such orbits are not evenly distributed across a band, but are grouped into collections with small gaps – called Kirkwood gaps – in between. These gaps are caused by the gravitational effects of Jupiter, which is now thought to have been responsible for the fact that no major planet coalesced from planetesimals in the gap between Mars and Jupiter.

Calculation shows that only in orbits well separated from the major planets can planetesimal debris collect into reasonably stable orbits. The gap between Mars and Jupiter is not the only one in the solar system. Theory predicts that there is another inside Mercury's orbit; whether asteroids ever collected there in the early days of the solar system is uncertain.

Some asteroids have been so influenced by Jupiter that they have been pulled into the same orbit around the Sun. They have not become Jovian satellites, circulating around the planet itself, but orbit the Sun at Bode's distance 52. These asteroids are gathered into two groups and are called the Trojans. Each group is separated from Jupiter, and imaginary

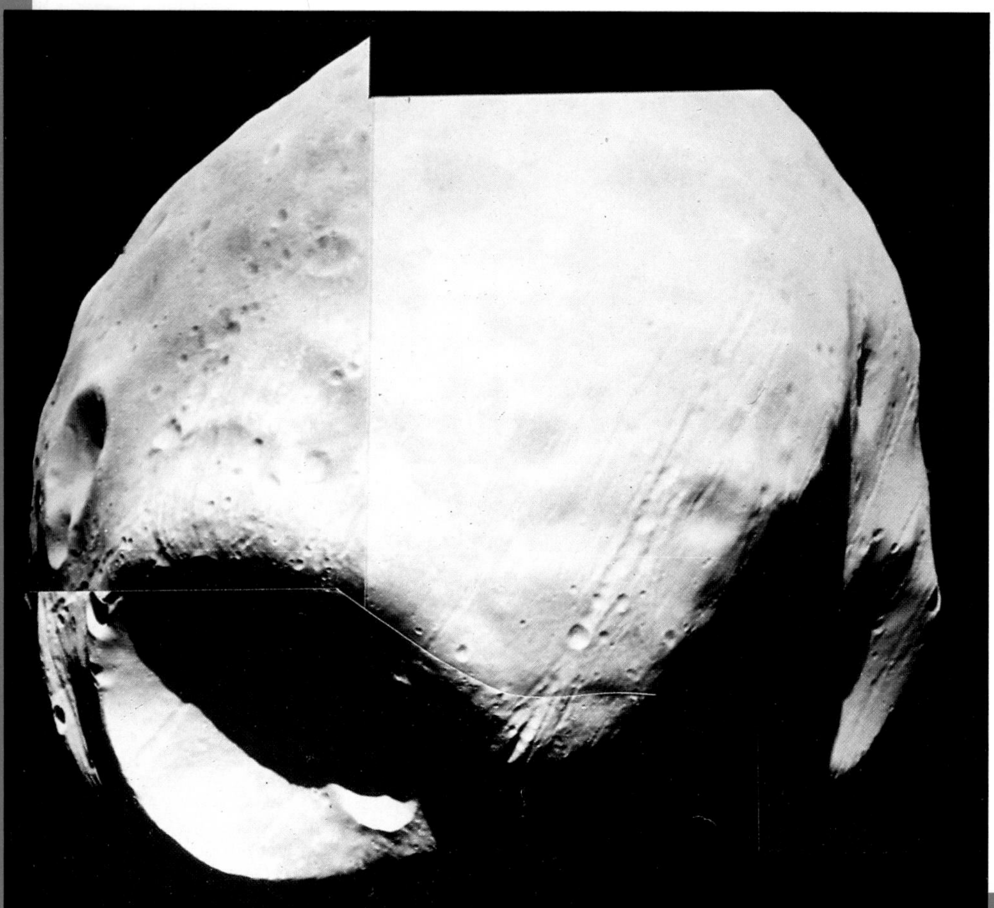

***Asteroids look like the moons of Mars*** *– small, potato-shaped and cratered. Phobos (below) is the larger of the Martian satellites, 28 km wide at its broadest. It is seen here in a mosaic of images made by Viking spacecraft. In fact, both Phobos and the other Martian satellite, Deimos, are probably asteroids captured by Mars.*

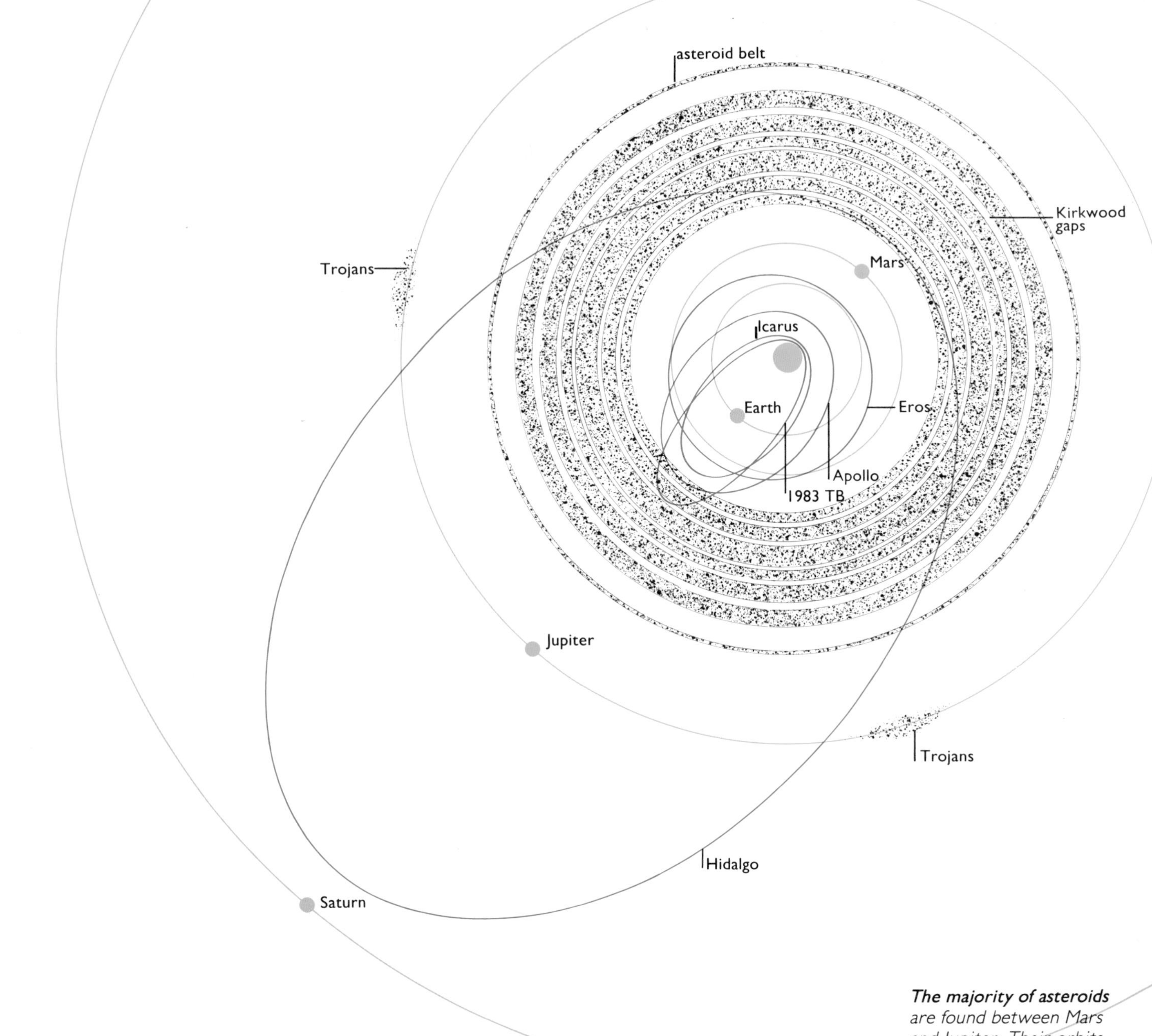


*The majority of asteroids* *are found between Mars and Jupiter. Their orbits are grouped into bands separated by so-called Kirkwood gaps, caused by the influence of Jupiter's gravitational field. Some asteroids, however, have eccentric orbits that cross those of the inner and outer planets. Icarus, for instance, comes close to the Sun, while Hidalgo's orbit nearly reaches out to that of Saturn. The Trojans are two groups of asteroids that orbit 60 degrees ahead of and behind Jupiter.*

lines drawn between Jupiter, a group and the Sun form an equilateral triangle in space. One group lies ahead of Jupiter and one behind.

Some 5 percent of the asteroids have quite eccentric orbits, which cross the orbits of Earth, Mars, Jupiter or Saturn; in the case of Chiron (not to be confused with Pluto's satellite Charon), the orbit crosses that of Uranus as well. Such orbit-crossing asteroids could crash into one of the terrestrial planets, but the chances are small.

Asteroids appear to be rocky lumps, and the great majority are thought to be potato-shaped or elongated bodies, just like the tiny Martian moons, Phobos and Deimos. And if Mars' moons are anything to go by, asteroids are probably pockmarked with dents and even small craters caused by collisions with smaller planetesimal debris within the solar system. Asteroids differ in colour, owing to the presence of various minerals and chemical substances which have water in their composition. This makes it clear that they have a wide variety of surface differences.

# COMETS

## • *Visitors from deep space*

***Cometary orbits around the Sun*** *are often exaggerated ellipses. Comets develop tails only when they get so close to the Sun that material is blown or boiled off them. Each comet has two tails, one consisting of dust and the other of plasma, or ionized gas. The dust tail is usually yellow, as it reflects sunlight. The ionized gas tail is often bluish. Tails reach their maximum brightness at the comet's perihelion, the point of closest approach to the Sun. At aphelion (when the comet is farthest from the Sun), comets are merely lifeless lumps of dust and ice.*

One of the most astonishing sights in the night sky is a large comet, with its bright head and long glowing tail. In ages when astrology held sway and celestial events were thought to presage events to come, comets were seen as signs predicting disaster; no wonder their occasional appearances struck terror into humankind. Not until the late 16th and 17th centuries were their paths studied with astronomical precision.

We now know that comets look spectacular only when their orbits carry them close to the Sun. When they are in the far reaches of their highly eccentric orbits, they are too small and reflect too little sunlight to be visible.

Comets can, in fact, be observed only in a small section of their orbits represented by their paths in the night sky. It was long a matter of debate as to whether they moved in straight lines or along curved paths, and if the latter, what kind of curves they might be. Independent confirmation of the elliptical orbits of comets finally came in 1758 when the bright comet of 1680 returned as predicted by Edmond Halley, whose calculations had been based on Newtonian theories.

Halley's Comet, as it has come to be called, has made many appearances since. On its last passage close to the Sun in 1986 it was investigated by a number of space probes, one of which penetrated the comet itself.

Comets are composed of material that could well have formed within the solar nebula. Each is an icy conglomerate of frozen gases and dust, just like a "dirty snowball". Halley's Comet is no exception. It is also probably fairly typical of other comets in that it has an elongated central nucleus – with the asteroidal potato shape – and is no more than 15 kilometres long, 8 kilometres wide and 8 kilometres thick. Its surface is dark and the whole object rotates once every 52 hours. This is a slow rotation and when the comet is close to the Sun, the sunward side becomes strongly heated and dust and ice evaporate from the nucleus. This happens to every comet as it approaches perihelion (the closest point in its orbit to the Sun).

The darkness of the Halley nucleus – it is darker than coal – was unexpected. Moreover, material does not evaporate from it as rapidly as would be expected if it were just an icy conglomerate. The comet loses material at a

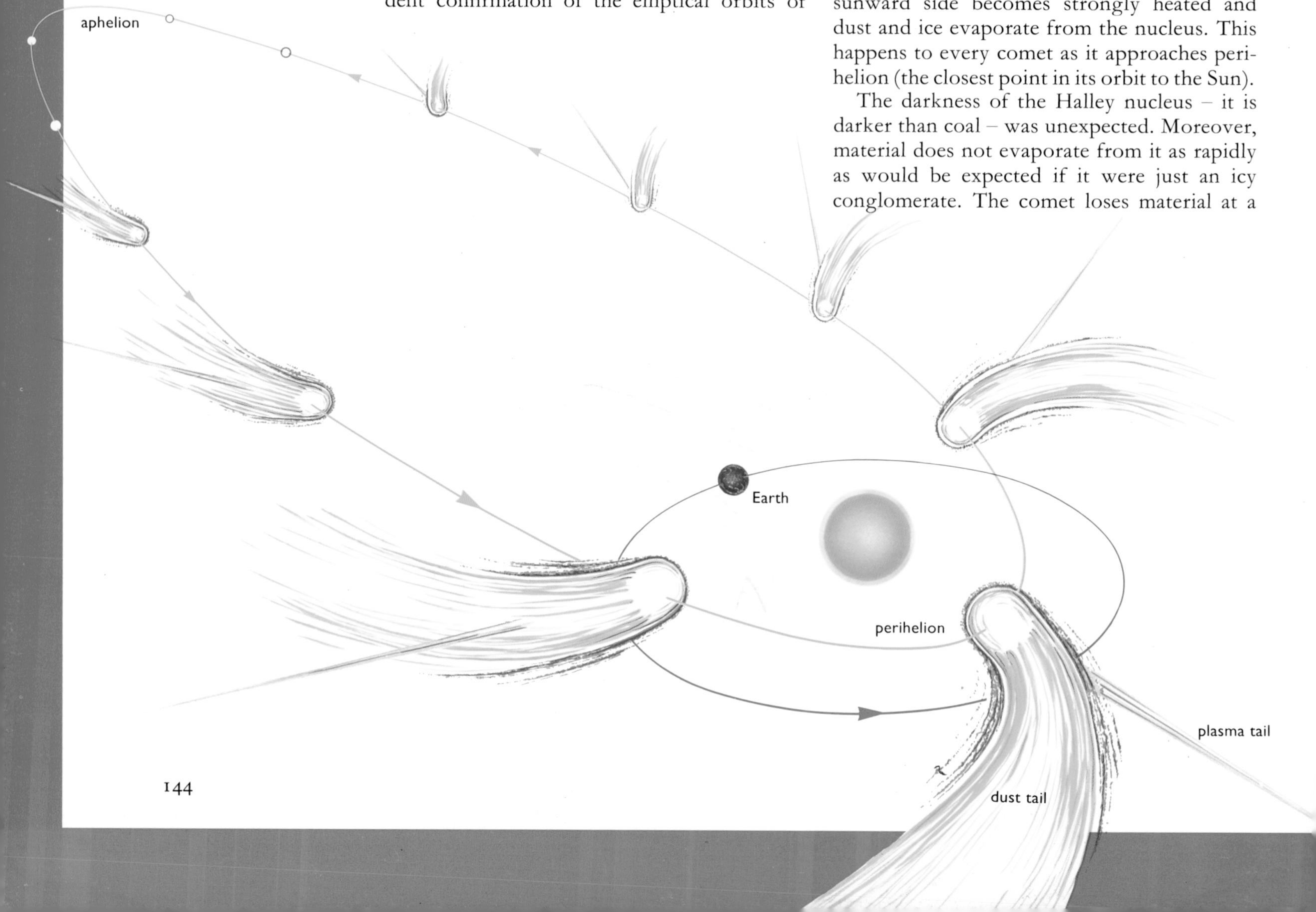

***The twin tails of Comet West*** *are shown in a false-colour image made in 1976. The upper, broader tail comprises dust from the comet's "dirty snowball" head. Dust is pushed outward from the head by the pressure of sunlight. The lower tail, consisting of ionized gas, is carried away from the comet by the solar wind.*

rate of only 15 tonnes per second, whereas ice would evaporate from 10 to 100 times faster.

From this evidence astronomers have concluded that the nucleus is covered with a mantle of dark porous material several centimetres thick. Such material could well be too heavy to blow off from the comet, or at least too heavy to reach escape velocity, so that it drops back before it gets very far. Indeed, only some parts of the surface of the nucleus poured material into space; other parts remained inert.

Nevertheless, Halley's Comet, like every other, loses some of its substance every time it passes perihelion, and some estimates suggest that Halley's ejected about 300 million tonnes at its 1986 apparition. This still leaves the 10 billion or more tonnes of the nucleus with enough material to sustain many more reappearances.

The materials that compose the nucleus of Halley's Comet consist of the expected mineral elements such as carbon, calcium, iron, magnesium, oxygen, potassium and silicon. Yet the spacecraft *Giotto* also showed that there were many light elements present, notably hydrogen and nitrogen. But the most prevalent material was a combination of both light and heavier elements to give organic molecules. The presence of these has led a few astronomers to suggest that Halley's Comet – and others – did not originate in the Oort cloud (p. 106) but came from interstellar space.

The tails of comets are generated by two mechanisms, both of which cause them always to point away from the Sun. In the first mechanism the pressure of radiation from the Sun pushes the dust outward from the comet's body to form a tail, especially as perihelion approaches. Dust particles from the comet form a large bright tail, which is usually slightly curved because the material first ejected lags behind that thrown out later as the comet increases speed when it moves ever closer to perihelion. Such dust tails reflect sunlight and appear yellowish in colour.

In the second mechanism ionized gases emitted from the comet are driven away in a straight tail by the solar wind. As the principal emission of radiation is due to ionized carbon monoxide, such gas tails appear blue in colour.

***The head of Halley's Comet****, in a computer-processed image made from photographs taken in 1910, appears to have a three-dimensional form. The bright arcs of gas in front of the head make up the coma, a glowing halo surrounding the nucleus. The coma is made of water molecules ionized by the Sun's ultraviolet radiation.*

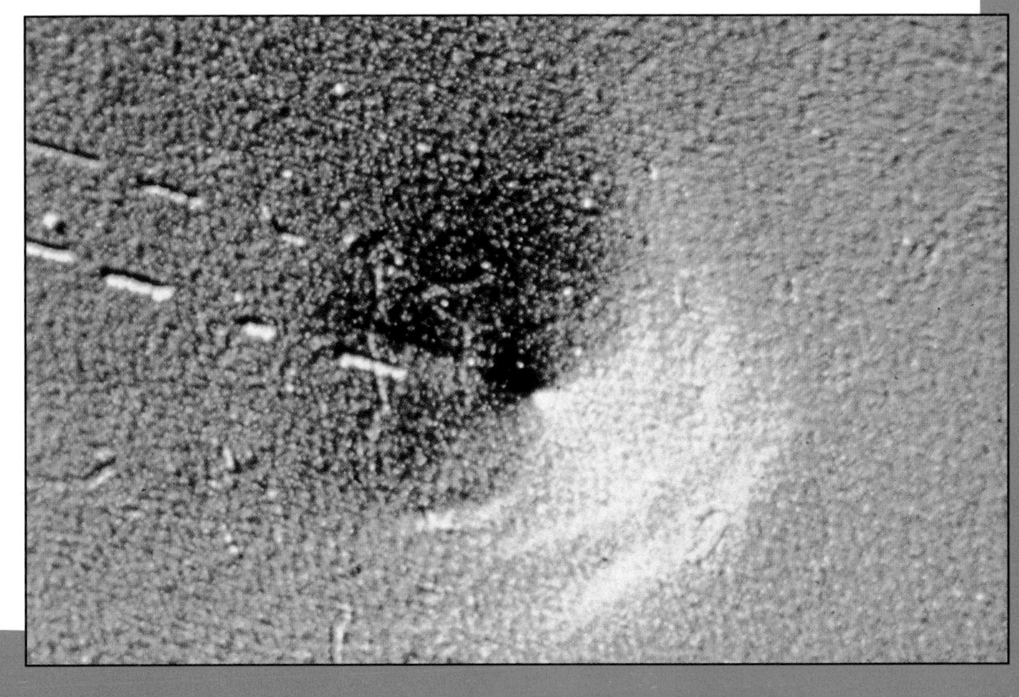

# METEORS

## • *Cosmic dust*

***Meteor Crater, Arizona,*** *is between 40,000 and 25,000 years old. The crater is well preserved because the climate in northeastern Arizona is extremely dry and there is therefore little erosion. The crater is 800 m across and 200 m deep. It is thought to have been made by the impact of an iron-rich meteorite, weighing thousands of tonnes, which largely vaporized because of the violence of the impact. The largest fragment that has been found weighs 635 kg.*

Falling or "shooting" stars are familiar to all civilizations but they are not stars in the true sense – they are meteors. Most meteors are, in fact, bits left behind by a comet. During a comet's orbit around the Sun it emits gas and dust and leaves a trail of material around its orbital path. Fortunately, comets seldom collide with our planet, but the Earth does periodically cut through cometary orbits.

When this happens, the trail of mineral and rock dust and larger debris is swept into the Earth's atmosphere where it becomes burned up by friction with air molecules. As a result, the molecules become ionized and emit radiation. Thus short-lived tails are produced by each piece of incoming material.

When the Earth crosses a cometary orbit the result is not just one or two meteors but often thousands, and such meteor "showers" can be spectacular events. The meteors all appear to come from one point in the sky called the radiant and the constellation in which this point happens to lie gives its name to the shower. Thus there are the Perseids with their radiant point in Perseus, the Leonids with a radiant in Leo, and so on. However, radiants are no more than an effect of perspective; they merely indicate the direction in the sky from which the meteor shower comes.

Not all meteors are the debris of comets. There are also sporadic meteors, which come from all directions in space and are dust and rocks, left over from the solar nebula, which are pursuing their own orbits around the Sun. During an average night a visual observer can see about 10 per hour.

These bits of debris enter the terrestrial atmosphere at velocities of anything between about 11 and 74 kilometres per second. The trails they leave depend upon the size of the incoming material, which may be no more than a dust particle or could be a large lump of rock. The lengths of the trails vary between about 7 and 20 kilometres.

Rarely, when an extremely large lump of rocky material enters the Earth's atmosphere the trail can be sufficiently bright to light up the countryside, and the meteor looks just like a glowing ball of fire.

Most meteoric material is completely burned up in the Earth's atmosphere but on occasions the ablation is incomplete, and a lump of rocky material actually lands on the ground. Examination of such meteorites provides a clue to the formation and composition of the asteroids from which they are probably derived.

Meteorites are best classified into three main types: metallic, lithosiderite and chondrite. There is also another minor group, the achondrite. Metallic meteorites are composed mostly of iron and nickel. The proportion of nickel varies; some contain less than 6 percent nickel, while others have between 8 and 20 percent. The metals are crystallized, and show that the asteroids from which they came underwent either slow cooling or sudden solidification.

The lithosiderites are made up of metallic material and silicates, which overlap each other. Such meteorites can be further classified into two types: mesosiderites and pallasites. In mesosiderites the silicates consist mainly of the minerals feldspar (an aluminium silicate) and pyroxenes (silicates which include iron, magnesium and calcium). This indicates that the asteroids that mesosiderites came from received their metallic components after they had solidified.

Pallasites contain much olivine (a silicate of iron and magnesium). They were probably formed by the intrusion of a metallic liquid between the core and the mainly olivine mantle of the asteroid from which they came.

The chondrites, or stony meteorites, contain small spherical particles, or "chondrules", which give them their name. These contain iron, feldspars, olivine and pyroxenes and are thus chemically rather similar to many terrestrial rocks. A subset of the chondrites, known as the carbonaceous chondrites, is a particular type composed of a mixture of carbon-rich crystals containing olivine, pyroxene, metals, glass and, uniquely, mica sheets.

Finally, there are the achondrites which, as their name implies, do not contain chondrules. They are very similar to lunar rocks.

Study of meteorites thus provides some indication as to the chemical nature of the solar nebula some 4.5 billion years ɔ. By that time iron and other heavy atoms had been synthesized, presumably within those early stars that had become supernovae and so distributed their contents into interstellar space.

***The Geminid meteor shower*** *occurs in the second week of December every year. The short diagonal streaks in this long-exposure photograph are star trails, while the longer ones belong to meteors. The Geminids peak at about 58 meteors per hour on the night of 14 December.*

# THE LIVING UNIVERSE

From the moment of the Big Bang, 15 thousand or more million years ago, the universe has evolved into the vast, expanding mass of galaxies we now observe, which stretches farther out into space than even the most sophisticated modern instruments can measure. In the immensity of the universe, our Earth is a mere speck of rock in orbit around a not very significant star. Yet the findings of modern science have led some astronomers and physicists to the view that the crowning glory of the whole cosmos is humankind.

In our solar system, only planet Earth supports life, and in this sense our planet certainly is unique. Only here, it seems, are conditions suitable for the evolution and existence of complex and intelligent life forms. Such life, based essentially on the versatile chemical attributes of the element carbon has, after millions of years, led to the astounding appearance of the self-reproducing, highly sophisticated, communicative species that is *Homo sapiens*.

The range of life existing on Earth is so vast and so complex that scientists have been led to speculate whether even the time span that has elapsed since the Big Bang has been sufficient to accommodate that evolution. An alternative explanation for the success of life on Earth might lie, for instance, in the arrival of complex molecules from outer space, which aided and speeded the evolutionary process. Painstaking study of meteorites and similar evidence now suggests that this is a definite possibility.

At the pinnacle of evolution, we human beings are endowed with powers that mark us out as special. Arguably the most important of these is our ability to communicate, not merely with spoken alphabetic languages, but through the universal symbolism of mathematics. But are we humans really the only intelligent beings in the universe? If we are not, then communication with alien civilizations is desirable – and, indeed, possible. Yet even if this were achieved, the problems of dialogue seem, at present, insuperable; for example, the time taken for messages to span the immense distances involved would impose the most daunting of limitations.

In considering the universe, and our place in it, we must also speculate about the evolution of the universe as a whole. Is our view too prosaic and parochial? What are the new laws of physics we must learn to unravel the mysteries of the actual creation, existence and future of the universe? And will these reveal that our ideas of the Big Bang and an expanding universe are, in fact, completely erroneous?

Evidence gleaned thus far suggests that our general picture seems to be correct. But the central part humankind appears to play in the scenario raises some more enthralling questions, most particularly whether the ability of humankind to observe and understand the universe has an effect on its reality.

As the sciences of physics and astronomy become more sophisticated, they tell us more about the far future of our universe. This may not be as simple or straightforward as was once thought. For example, black holes and the space they enclose present strange possibilities and provoke the speculation that the scientific outlook of today is still too limited in its concepts to allow full understanding of the universe.

***The lunar landing module* Eagle** *ascends from the surface of the Moon after the Apollo 11 mission in July 1969. For the first time, the human race had established a foothold on a world beyond its own. In the view of some scientists the appearance of intelligent life and its colonization of the universe was "designed into" the laws of the cosmos from the very beginning.*

# HOME PLANET

## • *Our living world*

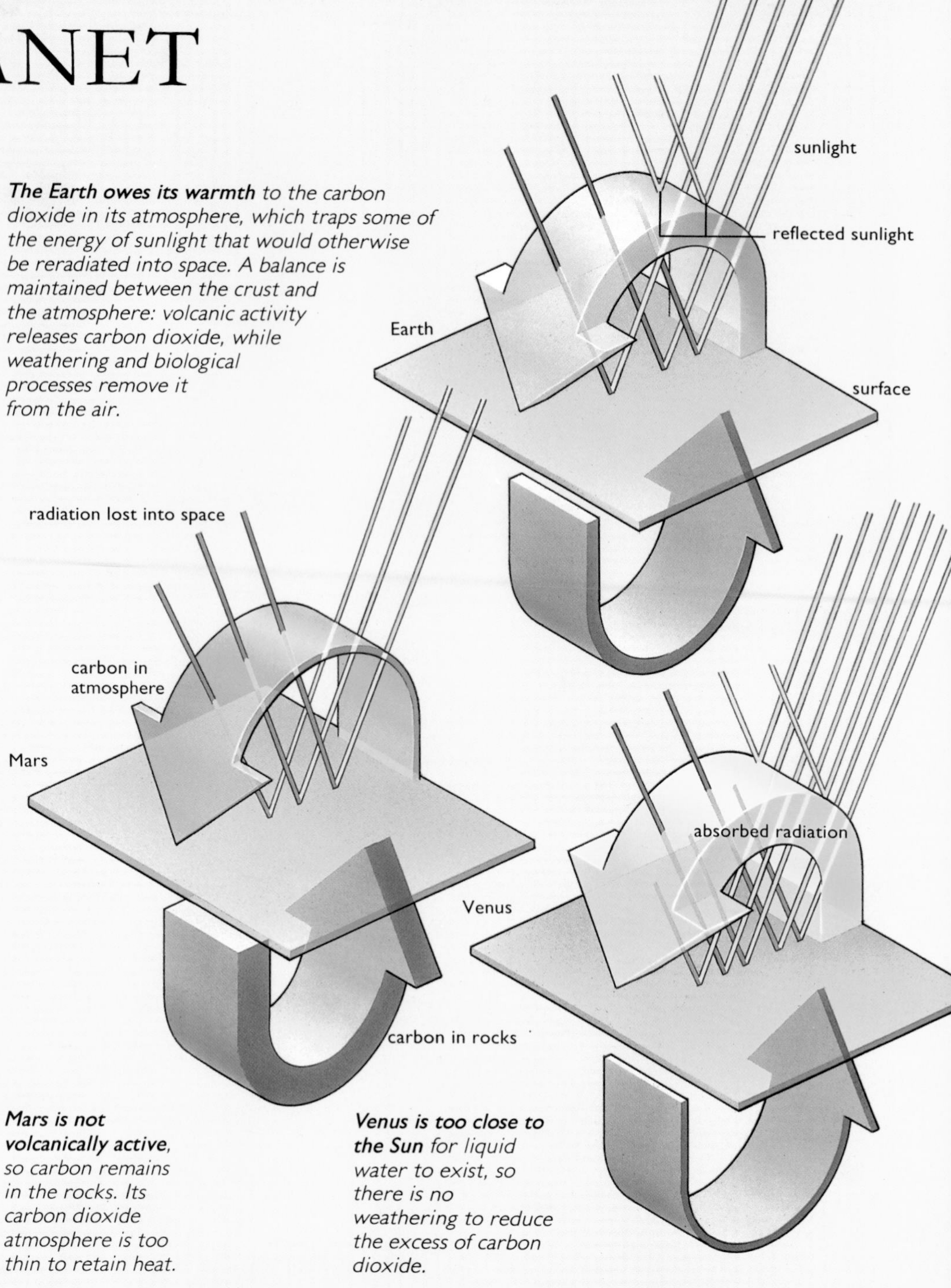


***The Earth owes its warmth*** *to the carbon dioxide in its atmosphere, which traps some of the energy of sunlight that would otherwise be reradiated into space. A balance is maintained between the crust and the atmosphere: volcanic activity releases carbon dioxide, while weathering and biological processes remove it from the air.*

***Mars is not volcanically active,*** *so carbon remains in the rocks. Its carbon dioxide atmosphere is too thin to retain heat.*

***Venus is too close to the Sun*** *for liquid water to exist, so there is no weathering to reduce the excess of carbon dioxide.*

Here on Earth, our planetary spaceship, we humans live and evolve – along with a vast complex of vegetable and animal life, from microscopic bacteria to huge elephants, and from minute viruses to giant sequoia trees. Although the Earth possesses, like the other terrestrial planets, a dense core surrounded by a mantle overlaid with an outer crust, it has this unique difference – life.

During the 20th century it has become clear that life exists on Earth only because conditions are right. It is neither too hot nor too cold, and potentially deadly radiation is kept at bay by the terrestrial atmosphere. Moreover, the chemical and biochemical environments are in perfect balance, to ensure not only the existence of living things but also their proliferation. This benign equilibrium has evolved, in tandem with life itself, over geologically long periods of time. Unbalance or destroy it, and the rich variety of life risks being diminished or even threatened with extinction.

Critically, the prevailing temperatures on Earth allow water – a vital ingredient of life – to exist in liquid form. Earth's average surface temperature is 15°C, well above freezing, and some 33°C higher than it would be if it were heated by solar radiation alone. The reason for this is that some heat radiated by the planet is trapped by the "blanket" of carbon dioxide and water vapour in the atmosphere and warms the sea and land below.

Yet if such warming increased, and continued unabated, this heating or greenhouse effect would run away with itself as it has on Venus (pp. 110–13). Consequently, the temperature would increase too much to allow life to continue. On Earth it seems possible that this galloping effect has been prevented by modification of Earth's environment by its living inhabitants. This forms the crux of the Gaia hypothesis.

The hypothesis was proposed in the late 1960s by the British atmospheric scientist James Lovelock who, with the American biologist Lynn Margulis and other colleagues, had begun to study the Earth's natural system of checks and balances. Looking at the whole Earth as a living organism, much as the Chinese had done in past times, Lovelock and Margulis pointed out that, despite continuing climatic change, the Earth enjoys climatic stability at any particular time. Carbon dioxide and water vapour help warm the surface, but there is no runaway effect because both are recycled in various ways.

Such recycling takes many forms. Sunlight, for instance, enables green plants, through the process of photosynthesis, to use carbon dioxide and water to manufacture carbohydrates – substances such as sugars, starch and cellulose. In exchange, oxygen is released as a waste product, and is used by animals for respiration.

When plants die and decay the carbon in their tissues combines with atmospheric oxygen; carbon dioxide is released, so promoting the greenhouse effect. Some of the carbon is, however, removed permanently. It then becomes locked into the shells of marine creatures as insoluble calcium carbonate and finds its way into sediments on the seabed. This, the Gaia hypothesis argues, prevents excess build-up of carbon.

***Photosynthesis by green plants*** *produces the oxygen that makes Earth's atmosphere unique in the solar system.*

Living things are also ultimately involved in the control of other cycles and in greenhouse effect regulation. For example, minute marine plants, or phytoplankton, are involved in the cycling of sulphur and iodine. Particles containing these elements are exuded by the plants and form the nuclei around which atmospheric water condenses to form clouds.

Clouds form as rising water vapour cools and condenses. And the higher the Earth's temperature, the greater the production of water vapour from the oceans that cover two-thirds of its surface. Being white, clouds reflect sunlight back into space, and so have a cooling effect. In due course, clouds bring rain, returning water to ground level.

Biological factors undoubtedly play a large part in environmental control on Earth. But the presence of humans on the planet – although we are part of the Earth's biology – introduces other factors. Two human-centred effects of current concern are the presence of "holes" in the ozone layer and global warming.

Ozone, a molecule composed of three oxygen atoms, is formed when solar ultraviolet light breaks down atmospheric oxygen. Most atmospheric ozone lies 13 to 24 kilometres above the Earth's surface. Crucially for life, ozone absorbs dangerous short-wavelength ultraviolet radiation which, if allowed to penetrate to the surface, causes deadly cancers in humans and other animals. It also threatens the existence of the phytoplankton, the plants crucial to element cycling. Additionally, food crops produce lower yields when exposed to ultraviolet radiation. Cooling of the upper atmosphere by ozone depletion is also likely to cause climatic change.

The destruction of ozone is caused mainly, it seems, by the commercial use of chlorofluorocarbons (CFCs), which are used in aerosols and as refrigerants. In cold air regions, for example over the Antarctic, the CFCs react with nitrous oxide in the atmosphere to form chlorine compounds. And one chlorine molecule can destroy 100,000 ozone molecules.

The use of CFCs is also thought to contribute to global warming – a long-term increase in the temperature of the Earth. Global warming appears to be accelerating, owing mainly to the widespread use of fossil fuels such as coal and oil. Burning these fuels releases carbon dioxide into the atmosphere, which in consequence traps more heat.

Unless carbon dioxide emissions are reduced it is estimated that by the middle of the 21st century the average temperature of our planet will have increased by between 1.5° and 4.5°C. This would lead to significant melting of land-based ice, and a rise in sea level of about 1 metre due largely to an expansion of the warmer upper layers of the sea. Parts of New York, London and Tokyo could be flooded, and small atolls such as the Maldives submerged totally.

The climatic changes accompanying such warming would have mixed effects. Parts of northern Canada and Scandinavia would become crop producers, while production would increase in central Australia and parts of Africa, China, India and South America as they became wetter. On the debit side, the grain-producing areas of America and the USSR would suffer droughts.

It remains to be seen whether the forces of Gaia will indeed "rescue" the planet from violent climatic change. What is irrefutable, however, is the fact that chemical and biological balance are crucial to the continuance of life on Earth. If chemical equilibrium were total – if there were no chemical cycling – Earth would be devoid of life.

Conditions observable on Earth also affect our opinions about the requirements for life to occur elsewhere in the universe. The Viking probes of 1975 made it clear, for example, that although Mars probably had surface water at one time, and may even have supported life forms in the distant past, it is barren and desiccated now. Perhaps a future generation of space explorers may find fossil remains and evidence of the processes that brought such life to an end.

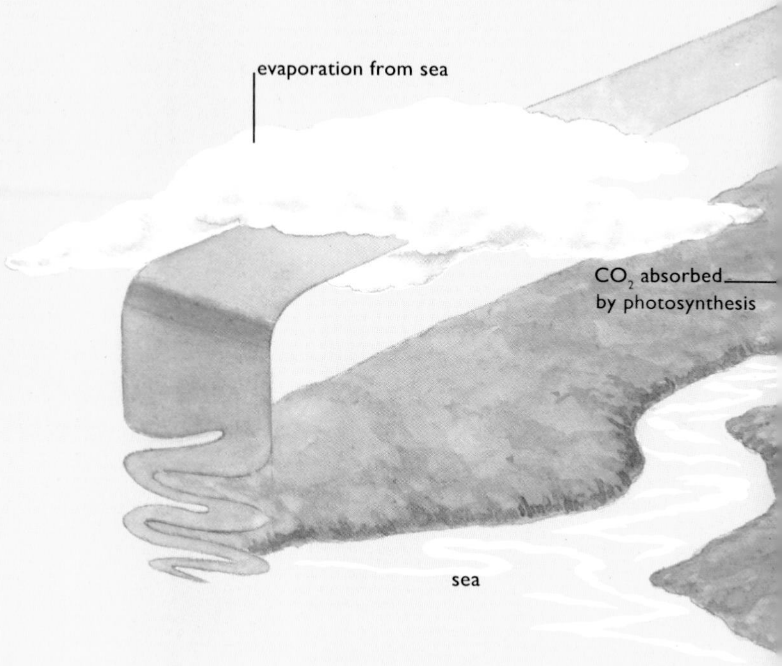

***Carbon and water are recycled*** *continually by natural processes. Water evaporates from the oceans and land, and is given off by the transpiration of plants. The water is released as rain, snow or sleet when air rises and cools. This water eventually flows into the sea, to resume the cycle.*

*Carbon dioxide gas is absorbed from the atmosphere by land plants, and dissolved carbon dioxide is absorbed by water plants, to be utilized in photosynthesis. When the plants die they decompose and return carbon dioxide to the atmosphere. Animals consume carbon in plants and release it in wastes or on death, when they decompose. Marine organisms form carbon-rich sediments when they die.*

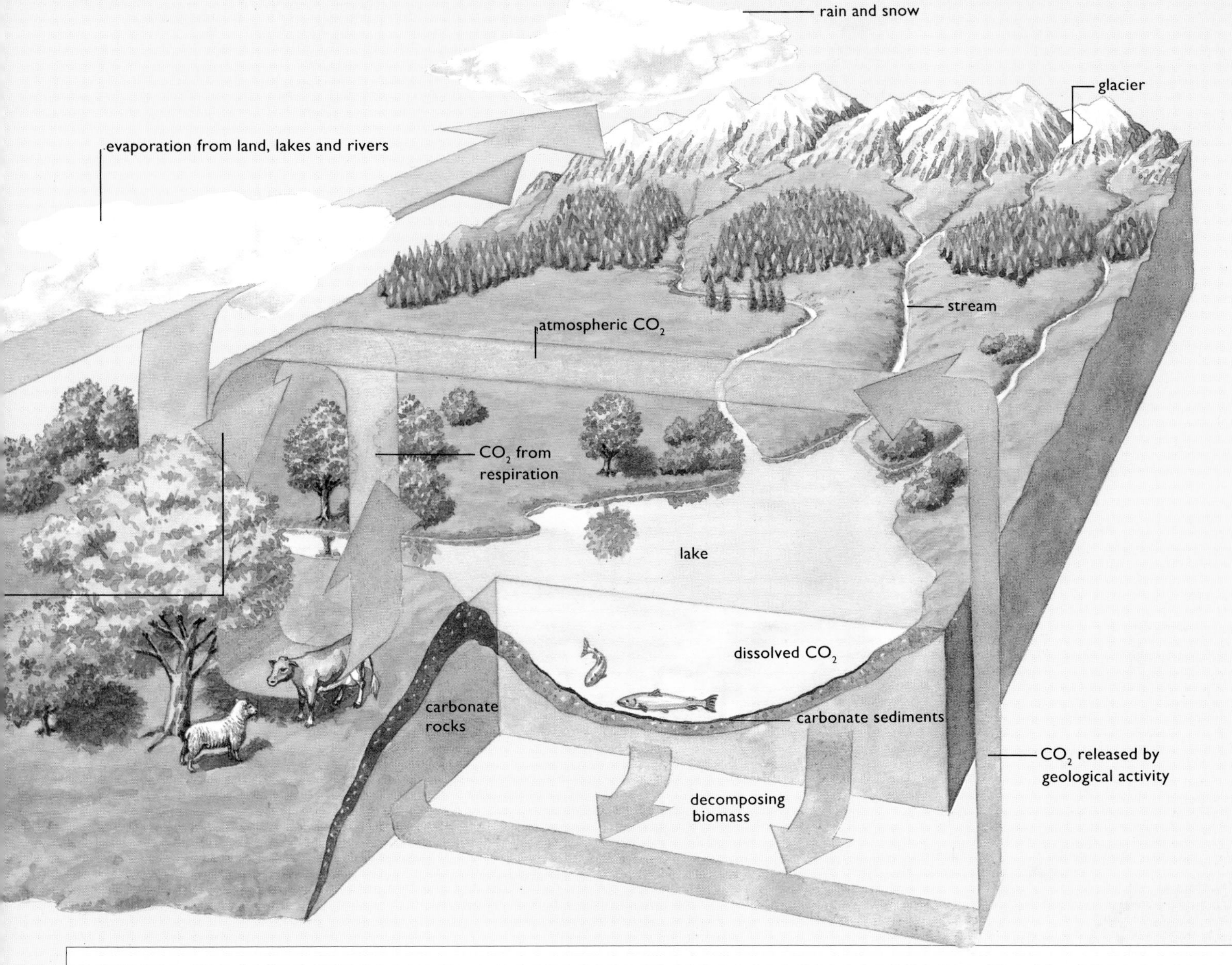


## Daisy worlds

Criticism of the Gaia hypothesis centres chiefly around the fact that the organisms involved need to "know" in advance what is required to keep the planet in balance. Lovelock countered this by designing a "Daisyworld" computer program for a hypothetical Earthlike world. The daisies were either white or black and began growing at 5°C, flourished at 20°C and stopped growing at 40°C.

At 5°C the black daisies would thrive best because they would absorb more sunlight and so become warmer than the soil. Ambient temperatures would slowly rise, causing the demise of the black daisies but allowing the white daisies to flourish because they could cool themselves by reflecting heat from their petals. Ambient temperatures would drop and in time the black daisies would again thrive.

Subsequently, daisies of ten different shades were studied; results showed that regulatory mechanisms continued, but in a more sensitive way.

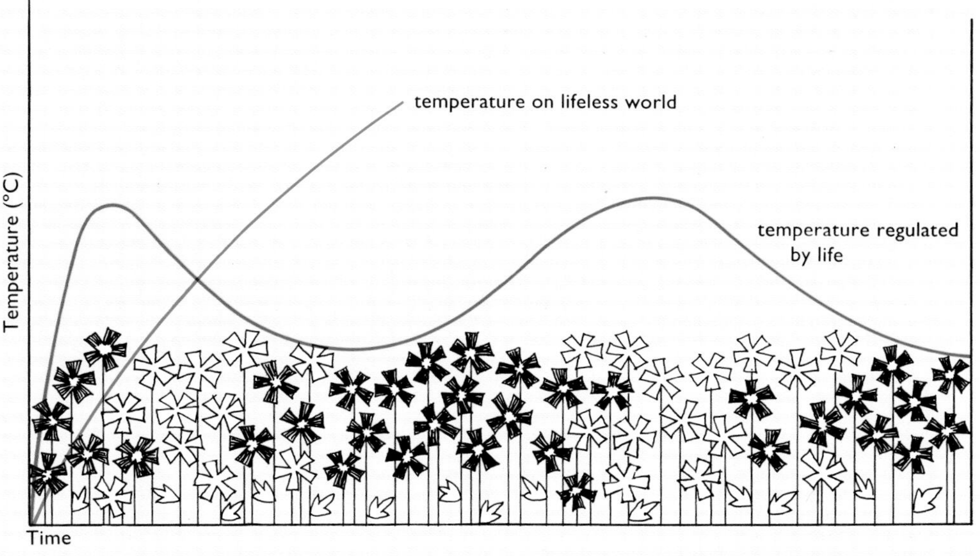

# THE NATURE OF LIFE

## • *Features of living organisms*

Life on Earth is all-pervading but elusive. It can be found from ocean floor to mountain peak, yet cannot be "made" in a laboratory. New life can begin in a test tube with a human egg fertilized *in vitro*, but this cannot be nurtured to viable babyhood outside the womb.

More than 2,500 years ago, the Greek philosopher Aristotle, who was also a fine biologist, stated "We mean by 'possessing life' that a thing can nourish itself and grow and decay". Aristotle was correct in that these are certainly recognizable attributes of living material but they are not the only ones.

Crystals, for instance, are non-living materials that display the behaviour he described. Suspend a piece of string in a strong solution of sugar and water and it will become covered in sugar crystals. The string will be "nourished" by the sugar and, in time, the crystals will multiply and grow until all the sugar has crystallized out.

Such growth and replication is certainly a characteristic of living material, but clearly not of living things alone. The sugar can be put back into solution, so that the crystals decay or "die", but this does not make them living. Clearly they lack other fundamental qualities.

Since nourishment, growth and decay are not alone sufficient to define life, other attributes must be required. Living organisms are defined by their ability to reproduce and repair. They must, in addition, be able to react to their environment – yet even here valid comparisons can be made with the behaviour of some inanimate objects.

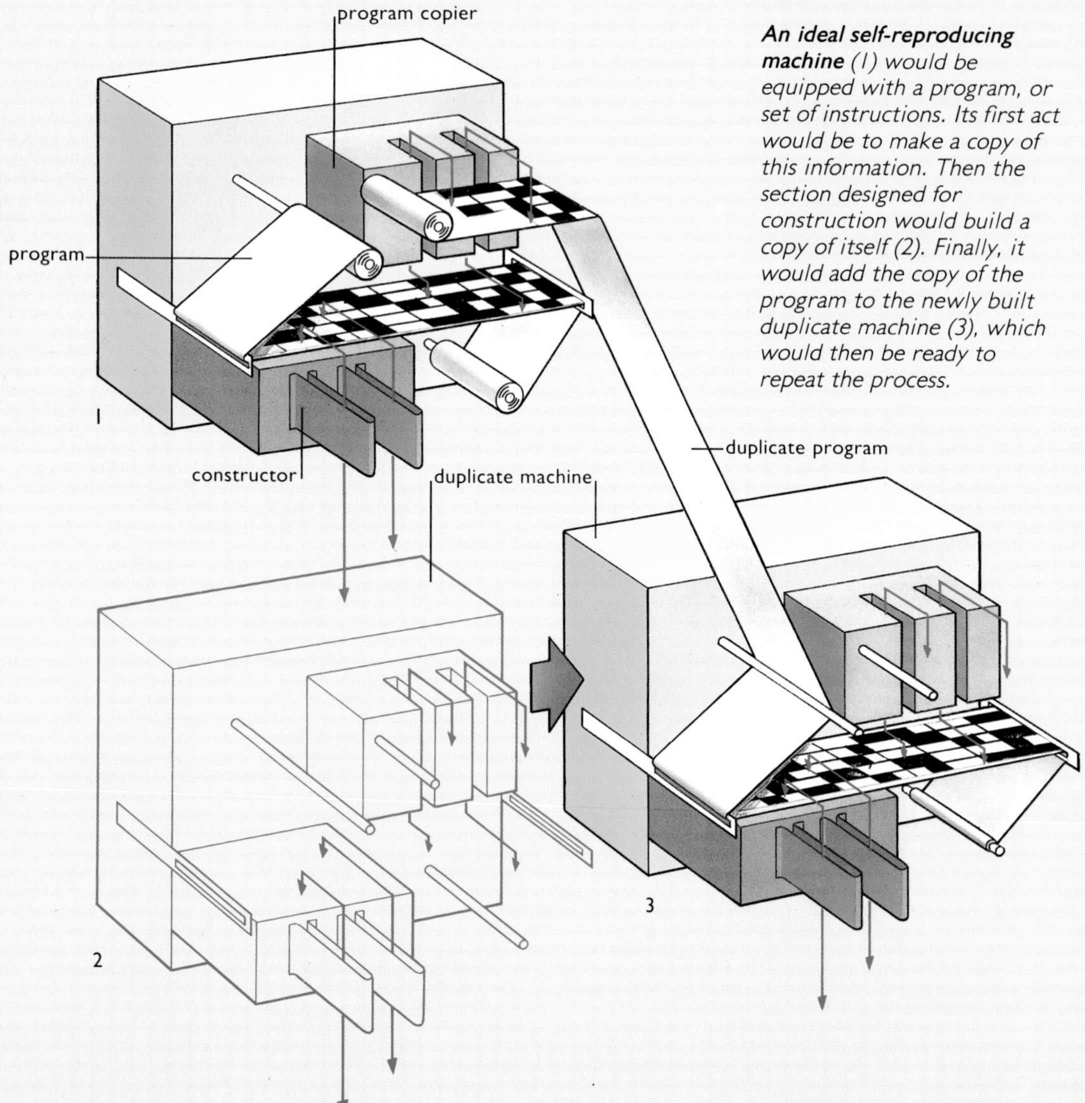


***An ideal self-reproducing machine*** *(1) would be equipped with a program, or set of instructions. Its first act would be to make a copy of this information. Then the section designed for construction would build a copy of itself (2). Finally, it would add the copy of the program to the newly built duplicate machine (3), which would then be ready to repeat the process.*

The Hungarian-born US mathematician John von Neumann, a pioneer in the development of modern computers, devised a mathematical representation of an idealized self-replicating machine. It is composed of two parts – a constructor and a computer program containing data and instructions. The constructor's actions depend on both the computer program and on the environment.

In order to operate, the self-replicating machine is dependent on being able to pick up and use the materials it has to hand. In favourable conditions the new "daughter" machine it produces has its own data bank and is self-reproducing. And because it uses raw materials available to it externally, the machine can be said to interact – albeit in a limited way – with its environment. The machine was not capable of self-repair, but this cannot be ruled out as impossible.

If such a machine is compared with a living organism, crucial differences emerge. First, all living matter is composed of cells which are complex constructs of simple biochemical materials. These cells can grow, divide and, in so doing, replicate themselves.

The essential difference between the processes in living organisms and non-living things lies in the way materials are used. The growth, replication and self-repair of living material occurs *inside* the body of an organism. Unlike a crystal or von Neumann's machine, which take in materials and use them as they are, a single cell can transform itself from the simplest of ingredients into a complex organism containing millions of cells. Once inside the crystal or machine, material is, by contrast, immutable. These inanimate objects cannot take in simple substances, then adapt and use them in order to replicate and multiply, in the way that cells do.

The power of self-repair in living things is remarkable. If a starfish loses an arm it can grow a new one. A deep skin wound can heal itself unaided. The human liver can regenerate even if reduced to less than a quarter of its normal size.

The way in which living things react

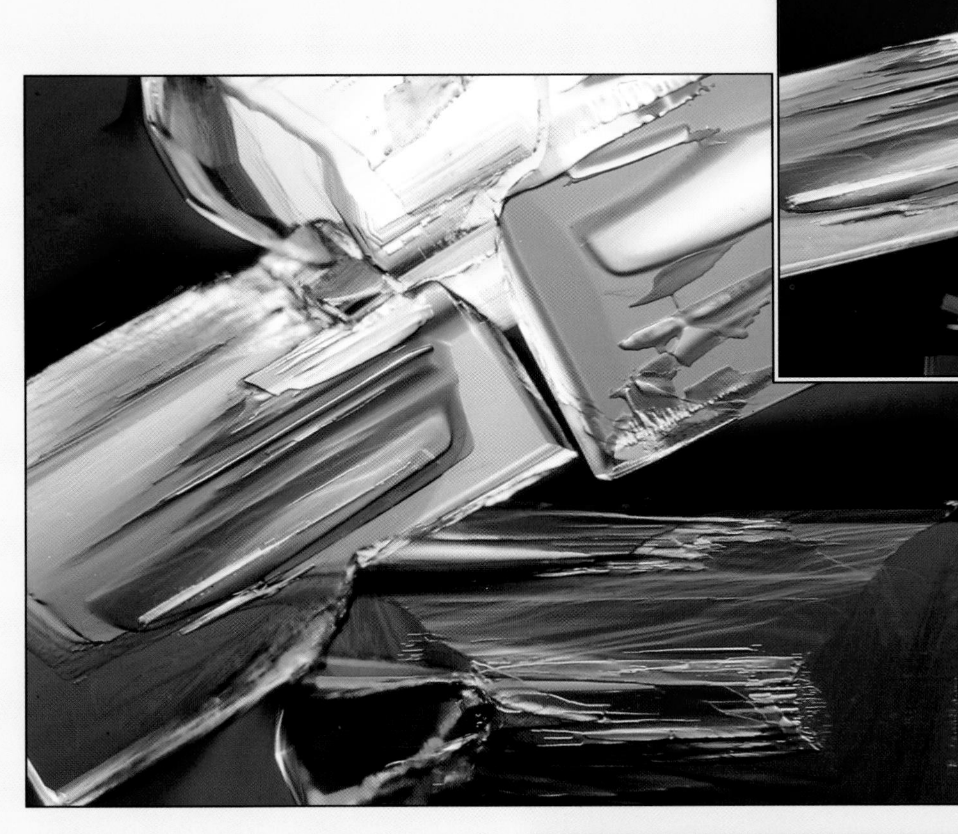

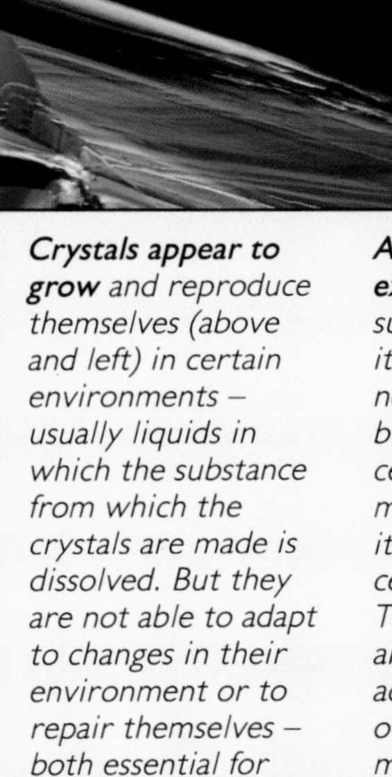

***Crystals appear to grow*** *and reproduce themselves (above and left) in certain environments – usually liquids in which the substance from which the crystals are made is dissolved. But they are not able to adapt to changes in their environment or to repair themselves – both essential for living organisms.*

***A virus cunningly exploits a host cell*** *to sustain and reproduce itself (below). Here new viral bodies are budding from such a cell. The virus cannot manufacture copies of itself but highjacks a cell for this purpose. The fact that it can arrange for – if not actually perform – its own reproduction means that it just qualifies as living.*

to their environment is equally varied. Bacteria, among the simplest of organisms, respond to changes in their environment – that is, their food supply – by subtle alterations in their internal chemical processes, something inanimate things cannot do.

Differences are yet more marked in highly evolved plants and animals. The Venus fly trap, for instance, is a carnivorous plant which eats insects that settle on it. By means of chemical "sense organs" it can detect the presence of food and react accordingly. In animals the senses of hearing, sight, touch, smell and taste all allow complex reactions to the surroundings, including the vital tasks of finding food and a suitable mating partner.

The acme of development is, arguably, the human being. Our complex reactions to the environment, our powers of memory and constructive thought, and our consciousness put us far above any other species. Certainly it may one day be possible to create robots capable of imitating some of these functions, but they will ultimately be of biological origin since they were designed by humans in the first instance.

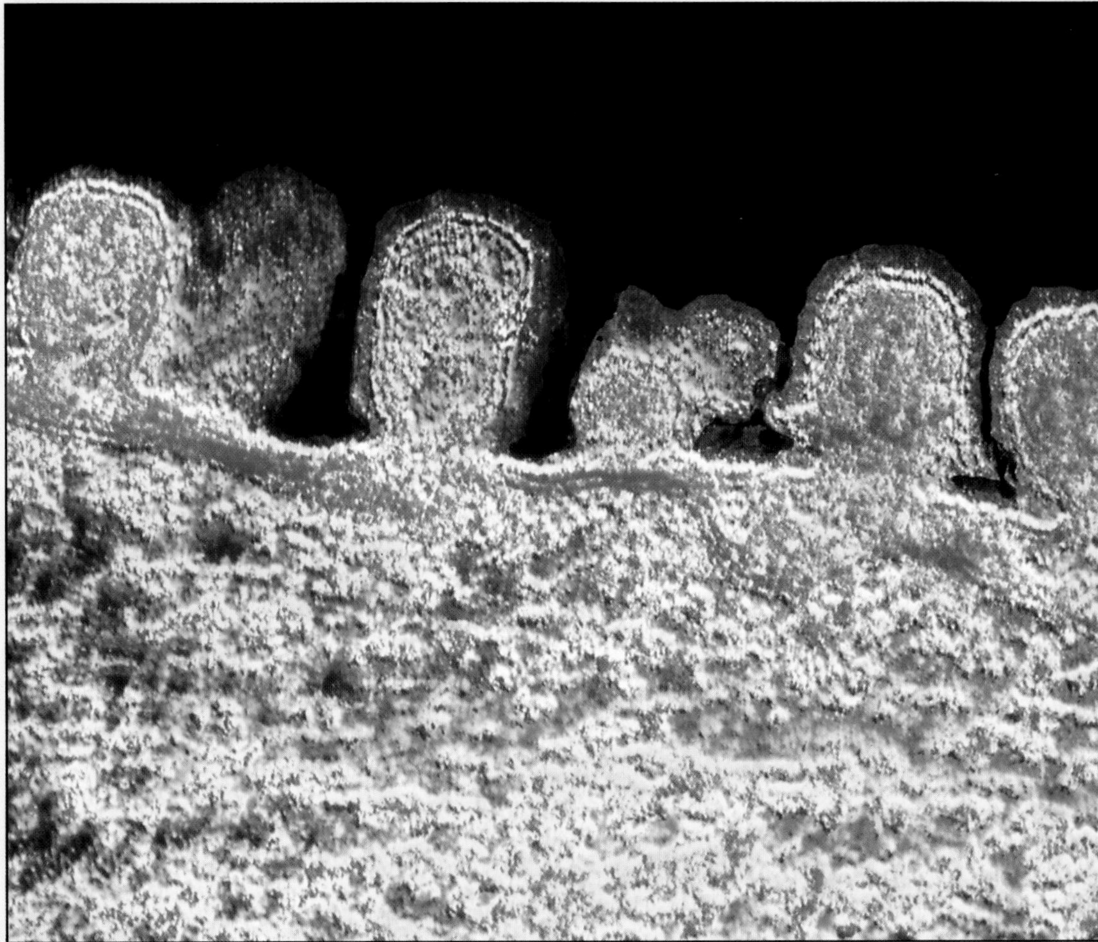

# CHEMISTRIES OF LIFE

## • *The basic building blocks*

All living things on Earth, from micro-organisms such as bacteria to the largest plants and animals, are based on the chemical element carbon. This seems to be because the carbon atom is unique in the way it links with other carbon atoms and with atoms of other elements, notably hydrogen.

Carbon atoms can join together by sharing electrons. Through such covalent bonding, a carbon atom can link itself to another carbon atom and, at the same time, join with other elements or compounds. The result is that carbon can form molecules that have a "ring" formation based on the linkage of six carbon atoms. Such molecules are essential to many life processes, including photosynthesis, and many are soluble in water, an attribute vital to earthly life.

As well as forming rings, carbon atoms can also combine with each other to make long chains or polymers. In nature, these polymers form the chemical backbone of complex organic molecules that are vital to the structure and maintenance of life, from the supportive walls of plant cells to insulin, the hormone humans need in order to metabolize sugars.

The versatility of carbon means that the temperature range of carbon-based life is enormous by earthly, if not by celestial, standards. It covers approximately 100°C – the range between the boiling and freezing points of water, the substance that makes up a high proportion of all living organisms. At the upper end of the range, some algae flourish at temperatures of 70° and 80°C, and some bacteria still grow actively at 95°C. In the laboratory they will survive above boiling point, at 104°C.

Many lichens, plants in which an alga and a fungus live in close association, can exist equally well in the heat of the Sahara or the cold of the Arctic. And some fishes with special blood can survive a few degrees below freezing. Outside Earth, low-temperature life might be possible, but because there would be little energy in such systems, evolution of complex life forms would be so slow that it would be outpaced by environmental change and thus could not survive for any length of time.

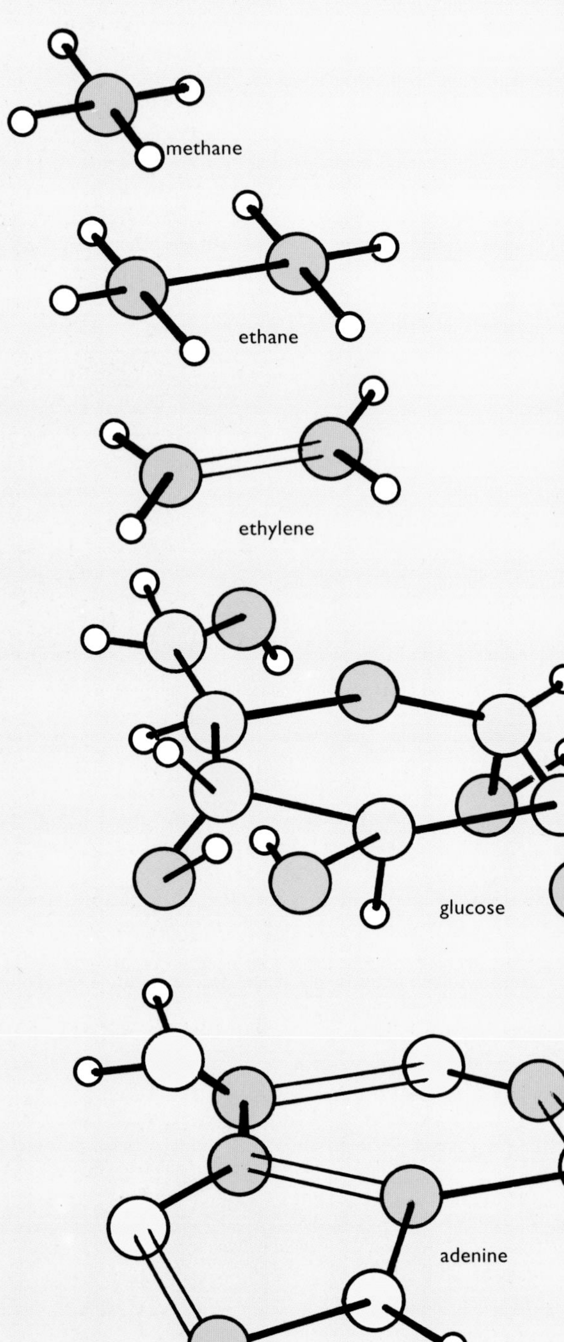


***Carbon is unique** in the variety of molecules it can form. They are typified by methane (one of the simplest), ethane and ethylene (in which a single or a double bond links two carbon atoms), glucose and adenine (containing carbon rings). Here carbon atoms are pink, hydrogen white, oxygen blue, nitrogen yellow.*

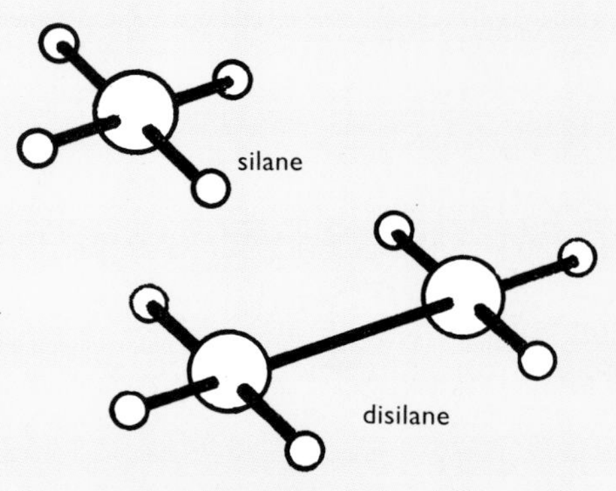


***Silicon resembles carbon** in its ability to form large chain molecules. The molecules of the compounds silane and disilane, for example, are analogous to those of methane and ethane. Such silicon compounds are, however, generally less stable than their carbon-containing counterparts.*

Short-wave radiation also puts limits on life. The complex molecules of living material are broken down or changed by such radiation, as, for example, when white skin becomes sun-tanned by ultraviolet light. The ultraviolet causes the production of the pigment melanin, to help prevent the radiation from penetrating the skin. But because such radiation is everywhere in space, life can exist only where it is protected by, for example, a planetary atmosphere.

On Earth, the element oxygen is an

*Tiny diatoms* (left), the base of the food chain in the oceans, have found a way to use the element silicon in their shells to harden them.

*On the sunless floors* of deep ocean trenches live these tube worms, 2 m long. They, and other shellfish, live off bacteria that extract energy from hot water, rich in hydrogen sulphide, that pours from thermal vents. Such communities are the only ones known that do not ultimately rely on photosynthesis for energy.

important constituent of living things and a part of the atmosphere. Atmospheric oxygen is released by green plants as a waste product of photosynthesis, and is used by animals for respiration. Without oxygen the range of life would be much restricted, but in other situations in the universe, could some other element take its place in living systems? Sulphur, for example, has been found to act in similar ways to oxygen, and provides the energy source for some bizarre deep-sea creatures.

Biochemists have thus considered the possibility of living systems based on the element silicon rather than on carbon. Although it cannot form ring-shaped or long molecules, silicon does combine readily with oxygen. Sand, or silica, is one of the commonest chemicals on Earth, and compounds of silica with calcium and aluminium are found in many terrestrial rocks.

These compounds are not, however, soluble in water, which makes them unsuitable as a basis for life. This is not to say that such a scenario is impossible. Some terrestrial bacteria can produce insoluble substances such as sulphur and process them perfectly adequately.

In general, silicon molecules are more resistant to heat than those based on carbon. Silicones – polymers with a central chain of alternate oxygen and silicon atoms used in lubricants and synthetic rubber – are stable up to 350°C. If silicon-based life were possible, the biochemists argue, it would probably be stable up to 250°C. If the atoms on the side of the silicone chain were more widely varied, it is conceivable that they could provide the basis for life elsewhere in the universe, where conditions prevailing on planets are hotter than on Earth.

Most recently, biochemists have considered alternatives to both carbon and silicon. Germanium, selenium and sulphur compounds have been suggested, but the restrictions seem to be much as they are for silicon.

# THE BEGINNING OF LIFE

## • *Building the essential molecules*

How life began on Earth remains a mystery, but it probably happened in one of two ways. Either it started when conditions were suitable for the synthesis of its basic chemicals, or it came about through the arrival of complex organic materials from space.

In the 1930s, the Russian scientist Aleksandr Oparin, and the Americans Melvin Calvin and Harold Urey, concluded that life could have begun in an atmosphere where there was no free oxygen – at an early stage in the Earth's history it was all combined with other chemical elements. Their supposition was upheld by the Chicago-based chemist Stanley Miller in a notable experiment that attempted to mimic conditions on Earth as they were 3.5 to 4 billion years ago.

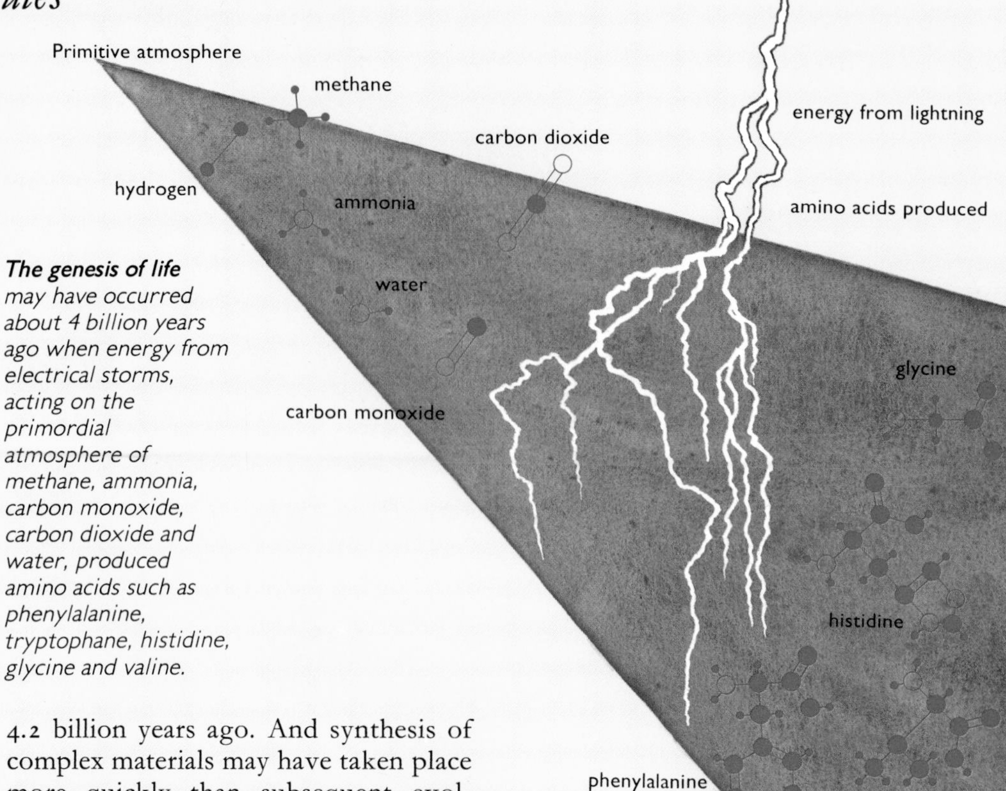


***The genesis of life*** *may have occurred about 4 billion years ago when energy from electrical storms, acting on the primordial atmosphere of methane, ammonia, carbon monoxide, carbon dioxide and water, produced amino acids such as phenylalanine, tryptophane, histidine, glycine and valine.*

Miller passed electric discharges through a mixture of ammonia, hydrogen and methane, in a flask containing distilled water. The chemicals represented the components of the early atmosphere, the electrical discharges flashes of lightning. After only one week the water had become deep red and, besides simple acids, contained amino acids – the organic molecules that are the building blocks of proteins, themselves a vital constituent of living things.

This chemical mixture represented the primordial soup, life's supposed starting place. It later became clear that other chemicals could also be transformed into fundamental organic molecules under similar conditions. These could have included sugars and nucleotides (components of the genetic material DNA).

Although plausible, this scenario raises a problem of time scale. The fossil record suggests that simple organisms such as bacteria and algae lived on Earth at least 3.2 billion years ago. Yet for all their biological simplicity, the chemical construction of these organisms is relatively complex. The first synthesis of the basic component chemicals here occurred no more than a billion years earlier. Was this enough time for the necessary evolutionary changes to take place?

Synthesis of amino acids may, of course, have happened even earlier than 4.2 billion years ago. And synthesis of complex materials may have taken place more quickly than subsequent evolutionary changes – at least as detected in the fossil record. Or maybe suitable organic materials arrived from space.

In the late 1960s, radio astronomers discovered the existence of organic molecules in dark nebulae. Since that time there has been a very real possibility that even more complex molecules may have arrived on Earth from space. Such molecules must have come from some scientifically plausible source, and must have survived the risk of destruction by short-wave radiation on its way to our planet. There are two possible ways in which the molecules might have travelled: in meteorites or in cometary dust.

Analysis of a meteorite that fell to Earth in 1969 has shown it to contain at least 74 amino acids. Sceptics suggested that the meteorite was contaminated after it fell, but in fact the 74 amino acids showed both similarities to and differences from those known on Earth. The atoms in the molecules of such amino acids can link with the internal carbon chain of the molecule in either a left-handed or a right-handed arrangement. Terrestrial amino acids are left-handed, but both left- and right-handed forms were found in the meteorite.

Some 70 million years ago, a giant comet deposited a mixture of meteors and dust into the solar system. Some 20,000 years later, a meteor from this comet landed at Stevns Klint on the coast of Denmark, an event pinpointed by evidence at the interface between rocks of the Cretaceous and Tertiary periods, which were laid down at this time. However, the Earth began to sweep up the comet's dust some 15,000 years before the meteor impact, and also for thousands of years after the collision.

It has been proposed that any amino acids within the meteorite would have been destroyed on impact, but that they would have been retained in the fine dust. This dust could have brought amino acids to Earth and, indeed, such compounds have been detected some tens of centimetres below the boundary between Cretaceous and Tertiary rocks.

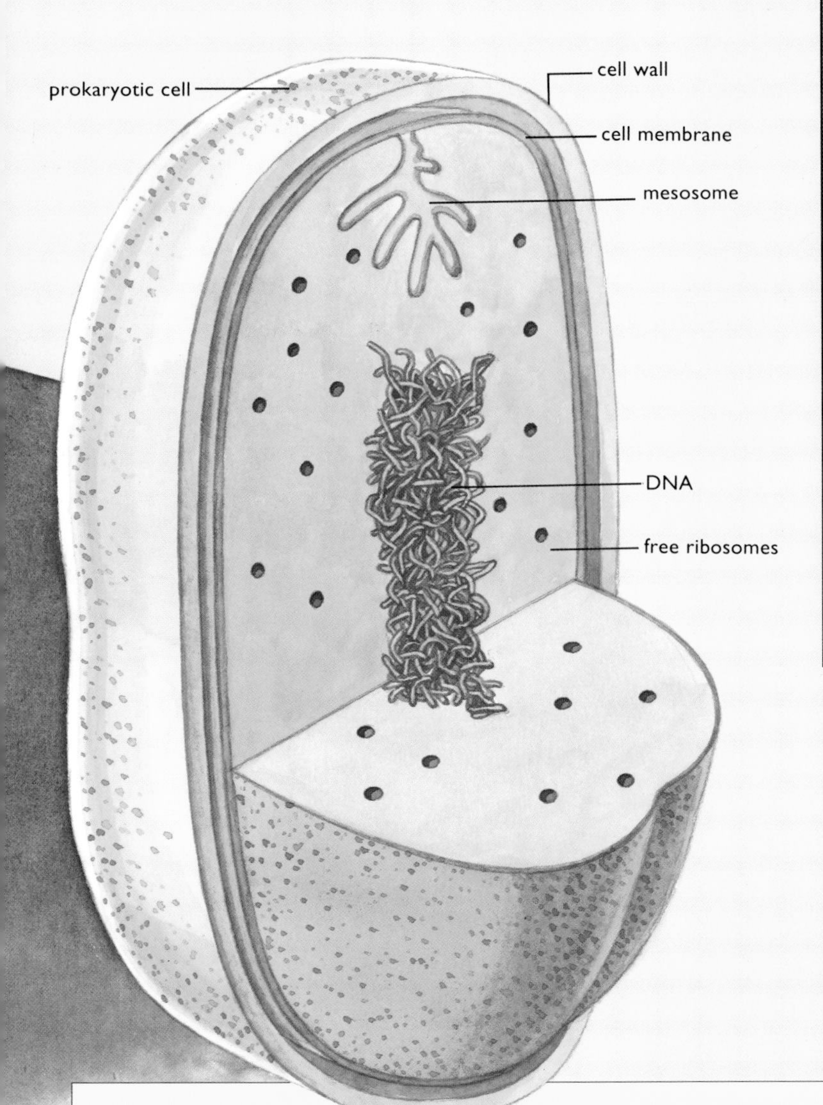


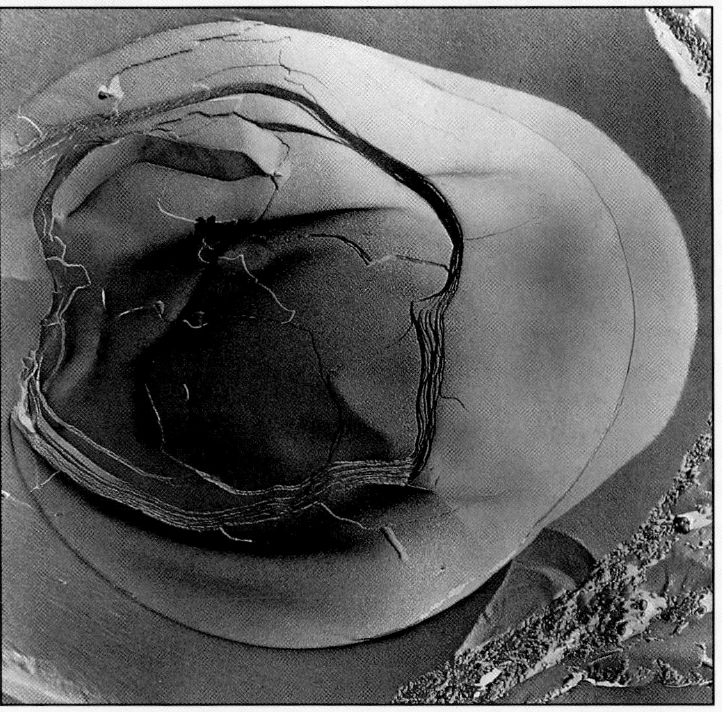

***The earliest evidence of life on Earth*** *appears in the fossil record about 3.8 billion years ago. It consisted of very simple single-celled organisms known as prokaryotic cells. This kind of cell does not have a clearly defined nucleus. Much of its chemical activity goes on at sites called mesosomes, which are highly folded openings in the cell membrane.*

***In the laboratory*** *scientists find that phospholipids – molecules with one end that can dissolve in water and one that can dissolve in fat – spontaneously form globular, multilayered structures, called liposomes, when placed in water (above). The outer wall of a liposome is a membrane, and in the evolution of life the appearance of cell membranes was as important as the evolution of DNA.*

### Molecules from space

When Halley's Comet (right) made its approach to Earth in 1986, observations from spacecraft showed that the central core of the comet's head was protected sufficiently for the substance formaldehyde to be safely tucked away within it. Such a substance could have formed the basis for the development of many other organic molecules. Evidence is gradually accumulating to show that the conditions in Halley's Comet might well occur in large meteors reaching Earth, and that organic substances might reach us in cometary "dust" swept up by the orbiting Earth.

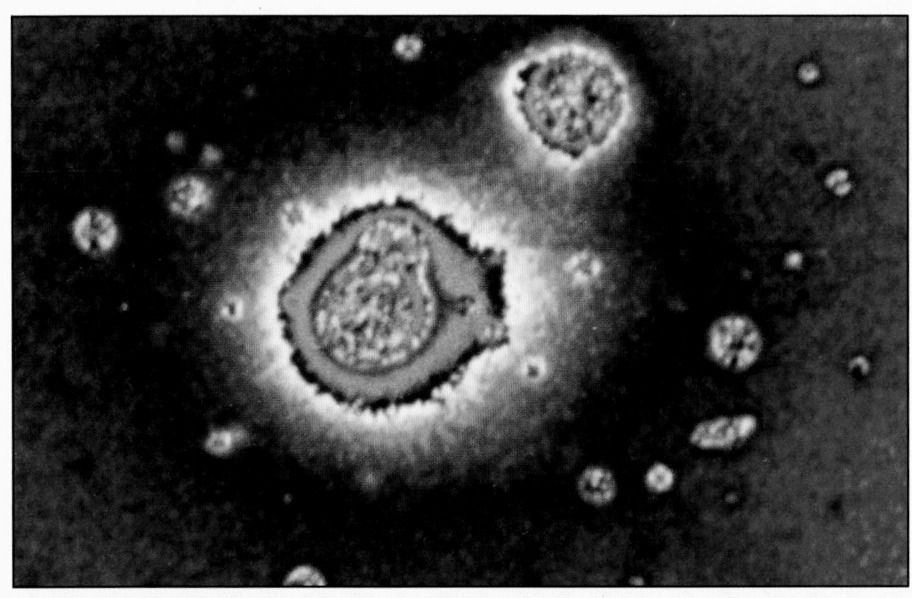

***Comets might incubate*** *the molecules of life, protected within the cometary nucleus.*

# MOLECULES OF LIFE

## • *The vital reactions*

Life's essential attributes come about through the biochemical properties of carbon-based molecules. These exist in a vast range of forms, but among the most significant are the amino acids, which are strung together and folded in three dimensions to form proteins. These amino acids are of 20 distinct chemical configurations. Their complex arrangements of atoms include not only carbon but also hydrogen, oxygen and nitrogen. Sulphur is another common constituent of proteins, as are iron and phosphorus.

Proteins fulfil major roles in the formation and maintenance of living materials. Structural proteins, for instance, form the building blocks of plant and animal cells. Other proteins control the chemical reactions that take place within and between cells. In this latter role they act as catalysts, facilitating and speeding the biochemical reactions on which the maintenance of life depends.

The catalytic proteins of living systems are known as enzymes, and each specific enzyme usually controls one particular biochemical reaction. This is achieved in an extraordinary way. The amino acid chain, which forms the protein's backbone, is so folded that it fits exactly the chemical on which the enzyme is to act.

Although there are only 20 amino acids, the variety of reactions they make possible is enormous because of the number of ways in which they can be arranged. A protein containing only 10 amino acid molecules would have 100 billion billion alternative forms of behaviour. In fact, real proteins never have fewer than several hundred amino acids in their chain, making the scope for biochemical activity virtually infinite.

As well as proteins, all living things contain substances known as nucleic acids, including deoxyribonucleic acid, DNA. This relatively simple, large, long-chain organic molecule is composed of chemical units, each containing a phosphate and a sugar. Each sugar is linked on one side to a base – a group of atoms typified by being soluble in water and reacting with acids to form salts. DNA contains bases of four chemical patterns; these are adenine, cytosine, guanine and thymine.

Each section of DNA – the phosphate, the sugar and the base – is known as a nucleotide, and the arrangement of the entire DNA molecule is a double helix. Although it may contain 300 million atoms, and if stretched out could measure up to a metre, it takes up little space because it is folded up on itself.

In most cells, DNA is concentrated in the chromosomes, rod-shaped bodies in the nucleus. Strung along the chromosomes are the genes, units of pure DNA in which the coded commands both for making and maintaining an entire organism – its genome – are carried.

It is the sequence of atoms in DNA that spells out the genetic code. Generally, DNA works so that one gene codes for the production of one enzyme, which then controls certain reactions within the cell. In addition, genes can act to control each other – by acting, for example, as "on" and "off" switches.

As well as making possible life's unique biochemical synthesis, DNA also allows for the vital processes of self-repair and self-replication. In the simplest form of reproduction – the creation of an exact copy of a cell – the double helix unwinds and a new copy of the complementary chain is added to it, so that two new identical chains are made before the cell divides. In sexual reproduction, the process is similar, but it is complicated by the mixing and exchange of genetic material.

While shuffling of the genetic pack brings about change in the appearance, or phenotype, of an individual, this is not due to any essential alteration in the genetic material. Genetic change, or mutation, is a permanent alteration in the nucleotide sequence in DNA.

Mutation, which can come about spontaneously, or be caused by environmental influences, such as radiation or certain drugs, can be deleterious to an organism, causing severe congenital deformity, for example. But it can also have beneficial effects, making an organism better adapted to its circumstances. Mutation allows for such adaptability and makes evolution possible.

While even single-celled bacteria can reproduce independently, this is not so for viruses, whose successful DNA replication depends critically on outside influences. A virus can reproduce only by making use of chemicals inside a suitable host – another living cell. For this it "highjacks" the cell's ribonucleic acid or RNA, a messenger molecule used in replication and protein synthesis that is very similar to DNA.

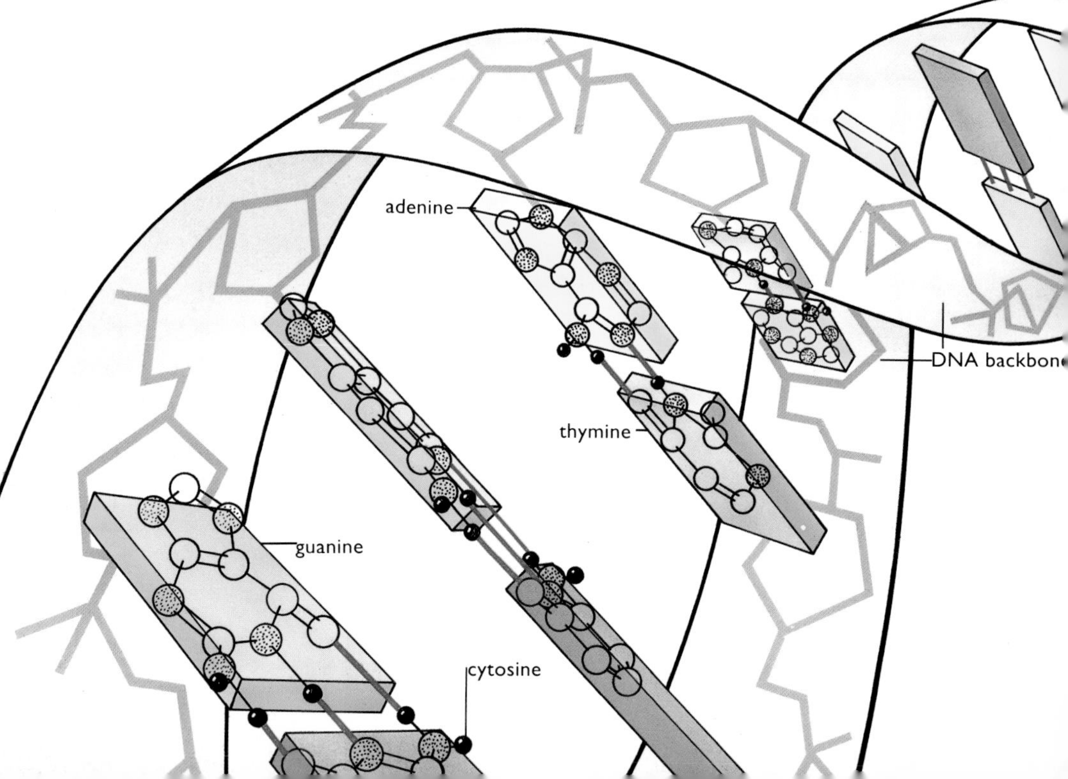

The virus is thus a kind of "hybrid" between the living and the non-living. Isolated viruses may cluster together, looking like clumps of inanimate crystals. Yet within them is enough DNA, the molecule of life, to provide sufficient coding for existence in favourable circumstances. It has even been speculated that viruses arrived on Earth from space. Whether or not this is true, they illustrate most graphically the thin dividing line at the edge of life.

***Cell replication** (right) involves passing on a precise copy of all genetic information to the offspring cells. As two nucleotide chains of the old DNA uncoil, new nucleotide chains are synthesized on their surfaces. Here, newly replicated chromosomes separate.*

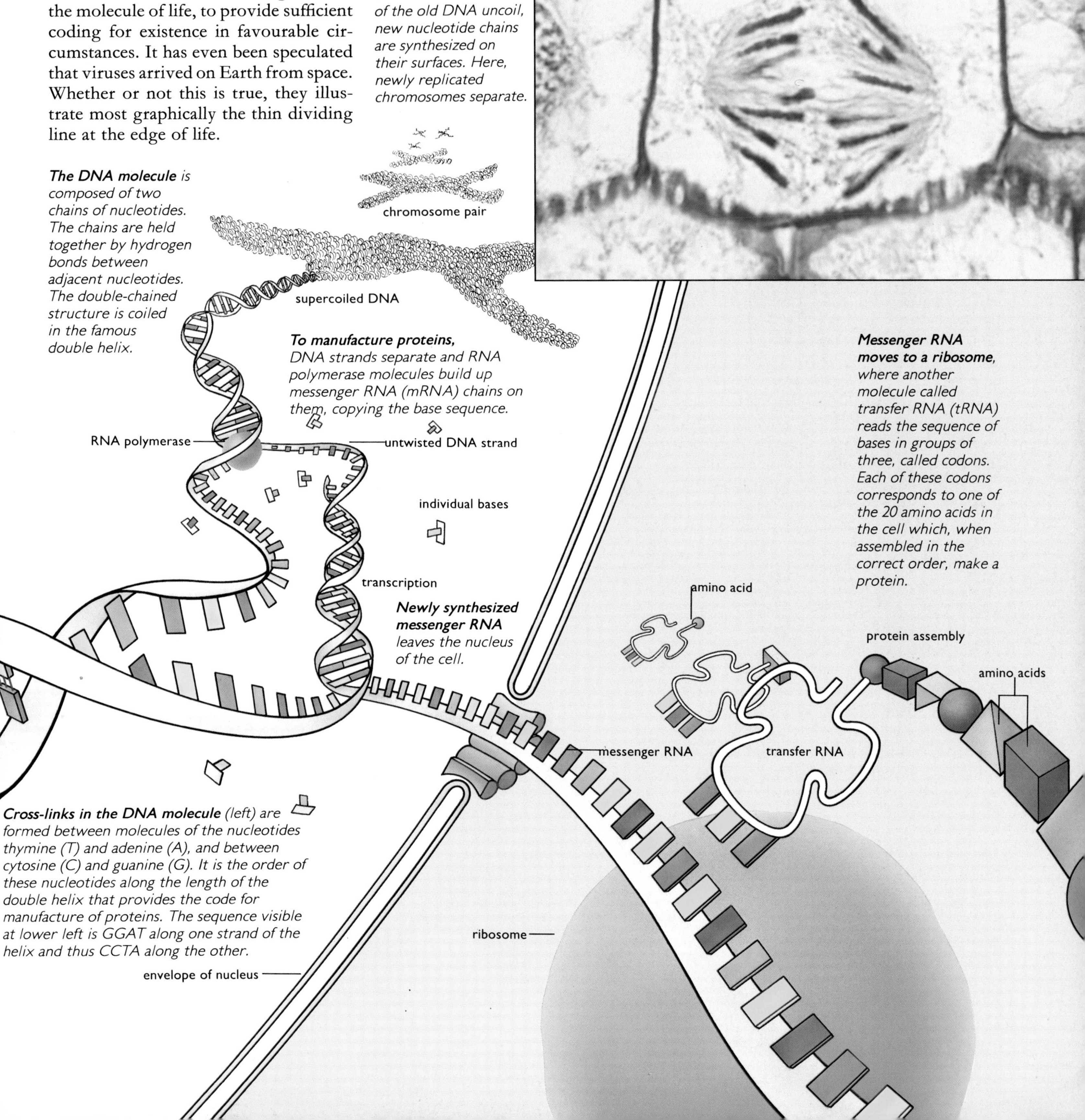


***The DNA molecule** is composed of two chains of nucleotides. The chains are held together by hydrogen bonds between adjacent nucleotides. The double-chained structure is coiled in the famous double helix.*

***To manufacture proteins,** DNA strands separate and RNA polymerase molecules build up messenger RNA (mRNA) chains on them, copying the base sequence.*

***Messenger RNA moves to a ribosome,** where another molecule called transfer RNA (tRNA) reads the sequence of bases in groups of three, called codons. Each of these codons corresponds to one of the 20 amino acids in the cell which, when assembled in the correct order, make a protein.*

***Newly synthesized messenger RNA** leaves the nucleus of the cell.*

***Cross-links in the DNA molecule** (left) are formed between molecules of the nucleotides thymine (T) and adenine (A), and between cytosine (C) and guanine (G). It is the order of these nucleotides along the length of the double helix that provides the code for manufacture of proteins. The sequence visible at lower left is GGAT along one strand of the helix and thus CCTA along the other.*

# MASTERING THE PLANET

## • *The scope of life*

The universal unit of life is the cell. Whether organized singly or in groups, cells are the substance of every living plant and animal on planet Earth. All living cells have a specific organization: essentially, variations are only matters of detail. Every cell has an outer covering, or membrane, which is, in effect, a sandwich of layers of protein and globules of fat; through it, substances pass in and out of the cell. In plants this membrane is covered with a cellulose sheath which provides rigidity.

The cell is filled with a jellylike substance, the cytoplasm, within which lie subcellular "particles", or organelles. These are involved in activities such as energy generation or the manufacture of fats or proteins. Many of these proteins act as catalysts, promoting other chemical reactions, either in the cell in which they are made or in adjacent cells. Still other organelles digest substances entering the cell, including noxious material.

At the heart of all cells lies a nucleus. Within it is housed the genetic material which, combined with proteins, forms the chromosomes. It is from here that the cell's activities are orchestrated; without a nucleus a cell will die.

Unlike those of animals, the cells of plants contain organelles known as chloroplasts. Within these are tiny packets of the green pigment chlorophyll that enables plants to use solar energy and build up carbohydrates in the process of photosynthesis.

The first cells to emerge from the primordial soup and to exist independently on Earth some 4 billion years ago were probably bacteria, which "fed" on organic molecules within their chemical-rich environment. From these probably evolved the first plants, similar to the single-celled algae of today.

As they photosynthesized, these primitive plants increased the oxygen content of the water in which they lived (oxygen is a waste product of photosynthesis). Only then was it possible for animal life, which depends on oxygen for existence, to come into being.

More elaborate forms of life took many millions of years to evolve. Gradually multicellular organisms developed, with cells specialized for particular tasks such as reproduction, and sensing and responding to the environment. By the Cambrian period, which began about 560 million years ago, complicated organisms such as large marine plants and animals had evolved.

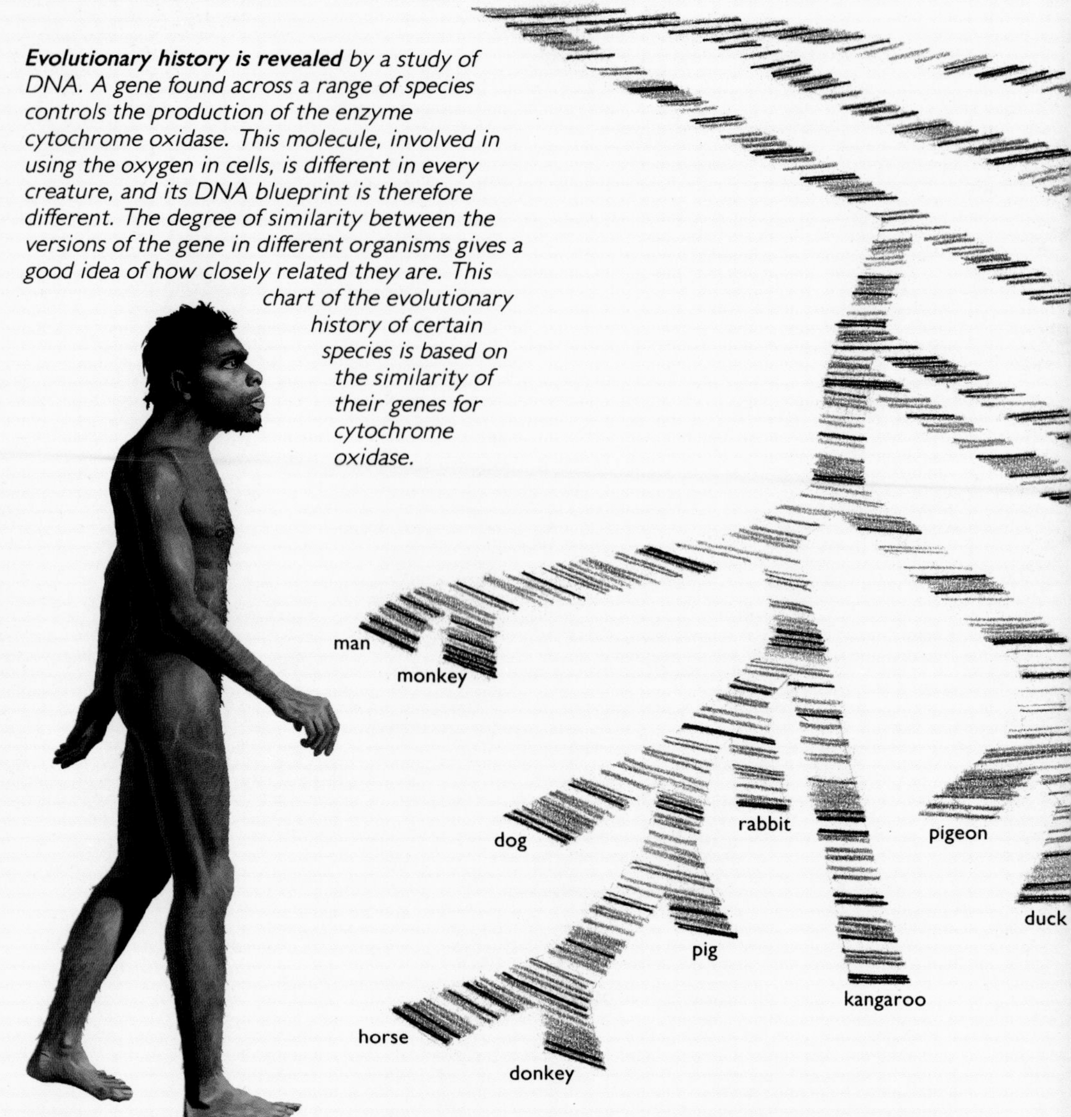


***Evolutionary history is revealed*** *by a study of DNA. A gene found across a range of species controls the production of the enzyme cytochrome oxidase. This molecule, involved in using the oxygen in cells, is different in every creature, and its DNA blueprint is therefore different. The degree of similarity between the versions of the gene in different organisms gives a good idea of how closely related they are. This chart of the evolutionary history of certain species is based on the similarity of their genes for cytochrome oxidase.*

As evolution progressed to a new era, the corals and the many-limbed trilobites flourished and proliferated, followed by the first fishes. Yet still the seas were the only theatres of life. The land masses of the Earth, which were joined together in one megacontinent, remained barren.

The first plants appeared on land some 400 million years ago and ushered in the Devonian period. They evolved into a range of types, from small mosses to giant tree ferns, and caused a change in conditions for life on Earth. Free oxygen was pumped into the atmosphere and land animals – insects, air-breathing lungfish and the first amphibians – could now populate the planet.

The Carboniferous period, which dawned some 345 million years ago, witnessed the appearance of the first reptiles and of the first winged insects. The latter evolved in tandem with the flowering plants, whose fertilization the insects made possible.

Successive periods of prehistoric life witnessed the gradual colonization by plants and animals of all the niches that the Earth has to offer. Significant events included, for example, the appearance of the first freshwater fishes, the rise of the

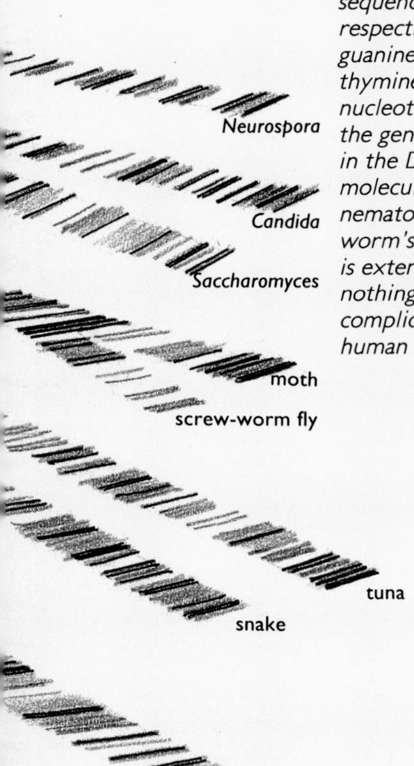


***Each band in these four-band groups*** *represents the sequence of, respectively, adenine, guanine, cytosine and thymine, the four nucleotides that form the genetic blueprint in the DNA molecules of a nematode worm. The worm's genetic code is extensive – yet nothing like as complicated as that of human beings.*

reptiles, the age of the dinosaurs, the flight of the first birds and – a turning point in evolution – the appearance of the mammals.

Only when the rule of the reptiles was ended some 65 million years ago – possibly by the collision of a huge comet with the Earth, which would have caused great climatic change – did the mammals gradually come into the ascendant. The early mammals were small, shrewlike creatures; but in time the group diversified to include the animals we know today, including hoofed mammals and the primates. Only about 4 million years ago did the first ancestors of humans walk on Earth.

Although comparatively small in size compared, for example, with the massive dinosaurs whose world they took over, the mammals have proved to be the great success story of evolution. This is because of the mammal's high degree of encephalization; that is, its high ratio of brain weight to body weight. This resulted in a level of intelligence far greater than that of any species yet seen.

Among mammals, humans show the most marked degree of encephalization, and this is reflected in every facet of human behaviour. But the phenomenon is not confined to primates. Whales and dolphins, for example, have a comparatively high degree of encephalization, but further development seems to have ceased some 20 to 30 million years ago. As a result, these animals have an intelligence probably similar to that of a dog.

Why this brain growth stopped among the large sea mammals is uncertain, but it may be linked to the fact that encephalization seems to demand a bodily system capable of producing a great deal of energy, that is, a very high metabolic rate. This is something which the dolphins and their allies seem incapable of achieving. And the same limitation was almost certainly a significant factor in the demise of the dinosaurs.

Encephalization brings with it a need for a long gestation period for offspring. Born helpless, they have much learning to do to "train" their large brains to function fully. This might be considered an evolutionary disadvantage – helpless young are easy prey – but against it is set the versatility and ability that a large

***Ants appear to show great intelligence*** *when they cooperate in the work of the colony. But individual ants are creatures of strictly limited abilities. These leaf-cutters, for example, are "programmed" for their tasks and are not capable of individual problem-solving. They show their limitations in the way they respond to errors: if, for example, one drops a leaf, it does not work out how to pick it up but starts again, going all the way back to the plant to chew off another.*

brain has to offer. And humans have used their brains to manipulate their environment to their own advantage. Science, technology and medicine make it possible for babies to have maximum protection from accident, disease and other threats to life.

Intelligence, as developed in *Homo sapiens*, has led to vast differences between the most advanced of the other mammals and the least developed of human societies. Apes may be adept at imitation, or even be able to "think through" the logic of placing a chair in the right place to reach an inaccessible banana, but this is the acme of their problem-solving abilities.

In addition, humans excel at communication. Certainly animals can and do communicate with each other, but in a much more restricted way than humankind. Some insects that live in communities – the social bees, for instance – are now well known for their ability to impart information. They do this by dancing around in particular patterns to tell other members of the hive where food is to be found. Although reasonably complex, this is all that the bees can "say" to each other. Ants also have certain elaborate communication systems, but again they are used only for specific and limited purposes.

Even among birds and mammals, communication is strictly limited and centred largely around the essentials of survival: avoiding predators, finding a mate and rearing young. But among many creatures this communication does involve smell, body language and, mainly in primates, facial expression.

These forms of communication are important to humans also, but what sets us apart from our fellow inhabitants of Earth – and possibly from all other inhabitants of the Galaxy and the universe – are our most remarkable and sophisticated powers of language, not only of speech but of abstract and technical language also. Indeed, the invention of specialized vocabularies for technical applications – of which mathematics is arguably the most elaborate and the most precise – has enabled humans to advance in a way denied to less intelligent creatures. And information to be communicated can be recorded permanently in writing, on audio tape and so on.

As with other human attributes, the use of language depends on a brain that has a large capacity for memory, as well as highly developed powers of learning, thinking and reasoning. Some 15 billion nerve cells, or neurons, in the brain alone are devoted to the tasks of processing information from the eyes, ears and other sense organs, and from the muscles and other internal parts that make the body "work".

Yet the brain cannot act alone. To operate as the control centre of the body it needs elaborate support mechanisms to help monitor the environment and to carry out the tasks it dictates. It is, thus, no accident that the human brain is a central processing unit, with connecting "wires" to every part of the body to receive and transmit information.

This wiring is the nervous system and its constituent neurons. Each neuron acts as a receiver, conductor and transmitter of nerve signals, items of electrical and chemical information which together provide all the necessary data for monitoring and controlling human activity, from walking and talking to designing a spacecraft.

The main nerve pathway out of the brain is the spinal cord, which has connecting nerves to all body parts, the nerves becoming finer and finer as they branch out to the various tissues and organs. The brain is fed by sensory messages coming in from the sense organs; motor messages are those carried outward, instructing tissues and organs to behave in a certain way.

The nervous system is also divided into two subsystems, one at the level of consciousness, the other involuntary, or autonomic. The advantage of this is that the autonomic system takes care of essential life-support activities, such as breathing and digestion, which we perform without thinking about them.

***The ability to solve problems** is one of the most astonishing of human faculties. The process often calls on complex concepts, diverse abilities and much background knowledge. Most mysterious of all, human beings often create new and original solutions to the problems they encounter.*

1

***Nicholas Copernicus** solved one of the greatest puzzles of his time (1) when he studied the mathematics of the movements of stars and planets. He sought a better account of them than the prevailing geocentric theory.*

2

3

***Solving such a complex problem** required deep knowledge of the known facts and previous theories, and the quest was strongly influenced by the philosophical and religious worldviews of Copernicus and the society of his day (2). The successful solution of the problem demanded abstract reasoning and mathematical calculation.*

***The solution arrived at by Copernicus** was that the Sun, and not the Earth, is the centre of the solar system (3). The theory was published in 1543, the year in which Copernicus died, but like any new idea it could not become the common property of humanity until it had survived the crucial process of critical discussion by his peers.*

# UNIVERSAL COMMUNICATION

## ● *Is there anyone out there?*

The ability to communicate in complex and sophisticated ways is one of the special attributes which sets human beings apart from other creatures on our planet. As well as making us unique, our powers of communication also provide us with the possibility of making contact with other living organisms in the universe, should there be any.

The prospects for such communication have been enhanced by the development, in the latter part of the 20th century, of the technology that allows communication across vast distances of space by means of radio. Only radio waves will suffice for this; even the most powerful light beams that could be produced would be too faint. The seriousness of scientists in their approach to interstellar communication is epitomized by the fact that when, in 1967, they discovered radio pulses emanating from space, radio astronomers conjectured that these might be coded messages from an alien civilization.

One of the chief protagonists of interstellar communication, the American astronomer Frank Drake, has devised two basic ways of making contact with civilizations whose languages may be totally different from those used on Earth by humans. The first is to send a message in the form of drawings or diagrams, and to include within it an astronomical explanation of the source from which it is being sent. The second, more sophisticated, technique involves the use of mathematics – the one language that may be universal.

The pictorial approach to communication has parallels with pictographic forms of writing, such as those used by the ancient Egyptians, and the style developed by the Chinese into non-alphabetic writing. This method was used in the message sent into space with *Pioneer 10* in 1973, designed by Drake and fellow American Carl Sagan.

A gold anodized plaque in the spacecraft contained drawings of a man and a woman, the man with his hand raised in a sign of peace and welcome. The figures stand against a silhouette of the spacecraft to provide a sense of scale. The plaque contained a diagram of atomic hydrogen (the simplest element), and a plan of the planets of the solar system, showing the path of the craft. Also on the plan was the position of the Sun in relation to the 14 pulsars then known. Symbols were provided representing the pulsation periods of each one. By comparing these with their values at the time the spacecraft is encountered, beings of another civilization should be able to calculate when it was launched.

For the Voyager probes, launched four years later, Drake and Sagan produced a video disc which contained both scientific information and material conveying the sights and sounds of Earth, including messages from the Secretary General of the United Nations and the American President, spoken information in a host of languages and music from Bach to rock-and-roll.

The second type of message uses binary arithmetic – the system used in all digital equipment, from computers to compact disc players, and even for relaying pictures back from space. The binary system "counts" using ones and zeros, and by convention 01 is number one, 10 represents the number two, 11 represents the number 3, 100 the number four, and so on. This may seem cumbersome, but it is just what the computer needs, for 1 and 0 can represent circuits that are switched on or off, respectively.

As an exercise, Frank Drake used the code to create a message which, if translated, could provide an intelligent alien with crude symbolic pictures of a human being, the solar system and the atomic structures of carbon and oxygen. All this was done with only 551 zeros and ones.

While we may be able to devise a message to send to an alien civilization, or to receive one from beings elsewhere in space, the possibility of interstellar dialogue seems remote, not least because of the vast distances involved. Even if, for example, there are planets orbiting Barnard's star, which is only 5.9 light-years away, and even if there is a highly developed alien civilization on one of them, any message sent will take almost six years to get there. Any reply will take six years to get back.

Yet we cannot expect that there is necessarily an intelligent civilization at that distance. If there were such a civilization within 100 light-years, any response to a message sent by one generation of humans would not be received until seven generations later. And to receive and understand the message the civilization must not only be intelligent but have reached at least the technological level of Earth.

***The language of modern telecommunications*** *is binary code, which employs only the two digits 0 and 1. For example, the Voyager 2 spacecraft (right) converted its images of Neptune's satellite Triton into binary numbers for transmission to Earth. The process is shown by the sequence of pictures below. The image (right) was divided into tiny squares called pixels. The brightness of each was represented by a number from 0, for white, to 255, for black (centre). The numbers were then converted into their binary equivalents, sequences of zeros and ones. These were transmitted as a stream of radio pulses (left); a white square is a "1" pulse, a black square is a "0" pulse. On Earth these pulses were used to reconstruct the original image.*

### Message to the stars

In 1973 this image (below) was transmitted into space by the giant radio telescope at Arecibo in Puerto Rico. It was aimed at M13, a globular cluster comprising thousands of stars. The message contained 1,679 "on" and "off" pulses, representing a sequence of white and black picture elements. These could be arranged in only two ways: as 23 rows of 73 elements, or as 73 rows of 23, but only the latter arrangement gives an intelligible picture.

The message begins with a demonstration of the way in which binary-code numbers are represented. There follows a list of the atomic numbers of the elements hydrogen, carbon, nitrogen, oxygen and phosphorus, which are essential to life on Earth. Other numbers give a rough idea of the composition of nucleotides, the building-blocks of DNA and RNA (pp. 160–61), and a picture of the DNA molecule is included. There are also pictures of a human being and of the Arecibo telescope together with their sizes, expressed in terms of the radio wavelength used.

***An encyclopedia in code:*** *Binary numbers from 10 to 1 (1). Atomic numbers of certain elements (2). Formulae for nucleotides (3). Number of nucleotides in human DNA (4). DNA molecule (5). Human figure (6). Height of human (7). World population (8). Solar system (9). Arecibo telescope (10). Size of telescope (11).*

# THE FUTURE OF THE UNIVERSE

## • *The influence of missing matter*

In every direction astronomers look they see galaxies and quasars moving away from us. It would be natural to assume that this motion will continue for ever, with the galaxies becoming more and more thinly spread. But the theory of relativity and the cosmology derived from it show that the universe can follow one of three possible courses in the far future.

The general theory of relativity describes space-time as having a certain curvature (pp. 18–19). The first possibility is that this curvature is greater than zero, which is a mathematical way of saying that space is closed – it curves around on itself and is finite, though unbounded. The universe will expand to a maximum, before converging again.

The second possibility is that the curvature is less than zero. In this instance space curves in a different way: it has a hyperbolic geometry (p. 18), and it is infinite. The galaxies spread apart, into infinity. The same thing happens if, thirdly, the curvature is zero, when space is Euclidean and again is infinite.

Astronomers would like to observe how expansion is changing with time. Only then could they predict with confidence what will happen to the universe. They would like to be able to measure how the scale factor of the universe – the relative separation of two arbitrary points – changes with time. The scale factor is related to the red shift, so if we measure the red shift of a galaxy or a quasar, we can in principle obtain the scale factor of the universe at the time the light was emitted. But with the extent of present knowledge, this cannot be done precisely.

Cosmologists can approach the problem from another direction, however. The universe of today is dominated by matter. The density of matter is 1,000 times that of radiation of all kinds, of which the chief component is the microwave background. This density has a profound effect on the future state of things. Calculations show that if the density of the universe is greater than a certain critical value, expansion will gradually slow down under the influence of gravity and then reverse. The universe will contract, pulling everything together, leading to a Big Crunch.

If, on the other hand, the density of the universe is less than the critical value, gravity will not be strong enough to bring the galaxies to a halt, and the universe will expand for ever.

If the density should be exactly equal to the critical value, then the galaxies will, in theory, come to a stop; but at an infinite time in the future. In effect, this is also a case of unending expansion.

The question is, therefore, how much matter the universe contains, and this is not easy to answer. Until recently it seemed there was not enough matter to reach the critical density. But recent research is altering the picture.

Studies of our own and other spiral galaxies show that their velocities of rotation can be accounted for only if they possess perhaps twice as much material as appears on even the longest-exposed photographs. The recent discovery of dim, low-mass stars in our own galaxy indicates where some, at least, of such so-called missing matter may be found.

Again, it is now clear that galaxies in clusters are embedded in intergalactic material, much of which can be detected only at non-visual wavelengths. This provides a substantial amount of the mass required to close the universe: probably five times as much material is present in a cluster as was originally believed. Furthermore, unknown numbers of black holes, of unknown mass, may exist in galaxies.

In addition, we are uncertain as to whether exotic undiscovered particles, such as WIMPs (weakly interacting massive particles), permeate space. And there is still the possibility that neutrinos, which are present throughout space, have a minute mass.

Thus within a few years we may discover that the density of matter in the universe is above the critical value. In this event, the future of the universe will be a return to a concentrated state similar to that in which it started – even possibly to another Big Bang.

**Neutrino detectors** *may throw light on the future of the universe. Neutrinos are ghostlike particles that swarm throughout space. They are so unreactive that they can travel through an entire planet without, in the vast majority of cases, interacting with any atoms on the way. Yet the number of neutrinos passing through us at each moment is so enormous that the tiny proportion of reactions adds up to a significant number of cases. This detector is at CERN, the European nuclear research centre, in Geneva. Each neutrino reaction releases a tiny flash of light, which is detected by one of the thousands of photomultiplier tubes on the exterior of the device. Experiments with such devices may reveal whether neutrinos have a mass. If so, it must be tiny. Nevertheless, it could be enough to help overcome the expansion of the universe and eventually turn it into a contraction.*

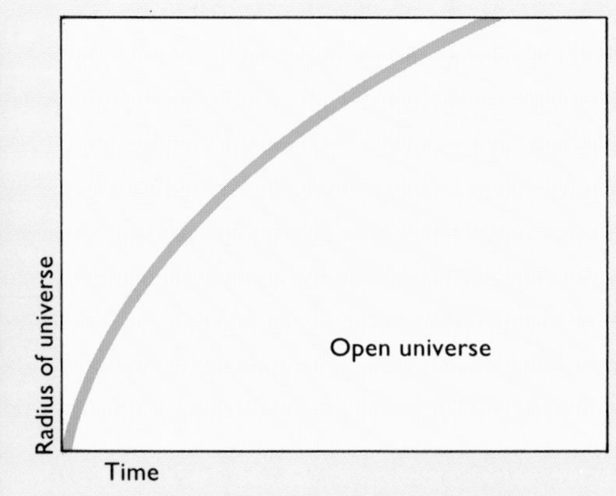


***In an open universe*** *the amount of matter present is not sufficient to stop expansion, since its gravitational self-attraction is not strong enough. If the matter we can directly observe, or detect by its gravitational effects, is the bulk of what exists, then the universe is in fact open and will expand for ever.*

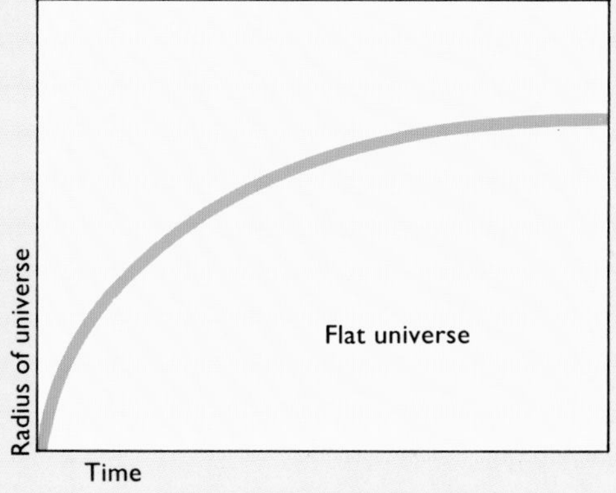


***In a flat universe*** *the density is at its critical value, exactly on the borderline between expansion and contraction. Such a universe is infinite, and Euclidean on the large scale; its expansion will never quite come to a halt. The modern Big Bang theory gives strong reasons for believing the actual universe may be flat.*

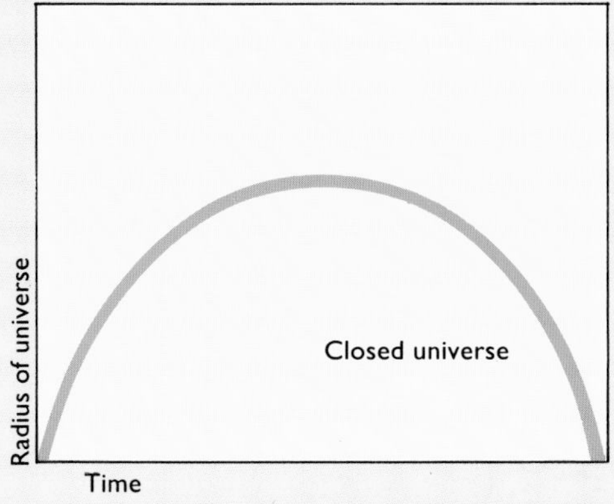


***A closed universe*** *is doomed to collapse on itself because its density is too high.*

# OBITUARY FOR THE UNIVERSE

## Death or rebirth?

If the universe is destined to expand eternally, the lines of the poet T.S. Eliot graphically echo the situation:

> This is the way the world ends,
> Not with a bang but a whimper.

Present knowledge of fundamental physics can enable us to foresee the history of matter over enormous intervals of time. But to discuss these remote times, it is necessary to use numbers many orders of magnitude larger than any that we have encountered so far. The lifetime of a star like the Sun is of the order of $10^{10}$ (10 billion) years. But the faintest, slowest-burning stars have lifetimes perhaps 10,000 times longer. So it will be in something like $10^{14}$ (100,000 billion) years that all stellar activity will be over, and there will be no stars left in the universe. The galaxies will consist of cold, dark matter.

In a further billion billion ($10^{18}$) years, the galaxies will be collapsing, for relativity theory predicts that in any system of orbiting bodies energy will be radiated away in the form of gravitational waves. A proportion of the matter in the galaxies could be swallowed up in ever-growing black holes at their centres.

If, as some physicists believe, the proton is not stable but decays after an extremely long lifetime, then matter itself will break down in the remote future. Protons will begin to vanish after not less than $10^{32}$ years from now, turning into lighter particles such as positrons or muons. All the atoms in the universe that have not already been swallowed by black holes will disappear, to be replaced by a sea of lighter particles and radiation.

If protons do decay in this way, the final end of the universe will be marked by the evaporation of black holes (p. 65). This will occur over a great range of time scales, for the rate at which a black hole vanishes depends on its mass. A black hole with 10 times the mass of the Sun will evaporate $10^{68}$ years from now. One that is 10 times more massive will last 1,000 times longer – until $10^{71}$ years from now. The giant black holes will take longer still; they will last for some $10^{90}$–$10^{100}$ years.

However, if protons do not decay, the situation will be different. After the vast time of $10^{1,600}$ years, white dwarfs will all collapse to become neutron stars, and a very long time after that – too great to be able to describe conveniently, even in the powers-of-10 notation we have been using – all neutron stars will coalesce to form black holes. The end will come when these eventually evaporate, yielding a featureless universe of radiation and particles.

Most people will regard this protracted fading of matter as a less attractive prospect than the alternative of a closed universe, in which gravity will have the final word and the end will be violent. With the continuing discoveries of missing mass in the universe, such an alternative seems a definite possibility. In this case the scenario is vastly different and the time-scale far shorter.

The actual time when expansion ceases and turns into collapse depends on the precise value of the Hubble constant, which is not known for certain. But after contraction has proceeded for billions of years, clusters of galaxies will begin to mingle something like a billion years before the Big Crunch. Hundreds of millions of years will then elapse before the galaxies themselves begin to merge.

The merging of galaxies will result in a single super-hypergalaxy, exerting an immense gravitational pull on its constituent stars. Within another million years the stars will approach each other so closely that the night sky will become as bright as the Sun. The temperature of space will rise until eventually it becomes hotter than the stars, which will explode. Black holes will grow apace in the hot, dense collapsing matter of the universe, and 100,000 years before the Big Crunch they will be forming at a catastrophic rate, sucking up everything around them.

The end of everything could be the collapse of the universe into a singularity – a single point of space and time, where density and temperature become infinite, and theories of physics become invalid. But there might, instead, be a sequel in which the conditions prevailing during the Big Bang are re-created, the four fundamental forces are reunited, and the universe returns to its original state, ready to expand again.

If such a "Big Bounce" occurs, the universe will gain a new lease of life, expanding until gravity again takes over and brings about another contraction. The life cycle of expansion and contraction will be repeated continually, in a bouncing universe that lasts for ever.

***1 If the universe should begin to contract** again at some remote future date, its evolution could resemble a film of its past history played in reverse. When the Big Crunch is only a few billion years away, the galaxies will be closer together than they are now, the sky will be brighter and the temperature of the universe will have risen.*

***5 The universe could conceivably be reborn*** *in another Big Bang following its collapse. If this occurs, then the universe would repeat the same cycle of expansion and collapse over and over again. In some theories, each cycle would be longer than the previous one.*

***4 In the last decade of the collapse*** *the black holes will begin to coalesce, until the entire universe is swallowed up inside a single supermassive black hole. This could mark the end of the universe and of time itself.*

***3 When the final collapse is perhaps only a few centuries away,*** *the temperature of the background radiation filling space rises to the point where it tears the stars apart. At the same time black holes will begin to swallow matter and radiation.*

***2 Eventually the galaxies will be separated*** *by distances comparable with their own diameters. They will interact strongly with each other and merge more frequently than they do now. But the increasing numbers of black holes within the galaxies will begin to make the process of collapse different from that of expansion.*

the behaviour of subatomic particles. His results were expressed in equations that turned out to be just like those of a vibrating string. It seemed, therefore, as if there were entities like elastic strings binding the nucleus together.

When the quark concept was developed, Veneziano's theory ran into difficulties and was shelved. However it did successfully describe gravitons, the messenger particles of gravity, so perhaps string theory was really a theory of gravity.

In the early 1970s other theories arose. One explained how quarks were bound together by messenger particles called gluons, which were later found to exist. In these theories symmetry (p. 28) played an important part. One type of symmetry that has been proposed is called supersymmetry. It unites the two great families of particles: bosons (particles that have whole-number spins, such as photons) and fermions (particles that have fractional spins, such as protons and electrons).

But supersymmetry involves many more dimensions than the four of Einstein's space-time. One theory, supergravity, calls for 11 dimensions – 10 of space and 1 of time. Another theory is a development of string theory taking supersymmetry into account. Called the theory of superstrings, it is the result of the work of the physicists John Schwarz, an American, and Michael Green, an Englishman.

In superstring theory, the mathematics shows that particles can be described as vibrations of open strings, or of closed, loop-shaped, ones. The size of open strings is roughly equal to the Planck length, a distance of only $10^{-32}$ millimetres, equivalent to a hundred-billion-billionth of the diameter of the atomic nucleus. The vibrations of open strings give rise to massless particles of spin 1, such as the photon. Open strings can close up to form loops, which give rise to other kinds of particles, including the massless spin-2 gravitons, which are yet to be observed.

Open strings and closed loops are combined in the heterotic or "cross-bred" superstring theory. According to this theory, vibrations moving around a loop in a clockwise direction are 10-dimensional, while those moving counterclockwise are 26-dimensional.

In relativity the trajectories of particles through space-time are called world lines (pp. 62–65). In superstring theory the fundamental strings and loops sweep out a two-dimensional surface in space-time, known as a world sheet, which is analogous to the film of a soap bubble. Interactions between strings and a shimmering motion of the world sheet account for the quantum behaviour of subatomic particles, and of messenger particles as well.

Since superstring theory requires 10 space-time dimensions and Einstein's space-time occupies 4, the remaining 6 must be rolled up, in the manner suggested by Klein. However, the theory does not yet explain why this should be so. Possibly, in the earliest moments of the Big Bang, all dimensions were rolled up and were equally important. For some reason, only three space dimensions have since unrolled to the colossal size of the present universe.

The rolled-up dimensions have extremely strong curvature. It must be measured in terms of the size of the strings, $10^{-32}$ millimetres. If a particle could travel at the speed of light through one of the extra dimensions and return to its starting point, it would be gone for a time less than the Planck time of $10^{-43}$ seconds. Its absence would never be noticed on the macroscopic scale.

Furthermore, according to the Heisenberg uncertainty principle (p. 25) such small distances can be probed only with extremely high energies. In fact, to explore distances of the order of the Planck length would require giving a particle energy of a magnitude that has not existed since the Big Bang.

An interesting consequence of superstring theory is that there may be a previously unsuspected type of matter, which can be detected solely by its gravitational effects. It has been called shadow matter, and could contribute to the missing mass that is believed to exist and that could, just possibly, turn the expansion of the universe into a contraction.

Superstring theory awaits confirmation. It is too soon to make sweeping judgments about it, but it holds the possibility that it is a step toward the most fundamental basis of physics: a TOE, or theory of everything.

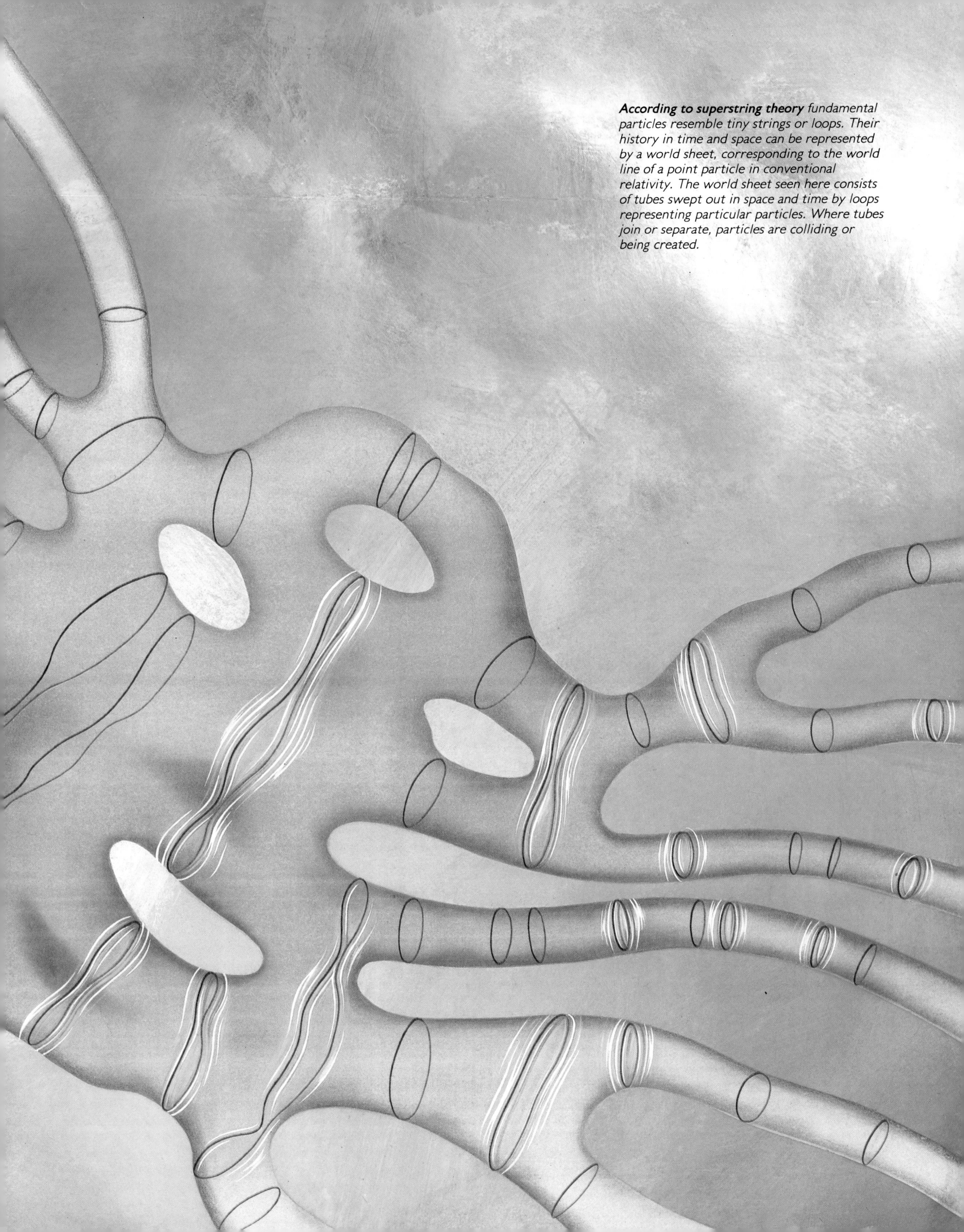

***According to superstring theory*** *fundamental particles resemble tiny strings or loops. Their history in time and space can be represented by a world sheet, corresponding to the world line of a point particle in conventional relativity. The world sheet seen here consists of tubes swept out in space and time by loops representing particular particles. Where tubes join or separate, particles are colliding or being created.*

# QUANTUM SPACE

## • *Fabric of the microscopic world*

The science of today confronts a barrier in its attempts to get back to the very start of the Big Bang, before the crucial Planck time of $10^{-43}$ seconds. The known laws of physics break down and become unusable under the extreme conditions of space and time then prevailing. Theorists are struggling to extend those laws or develop new ones, gaining new insights as they do so.

Another line of attack is the study of gravitational radiation. Relativity predicts the existence of gravity waves, and quantum theory predicts that, like all other kinds of wave, these should in some circumstances appear in the guise of particles. These conjectured messenger particles are called gravitons. The experimental detection of gravity waves or gravitons would be a crucial step in the unification of gravity and the other fundamental forces.

Since the 1960s a number of attempts to observe such waves have been made. The most famous are those of the American physicist Joseph Weber, who set up a huge bar of pure aluminium, weighing 4 tonnes. Such a lump of metal would be too massive to respond significantly to ordinary, local disturbances, such as traffic vibrations, or seismic tremors. But it would be squeezed and stretched by passing gravitational waves, which would make it ring like a bell, even though its distortion would amount to less than the size of an atomic nucleus.

Weber fitted the bar with extraordinarily sensitive detectors. Early on he did get what seemed to be a positive result, but it was later ruled out as a false alarm, and the instrument seems to have recorded no gravity wave events since.

Those physicists who have not been discouraged by the immense difficulties of detection have been devising new methods. In one of these, the detector comprises two beams from a laser. Each beam is directed along a stainless steel pipe, 3 kilometres long, at the end of which it is reflected by a mirror mounted on a massive lump of metal to keep it highly stable. To magnify the expected effects the laser beams are reflected to and fro 50 times, making the length of each arm equivalent to 150 kilometres. Eventually the beams meet and their waves interfere with each other.

A gravity wave will stretch and squeeze space as it passes, briefly altering the distances traversed by the light beams. The interference pattern, consisting of a pattern of bright and dark lines, will momentarily shift.

Such instruments may be built at Glasgow in Scotland and at Garching in Germany. If a local disturbance affects one, the other can be expected to remain stable, while a gravity wave will affect all suitable detectors. They will register its passing at different times, which will give information about the direction from which the wave comes.

But observations relevant to the existence of gravity waves have already been made. The pulsar PSR 1913 + 16 is part of a binary system – that is, it is orbiting another star, taking about $7\frac{3}{4}$ hours to do so. It was calculated that it should be losing energy by the emission of gravity waves, with the two stars falling together. In 1974 a lengthening of the pulsar's orbital period of 7.5 millionths of a second per annum was detected, and it seems extremely likely that this energy loss is due to the predicted gravitational radiation.

If the messenger particles of gravity, the gravitons, exist, they will be subject to the same uncertainty relations as all other particles (p. 25). The time and place of their absorption or emission can be specified only to a certain degree of precision. As gravity distorts space, so the presence of gravitons will cause space to curve around them. The quantum uncertainty will make the curvature fluctuate: space-time can be thought of as rippling because of the presence of the gravitons. But such rippling will be minute: compared with electromagnetism or the nuclear forces, gravity is an extremely weak force. Even a body as massive as the Sun deflects starlight passing close to it by only a tiny amount.

The uncertainty principle also means that for a split second particles can borrow quantities of energy. The amount to be borrowed and the duration of the loan depend on the strength of the force carried by the particles. In the case of gravitons, which carry the very weak force of gravity, the loan can only be for a very short time indeed – no more than the Planck time of about $10^{-43}$ seconds.

In that short time a particle, travelling

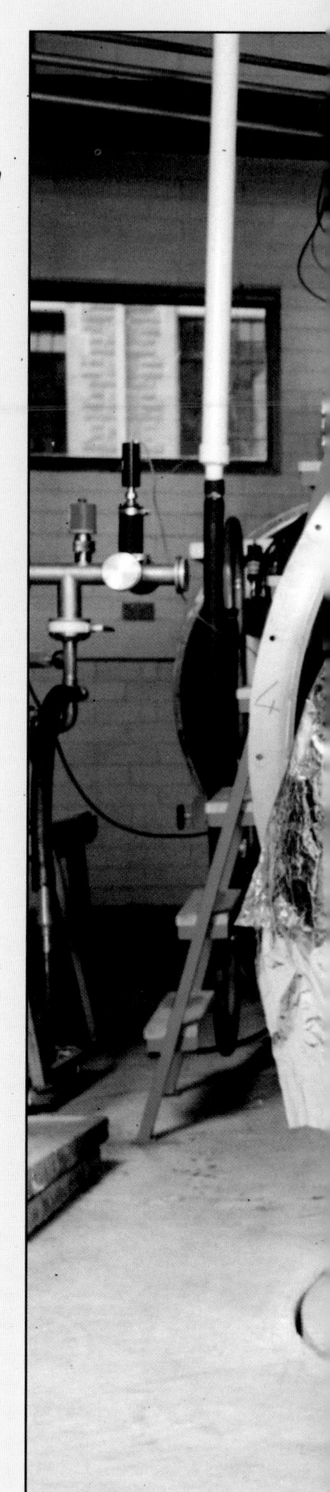

***A gravity wave detector*** *designed by the pioneer in this field, Joseph Weber. Located at the University of Western Australia, it consists of a niobium bar that would "ring" like a bell for a long time after a gravitational wave had passed through it. Although such vibrations would be smaller than the nucleus of an atom, they could be distinguished from the background noise of mechanical and thermal vibrations. It is claimed that bar antennae of this type detected gravitational waves from the supernova observed in the Large Magellanic Cloud in February 1987.*

even at the speed of light, can travel only about $10^{-32}$ millimetres, the Planck length. So it is on a scale smaller than this that we encounter ripples in space-time. And such ripples make our present theories invalid for the very earliest moments of the Big Bang, when the universe was still of this microscopic size.

All this indicates that, when it proves possible to extend physics into this so far forbidden area, we shall find that space itself is no longer continuous, but quantized – divided into elementary units. It may be like a sponge, displaying a discrete structure of ultramicroscopic dimensions. According to the American physicist John Wheeler, we shall even find "wormholes" leading from one part of space to another. Perhaps even past and future will no longer be so clearly distinguishable in this new microcosm.

# BIG BANG IN QUESTION

## • *Challenges to orthodoxy*

The problems encountered in trying to probe the earliest moments of the universe would become irrelevant if the Big Bang theory were to be abandoned. Although it appears to fit in extremely well with nuclear physics and with the universe as observed from Earth, a few scientists question its validity.

In the late 1940s the British astrophysicist Fred Hoyle and his associates Hermann Bondi and Thomas Gold proposed their famous steady-state theory of the universe. This suggested that the universe always looks the same, from any viewpoint and at any time. Thus, although galaxies are born, evolve and move away from each other, they are continually replaced by newly created matter, in the form of hydrogen gas, which evolves into galaxies and stars in due course. Such a universe has no beginning and no end.

After many years in which the steady-state theory has lain fallow, Hoyle is once again proposing it as a serious contender in cosmology. He claims the theory can now explain the observed abundances of deuterium, hydrogen and helium, which are so successfully explained by the Big Bang theory.

One of the strongest arguments in favour of the Big Bang view of the universe is the presence of the microwave background. This is regarded as being left over from the Big Bang itself, a cooled remnant of the hot fireball that gave birth to the universe. Yet Hoyle has a new suggestion: that the background radiation is due to comparatively recent events – supernovae. Astronomers agree that it is in such explosions that the heavier atoms in the universe – particularly iron – are formed. In Hoyle's new suggestion the iron atoms go on to form long, narrow "whiskers".

If the vapours of metals are slowly cooled, most crystallize into metal whiskers. Typically they have a thickness of two millionths of a millimetre and a length of no more than 1 millimetre. Hoyle says that iron whiskers of this size, floating in interstellar space, would absorb infrared and short-wavelength radio waves and re-emit them with the spectrum of wavelengths of the background radiation.

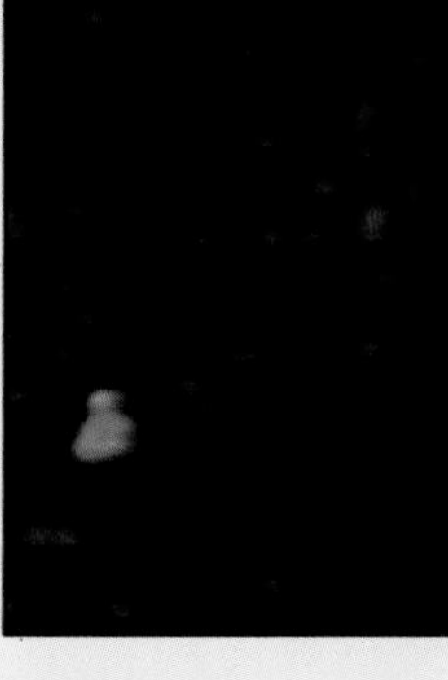

***The quasar Markarian 205,*** *which appears green in this image, seems to be physically linked with the galaxy NGC 4319 above it; faint signs of a bridge of gas can be seen here. Yet the spectrum of Markarian 205 shows a red shift 10 times greater than that of the galaxy, which according to most astronomers would indicate a correspondingly greater distance. Opponents of the Big Bang theory suggest that in such cases the active object is being ejected from the associated galaxy at high speed, giving rise to the high red shift.*

What is more, observations of the pulsar lying at the heart of the Crab Nebula – the scene of the supernova explosion in AD 1054 – show a drop in the radiation in the bands that metallic filaments would be expected to absorb. This could be due, Hoyle says, to the presence of such iron needles.

One difficulty for the steady-state theory is posed by quasars. When astronomers observe objects in the remote

***The steady-state theory*** *suggests that the universe always looks the same. As the galaxies move farther apart, new matter is continually created, so the universe does not thin out with the passage of time. The diagram shows part of the universe expanding and new material forming within it, thus keeping the average density constant.*

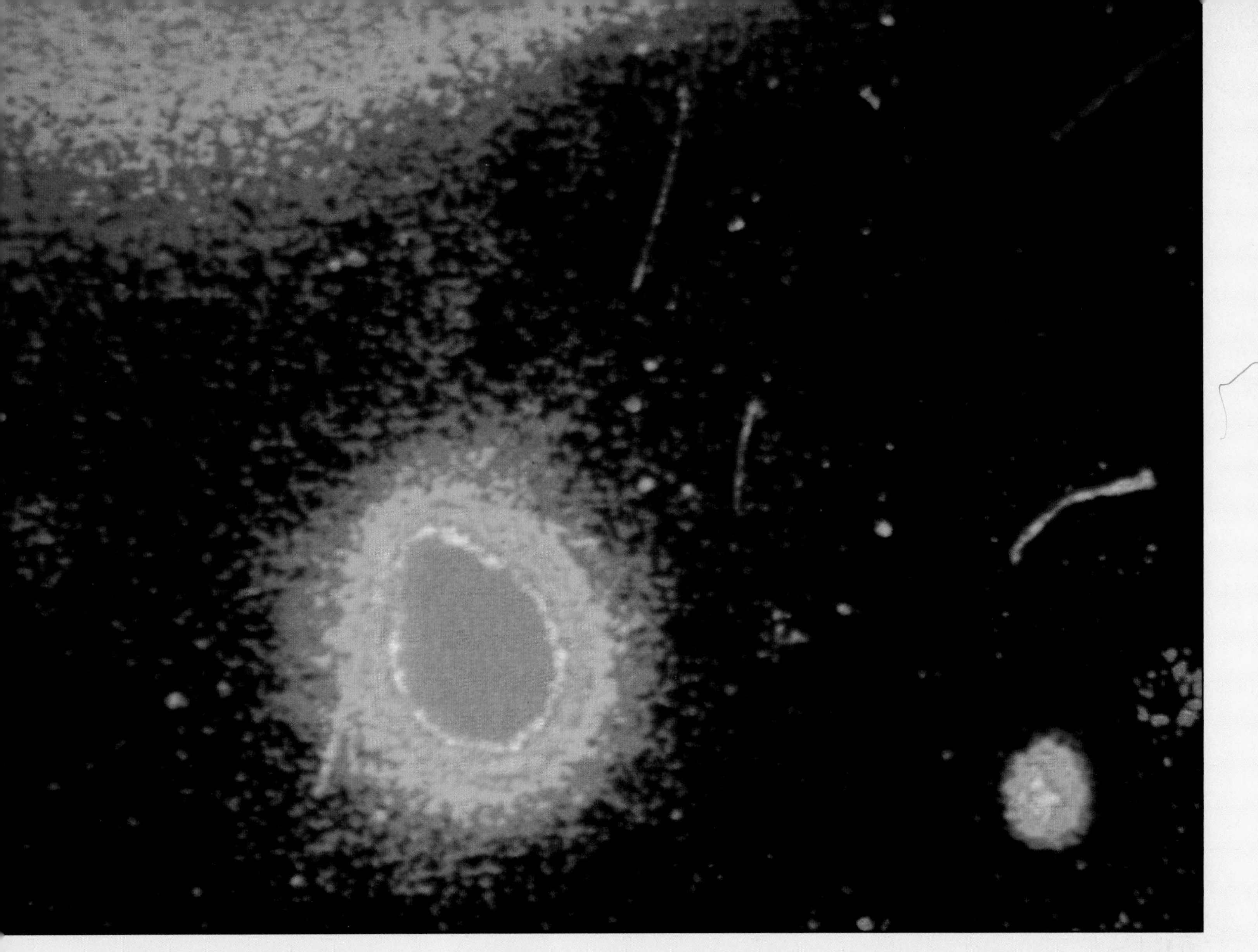

universe, they look backward to the time when light left them. According to the steady-state theory such distant regions should look the same as nearby ones. But it seems that quasars are more numerous at extreme ("cosmological") distances, and hence earlier epochs.

One line of attack on this problem questions the nature of quasars. Because Hubble's law tells us that distant objects move away from us at velocities that increase with distance, quasar red shifts are normally interpreted in the same way. They are so large, it is assumed, because the objects emitting them are so distant. This view is attacked by the American astronomer Halton Arp, among others.

Arp claims that many photographs show quasars that are close to, and apparently connected with, ordinary galaxies. Yet the companion galaxies do not show large red shifts, which they would do if they were at the distances claimed for the quasars. Arp interprets the quasars as lumps of matter ejected by the galaxies at very high speeds, thus giving rise to large red shifts.

Most astronomers believe that such apparently close associations are due to quasars happening to lie almost in the same line of sight as the galaxies: the alignment is purely fortuitous, they say. Not so the opposers of the Big Bang, who claim there are too many alignments to have occurred by chance.

It is also said that many alignments are optical illusions, caused by gravitational lensing. Those few taking Arp's side retort that in many cases, the quasars are too far from the associated galaxies to be caused in this way. They believe, too, that quasars are associated with most galaxies (though most are unobserved), whereas lensing must be very rare.

In short, they believe the quasars we observe are comparatively nearby. Hence, on the large scale they may be distributed uniformly through the universe, as the steady-state theory requires.

Arp's alternative explanation removes the necessity of believing in a vast output of energy from quasars. If quasars are relatively close, their apparent brightness would not indicate the emission of prodigious amounts of energy. Thus they would not require the special conditions of galaxy formation that are an integral part of the Big Bang theory.

However, the views of Arp and his sympathizers do not seem to explain why all quasars display only red shifts. Blue shifts are not observed, yet if quasars consist of material ejected from galaxies, one would expect to see as many moving toward Earth as away from it, and hence as many blue shifts as red shifts.

With the difficulties in Arp's position, and the successes of the Big Bang theory, most astronomers remain unconvinced by the latter theory's rivals.

# THE ANTHROPIC PRINCIPLE

## • *A universe designed for human beings?*

For most of humankind's history the Earth was thought of as the centre of the universe. This was part of a general outlook shared by all civilizations, each of which put itself at the centre of things. To the ancient Egyptians, their country was the centre of the world, and their universe was long and narrow, like Egypt itself. To the people of ancient Mesopotamia, whose country spread over an area that was more nearly circular, the heavens were a dome.

With the coming of the great civilization of ancient Greece, various rival cosmologies were proposed. But the dominant conception of the universe was as a sphere, with the Earth, also spherical, immovable at its centre. This view, coupled with an Earth-centred mathematical analysis of planetary motions, seemed so incontrovertible that the geocentric concept was accepted by all scholars for well over 1,800 years. So for a very long time people were brought up in the belief that the Earth was the centre of the universe, with humankind as its crowning glory. *Homo sapiens* was the centrepiece of creation, the lord of an anthropocentric universe.

In 1543 Nicholas Copernicus, a Polish ecclesiastical administrator, proposed a new mathematical picture of the universe, with the Sun, not the Earth, at its centre. Later studies by Kepler, Galileo and Newton supported this view, so that within a little over a century humankind was dethroned from its prime position to that of the inhabitant of a not very large planet, in orbit around what later turned out to be a not very significant star.

This new view came at a time of intellectual revolution. Modern science was becoming established, with its insistence on observational and experimental evidence, backed up by mathematical analysis, as the touchstone for any theory. Humanity's demotion to an insignificant place seemed to fit in perfectly with this mechanistic universe governed by cold, impersonal laws.

During the 20th century our concepts of the laws of physics have suffered a severe shock with the advent of quantum theory. Now particles are no longer to be regarded as solid, with definite positions and motions, but are best described as waves of probability. What is more, it has become clear that in observing such particles, the very act of observation has its effect on the particle itself. So now the mechanistic universe of the previous centuries of modern science has broken down – at least in the microscopic world of subatomic particles. We live in a universe that has elements of randomness.

But underlying the new universe of

**Our place in the universe**

It was natural for early astronomers to imagine a universe with our viewpoint, the Earth, at its centre. Religious authorities, too, felt that humanity, having been created by God, must be at the centre. The Greek astronomer Aristarchus thought the planets might orbit the Sun, but this idea attracted little support in his day.

Then, in the early 1500s, the Polish astronomer Nicholas Copernicus demonstrated that a system with the Sun at

***An 18th-century orrery,** or mechanical model of the solar system, epitomized the new world picture in which the Earth had been dethroned from its central position.*

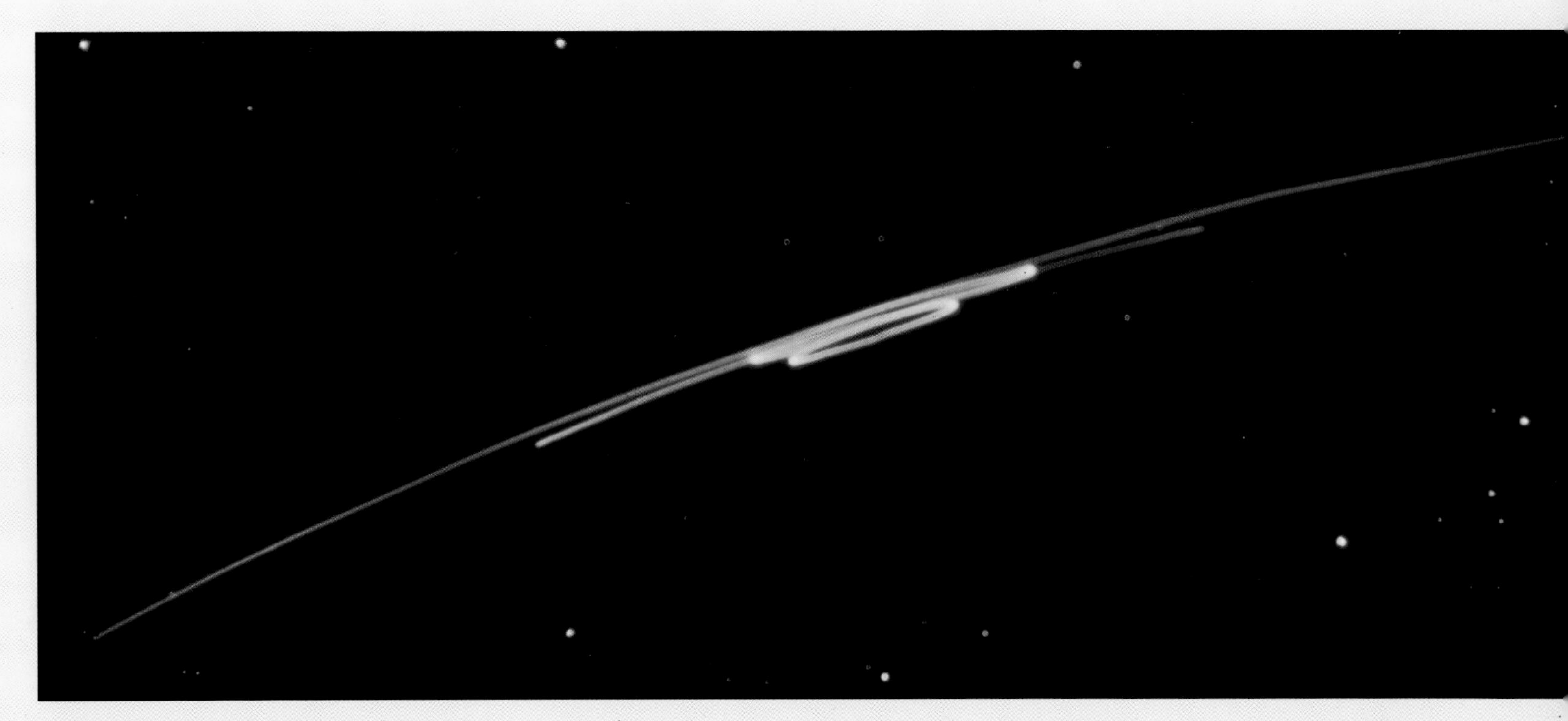

20th-century physics and astronomy is a strange pattern linked with the four fundamental forces and some basic quantities associated with them. These basic quantities include the masses of particles such as the proton and electron, their electric charges, and the strengths of gravitation, electromagnetism and the strong and weak nuclear forces. They also include Planck's constant, which sets the scale of quantum effects, and the so-called fine-structure constant, which describes how an electron behaves in an electric field.

The numerical values of these depend on our choices of units and therefore have no intrinsic significance; but pure numbers not dependent on choices of units can be obtained by considering various combinations of these quantities. The results are surprising.

They include, for example, the so-called large-number coincidences. The electrical force between the proton and electron in a hydrogen atom is $10^{39}$ times as strong as the gravitational force between them. This enormous ratio is almost the same as that between the size of the observable universe and the size of an electron, about $10^{40}$. Furthermore, $10^{40}$ multiplied by itself equals $10^{80}$, which is the order of magnitude of the number of atoms in the observable universe.

Furthermore, as the British scientist Martin Rees has pointed out, the lifetime of a star is related to the time it takes a photon to struggle out from the central regions to the surface. This is because the lifetime depends on the star's mass, and a link is found between the gravitational attraction of the star (and hence its mass) and the time that a photon takes to cross an atom. This link can be expressed by a pure number – which once again comes out to be $10^{39}$, whatever the size of the star.

Since these numbers are so vast, there is a great range of values that they could have taken. That they should have turned out so close to each other can scarcely be a coincidence. These relationships seem to be a fundamental characteristic of the universe. They suggest a basic order to the universe that is poorly understood as yet.

There are other sorts of coincidence among the fundamental characteristics of the universe. They involve the relative "tuning" of the fundamental forces and are crucial to our own existence. For example, we live in a universe that is expanding at a rate that was set in the Big Bang. If gravity had been only a little stronger, it would have taken over early on and made the new universe collapse;

***Mars sometimes seems to exhibit retrograde motion*** *when viewed from the Earth – it loops backward. Copernicus believed that real motion like this was unlikely, and that the effect must be caused by the Earth passing Mars, both planets being in relatively simple orbits around the Sun.*

its centre would provide a simpler and more acceptable explanation for the relative motions of the planets. We now know that his scheme is in principle correct and that the Earth is not at the centre of our planetary system. The solar system is also nowhere near the centre of our galaxy.

But the idea that the Earth has a special importance is once again gaining favour; some scientists now believe that the universe is of a kind that favours the development of living things, and the only place known to support life is the Earth.

if it had been weaker, the expansion would have become a runaway process and galaxies and stars would not have had time to form. If either of these alternatives had occurred there would never have been an Earth on which life could evolve.

Again, if other basic ratios and constants of nature were different, the universe would not be favourable to our existence. If, for instance, the strong and weak forces were slightly stronger in relation to electromagnetism than they are, hydrogen would not exist in its ordinary form. This would mean that heavier elements, such as carbon and oxygen, would never have come into being, and there could have been no living things.

Yet again, if the weak nuclear force were of slightly different strength, supernova explosions would not occur. So some sources of heavy chemical elements would be denied to the universe, for it is in supernova explosions that the heavier elements are scattered through interstellar space, to be incorporated later in planetary systems.

The strength of gravity is crucial in another way to the existence of life in the universe. If it were weaker, then it could not crush the material in a star the size of the Sun strongly enough to ignite thermonuclear reactions. Only very massive stars could shine by nuclear processes, and such stars would probably have lifetimes too short for any evolution of life to occur.

These hints that the universe has been arranged to favour the appearance and survival of life have been taken very seriously by certain scientists. Prominent among them is the astronomer Brandon Carter. He has pointed out that the time it has taken for the evolution of *Homo sapiens* from the first appearance of life on Earth is about four billion years. He also claims that the average period for any evolution of this kind should be much longer – longer in fact than the 10 billion years that is the lifetime of a Sun-sized star, and longer still than the period during which conditions favourable to life could exist on the Earth.

If this is so, then it would seem that intelligent life appeared on the Earth despite its being a highly improbable event. In 1974 this led Carter to propose what he called the anthropic principle: namely that our universe reflects the particular viewpoint of our own species. His bald statement has been the subject of intense study, and several related theses have been distinguished.

The weak anthropic principle states that observed features of the universe must be restricted by the requirement that carbon-based life can evolve and that there must have been sufficient time for it to do so.

Some scientists would go further and say that the universe was somehow arranged to bring humankind into existence. Carter has named this idea the strong anthropic principle. It states that the universe must have those properties that *allow* intelligent life to develop. And some go further still and say that the universe must be such that intelligent life *will* appear.

The word "must" has triggered fierce controversy because it introduces a principle that appears to lie outside science: that the universe was designed with such a purpose in mind. Some claim that this is a metaphysical concept, and science eschews metaphysics. Those who favour it claim otherwise, because they say that the presence of humankind really is exceptional.

Certainly, the universe has very specific properties that have allowed humankind to evolve. Given that it has these properties, the evolution of human beings appears to be a foregone conclusion.

What is more, humankind has unique abilities. Other animals have evolved on the Earth, but humans are the only species that can formulate the laws of physics and thus understand the nature of the universe.

The fact that this can be done is astounding. Why should the universe be comprehensible to us? Is it just a fantastic coincidence, or is there some deep reason why it has turned out that way? If there is such a reason, then we do occupy a very special place in the universe.

If the human species does have a privileged position, is it alone in its preeminence? Are there other civilizations out in space, of beings that can also understand the universe? We do not know. But as discussions on communication with alien intelligences show (pp. 166–67), there is no evidence that our messages have reached anyone, nor have any been received on Earth. After all, radio and television signals have been travelling outward from the Earth for many years now, yet we have received no reaction from space.

Possibly this is because not enough

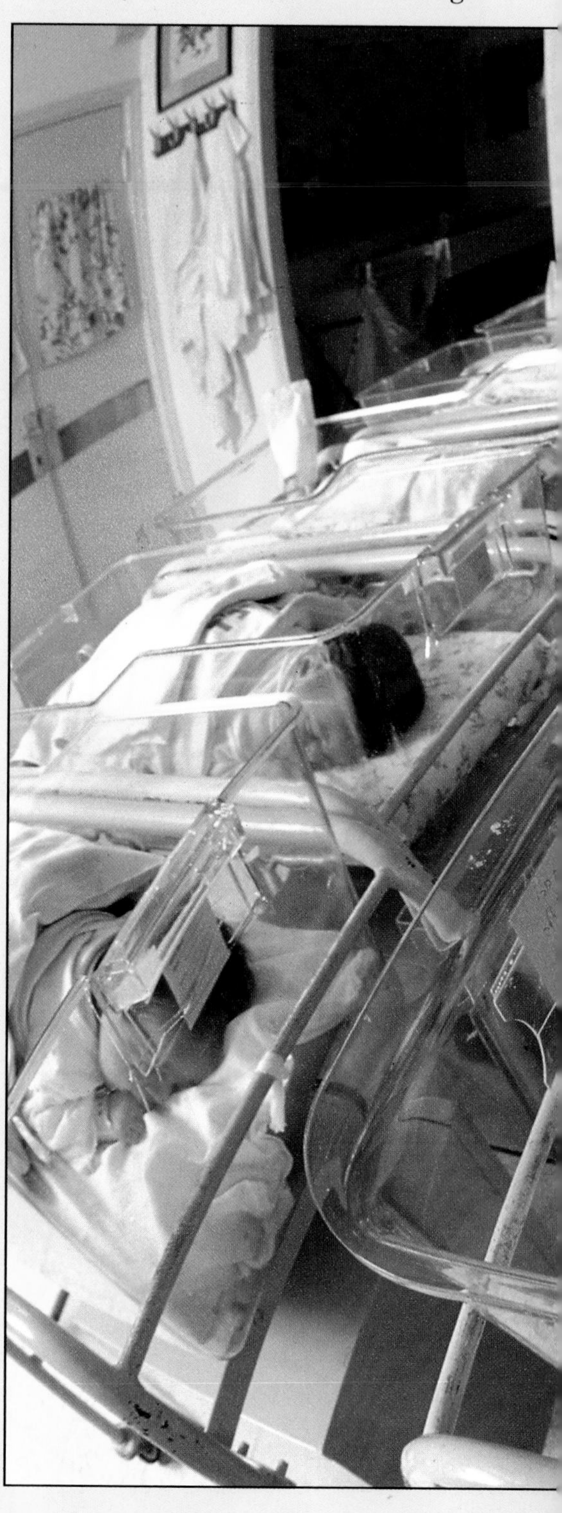

time has elapsed: it may require thousands of years before such signals can reach the nearest star system that is the home of intelligent beings that could respond in kind. However, the American scientist Frank Tipler thinks otherwise. He argues that within a billion years of its appearance any intelligent species would be capable of colonizing the Galaxy. If humankind were not unique, these others should have reached here by now.

If he is correct, then we return to a belief in the unique position of human beings in the universe. It is different from older views in that humankind is not now taken to be at the physical centre of the universe. But it resembles them in that our species is restored to a pre-eminent place in creation.

***These children are more than six billion years ahead of their time,*** *according to Brandon Carter, a proponent of the anthropic principle. That is how much longer he says it should have taken life to reach such a stage of development. He therefore concludes that the universe has been deliberately arranged to favour the development of life. Some supporters of this theory think the universe has been arranged specifically to favour one species – human beings.*

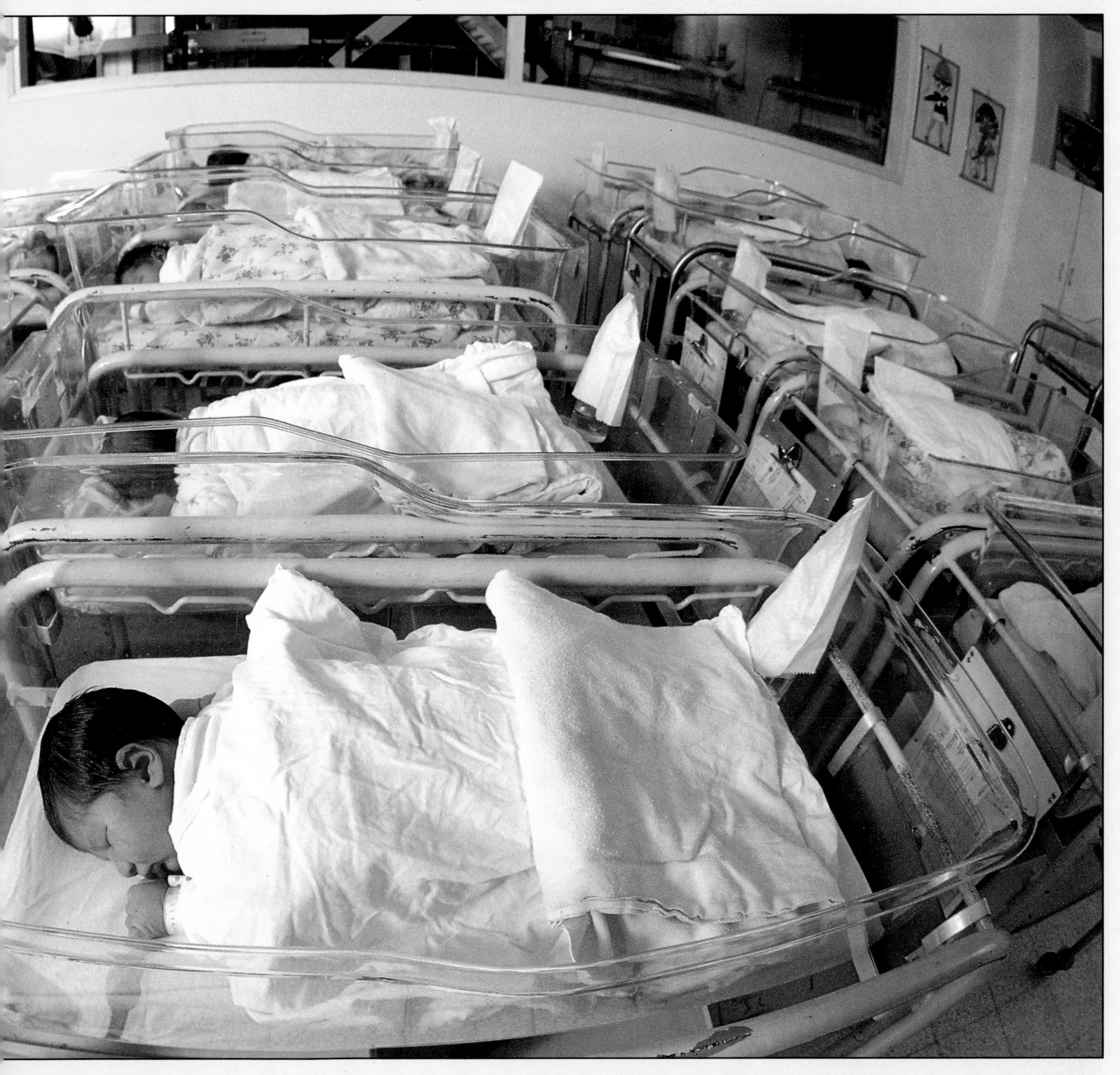

# CREATING THE UNIVERSE

● *Is reality in the eye of the beholder?*

The anthropic principle brings us face to face with the problem of the nature of the universe and our place in it. The universe now has to be viewed not as a clockwork mechanism, as philosophers once believed, but as containing an element of chance. It has to be visualized, furthermore, as a blurred picture, with its particles resembling waves or even ripples rather than points. But more than this – human beings, as observers, seem to play a more important role in the universe than was formerly imagined.

A scientist can be sure that an electron is at a certain place only if it produces some appropriate event there – if, say, it hits a television screen and produces a bright spot. Until the intervention of the observer it cannot be said even to have a definite position.

This realization has led some physicists to revive a view once expressed by Bishop George Berkeley, a younger contemporary of Isaac Newton, who said that "... all those bodies which compose the mighty frame of the world ... have not any subsistence without a mind."

The physicist can pursue this train of thought further by an experiment in which a beam of photons is split into two overlapping beams that interfere with each other. Letting these beams fall on a screen produces an interference pattern, displaying alternate light and dark bands. The bright areas are caused by photon waves reinforcing each other, the dark ones by waves cancelling each other out.

Suppose the original beam is now dimmed so that the number of photons is drastically reduced until, finally, only one photon passes through the apparatus at a time. Even in these conditions interference still takes place. Although no one can say where a given photon will go, it is more likely to go to a part of the screen where a light band falls, and to avoid one where a dark band falls. If a large number of such events occur, the interference pattern will build up. Although there is only one photon in the apparatus at a time, each seems to interfere with itself – as if it were present in both beams.

If physicists probe further by adding detectors, they can discover which of the two possible paths a given photon has taken, but at the price of destroying the interference pattern. Their intrusion changes things. By compelling the photon to behave in a particlelike way, they prevent it from exhibiting wavelike behaviour. Or to put it another way, only when the particle is detected – observed – does it become fully a particle, or fully real. This is why some physicists say that the universe depends on our observations. And since everything we observe, from molecules at one end of the scale to stars and galaxies at the other, is composed of subatomic particles, this argument must embrace the entire universe.

In the 1930s, the British astrophysicist Arthur Eddington remarked that when we examine the universe we find a pattern of footprints, and when we study these, we find they are the footprints of humankind: our theories of the universe have characteristics that stem from the fact that they are human constructs.

But some quantum scientists are now saying more: that the universe exists only because we are observing it. The strong anthropic principle states that the universe is such that it must give rise to intelligent life; but according to the American physicist John Wheeler, quantum physics shows that without participating observers there would be no universe. Observers are necessary to bring the universe into being.

If this participatory anthropic principle, as it is termed, is true, then our

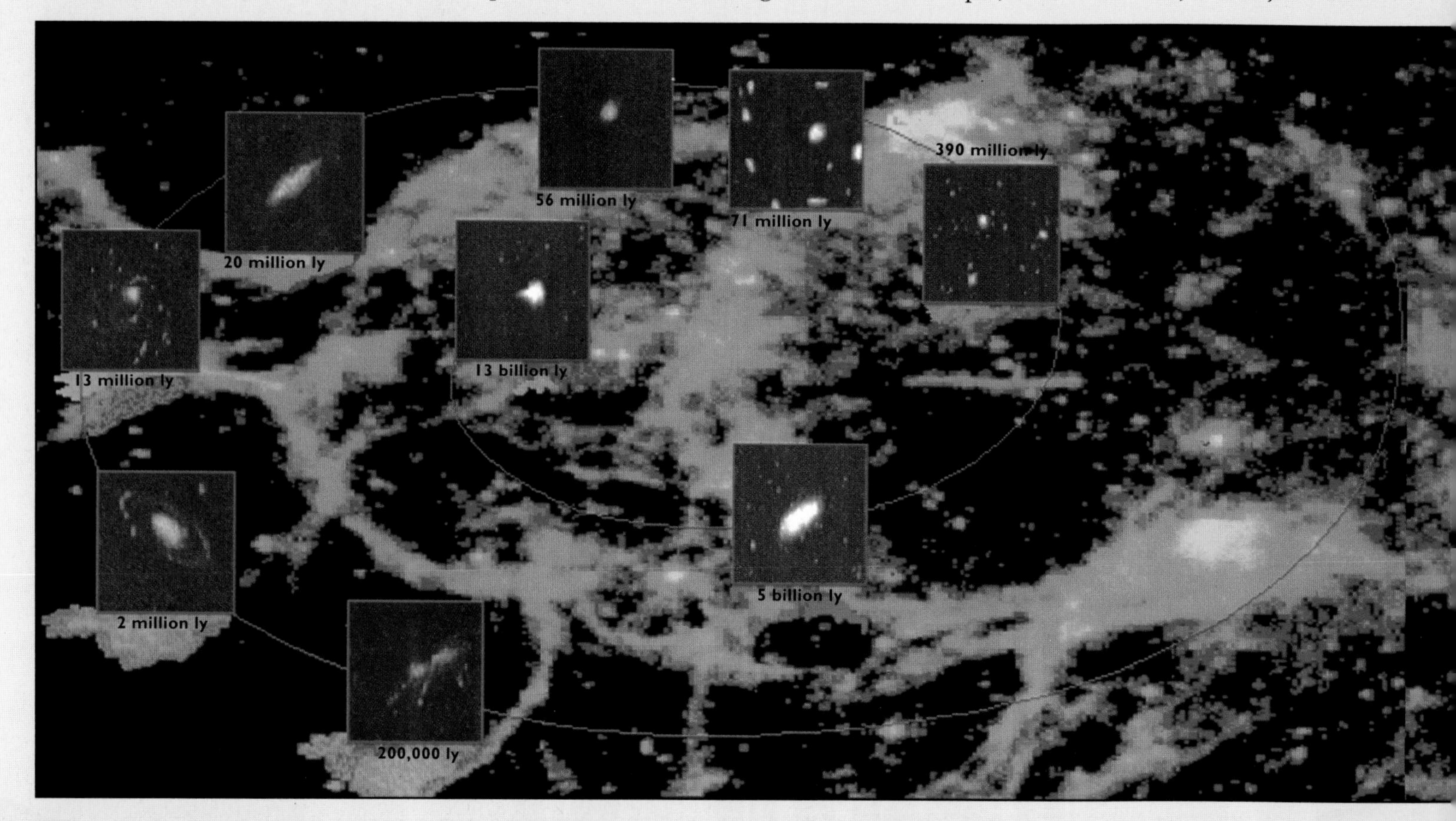

observations affect things in the past as well as in the present. When, say, a distant quasar is observed, its photons are projected into full reality in the moment that they are seen. In the preceding billions of years, they did not fully exist, and neither did the quasar – in respect of some of its properties, at least. Everything in the universe depends on intelligence comprehending it: without that, there is nothing.

With this view, the future of the universe poses a grave problem, for it seems that humankind must always be present to guarantee its existence. The American physicist Frank Tipler has tackled this question. He feels certain that no other intelligent beings exist beyond the Earth, but he points out that the presence of humankind is not essential to satisfy the participatory anthropic principle; what is necessary is the continued existence of a high level of intelligence. His solution is therefore to distribute copies of human intelligence throughout the universe.

Tipler suggests that the mind is essentially a computer program. At present it is housed in a particular kind of computer, the human body, but, he says, it is the program that is important. He looks forward to the creation of computers that are copies of ourselves. Then, even if humanity is swallowed up when the Sun becomes a red giant, copies of human intelligence will contemplate the universe for ever – or until the Big Crunch. Thus the participatory anthropic principle will continue to be satisfied.

As might be expected, this principle does not commend itself to many scientists. It seems to most to be extravagantly anthropocentric, for it makes humankind not merely the centre of all things, but the measure, and even the creator, of the universe.

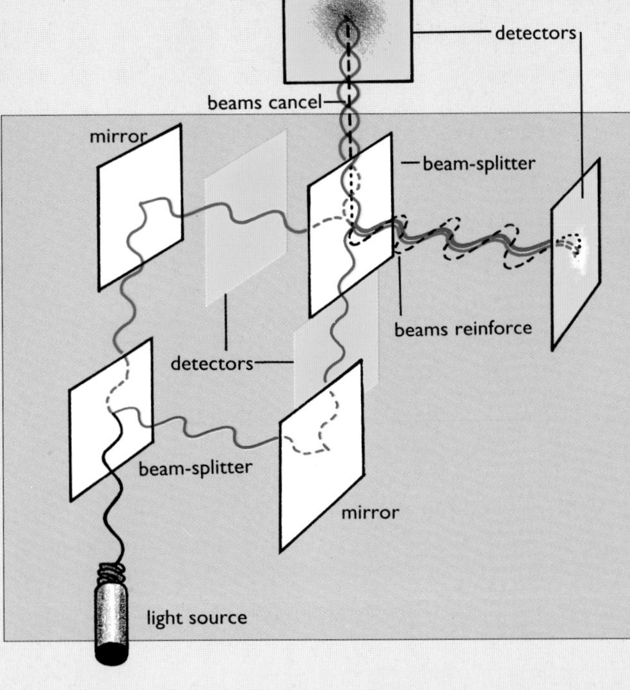


***An observer can impose*** *particle or wave behaviour on light. Here a beam that is split, recombined and only then observed, interferes with itself, so no light reaches the top detector. But if detectors are inserted before the interference occurs, they show which path each photon takes. This forces the light to behave as particles and destroys the interference effects.*

***To look into space*** *(below left) is to look back in time (below). We see the Small Magellanic Cloud as it was 200,000 years ago, when* Homo sapiens *was emerging; the Andromeda galaxy as it was when earlier types of human lived, 2 million years ago; galaxies beyond the Local Group as they were 13–20 million years ago, when early apes flourished. The light from galaxy M77 set out 56 million years ago, when dinosaurs were not long extinct, and from the Virgo cluster 71 million years ago, when they ruled the Earth. We see the Coma cluster as it was when animals had not yet left the sea, 390 million years ago. Light from some galaxies set out at the solar system's birth, 5 billion years ago; and from some quasars 13 billion years ago, as our galaxy formed.*

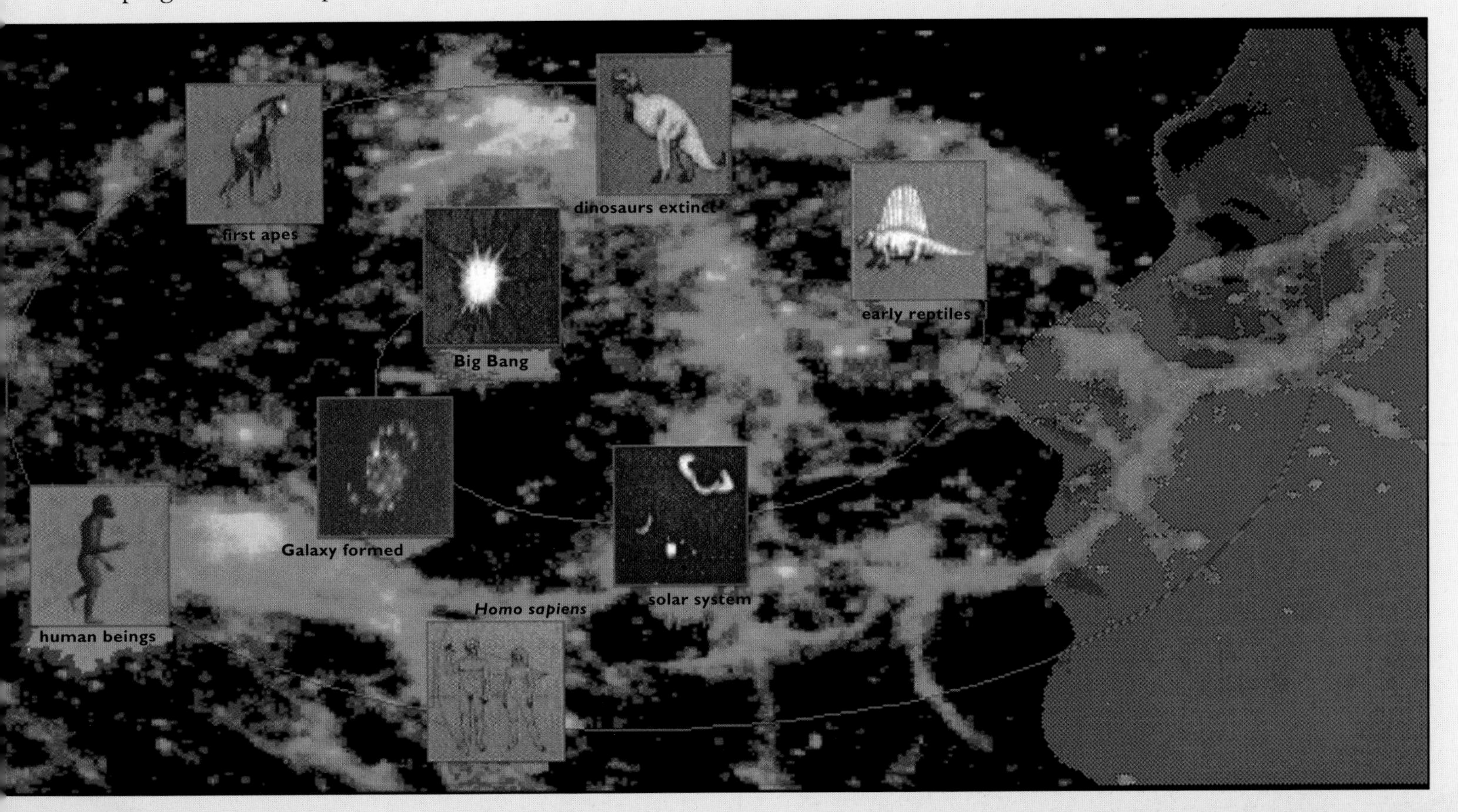

# FIRST AND LAST THINGS

## • *New universes rise from the ashes of the old*

The history of humanity's awareness of the universe is essentially one of widening horizons. Today we survey a universe that extends enormous distances into space and aeons into the past and future, and is of a complexity far exceeding that imagined by even the most advanced of the Greek philosophers. Yet could our picture of the world be still as parochial in its way as those of earlier civilizations? We have now amassed impressive evidence to show that we live in a universe expanding from a Big Bang, but is that the sum total of what exists?

The Indian scientist Jayant Narlikar has suggested otherwise. He believes that the universe we observe may be only one of many such expanding universes within a vastly larger space. The Narlikar hyper-universe can be likened to a gigantic container of bubbling liquid, with our own universe as one of the bubbles, and the other bubbles being universes in their own right.

This is not the only kind of grand scheme, embracing universes wider than the one we observe, that has been proposed. Others make use of the idea that through a black hole it might be possible to reach another, totally different, region of space. Mathematics shows that, in theory at least, a black hole could be connected to another region of space-time by a "wormhole", an extremely thin, convoluted neck or tube passing through other dimensions. The other end of the wormhole would be a white hole, spewing out material rather than sucking it in.

Such a wormhole would collapse as soon as it was formed unless a special quantum condition prevailed – the creation within it of negative energy. Negative energy, first predicted by the English physicist Paul Dirac in the late 1920s, may sound strange, but the notion is realistic enough: it led to the discovery of the positron in 1932 and is related to the existence of antiparticles in general. Wormholes, if they exist, would possess some strange properties: not least that in some circumstances they would permit objects falling into them to travel backward in time.

The American physicist Lee Smolin has extended ideas about black holes and wormholes to the question of the birth and death of the universe. The interior of a black hole can be imagined as "pinched off" from the main body of space-time by a tiny neck, akin to a wormhole. But from the inside a black hole would appear to be an expanding space – in fact, an expanding universe.

A black hole will not always contain a long-lived expanding universe: many black holes "evaporate" in a comparatively short time (p. 65). But Smolin has shown that, if the hole exists for long enough, it will give rise to an expanding universe.

The formation of such a new universe will be violent, and the result may be that some physical processes and fundamental "constants" will be modified. Rather as each new generation of living things varies randomly within narrow limits because of sudden genetic changes in DNA, so the physics of newborn universes will vary randomly. For example, electrons could have slightly different masses in different universes.

Some universes would be evolutionarily successful, lasting long enough for many black holes to form – by the death of massive stars, or the coalescence of stars in galactic nuclei, and so on. These black holes would each give rise to a further universe. Smolin calculates that in any long-lived universe the masses of protons and neutrons, for instance, will be almost equal to each other, as is the case in our universe. So such universes will not be very different from our own.

The outcome of all this is that the successful universes would be those with such conditions as would allow full development, including the production of life and the evolution of intelligence. So we should not be surprised that the anthropic principle prevails in our own universe, which seems adapted to the requirements of life: this is merely the effect of selection among a vast number of universes, of which by far the greater number were unsuccessful.

***A wormhole passing through higher dimensions*** *could in theory link different regions of space-time. One mouth of the wormhole would be a black hole, into which matter and energy are drawn, while the other would be a white hole, a hypothetical entity from which matter and energy would pour out. The connecting region, the wormhole, would be unobservable from the outside universe.*

In this picture we do not have to be disturbed by the problem of a first cause to explain the Big Bang – the latter was merely birth from a previously existing universe. Smolin's universe in the large will last for ever, propagating itself by giving birth to new universes.

Only further research can show how much truth there is in this highly speculative proposal, and, indeed, in the numerous others that will doubtless be proposed. But perhaps with Smolin's prospect of self-propagating universes born from the ashes of the old ones, we have come full circle, returning to some very early ideas, although these are now backed by scientific calculations. Only time will tell.

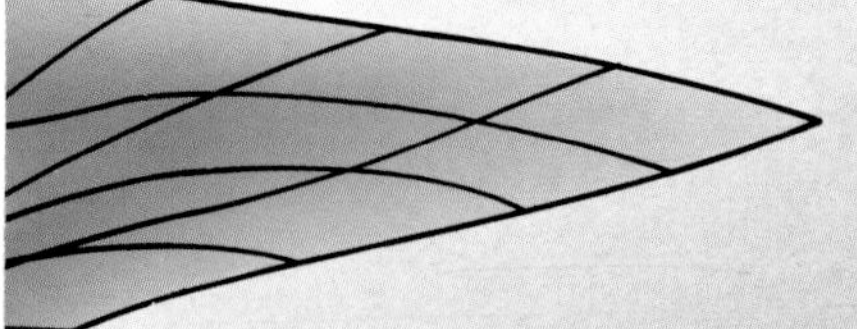


***In principle*** *it would be possible for matter passing through the wormhole to travel backward in time, emerging at an earlier time than that at which it entered. It has also been suggested that a black hole and a section of the interior of a wormhole could develop into a "baby universe" – a cosmos in its own right. Thus every universe could spawn others, throughout eternity.*

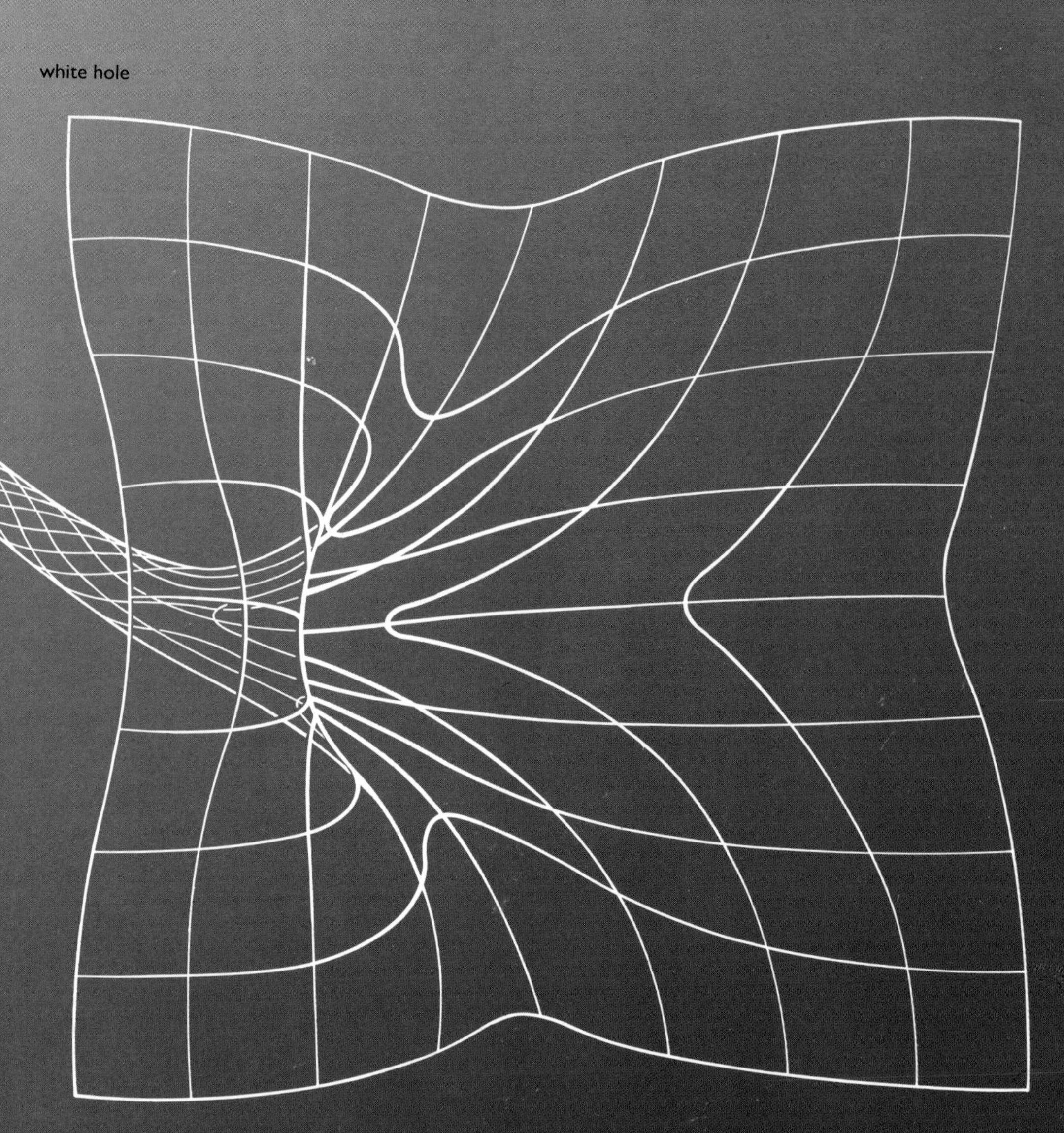

## Coriolis force
The deflection of a body moving above the surface of the Earth, caused by it's rotation. In the northern hemisphere it causes a clockwise circulation of air around an area of high pressure and counterclockwise circulation around an area of low pressure. It has the opposite effect in the southern hemisphere.

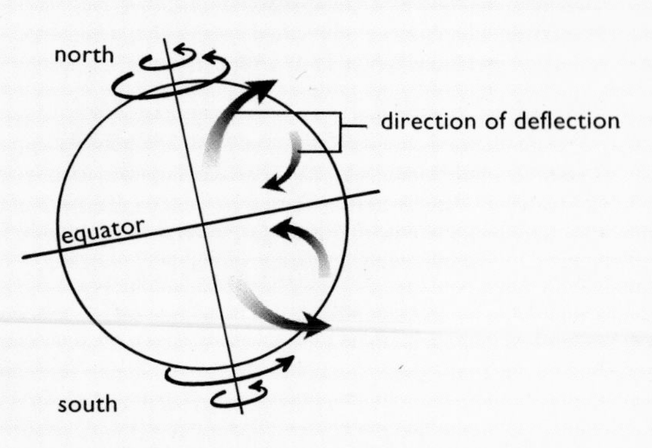


## corona
The outer atmosphere of the Sun. Extending some 10,000 km above the photosphere, it is extremely tenuous, less than one thousand-billionth of the density of the Earth's atmosphere. Composed of very energetic electrons, it reaches a temperature of some 2 million degrees at a height of some 75,000 km. Visible only at total solar eclipses or with special instruments, it also emits extreme ultraviolet and X-radiation.

## cosmic background radiation
Microwave radiation peaking at a wavelength of 1 mm, which is visible at the same intensity all over the sky. It is taken to be the cooled remnant of the primeval fireball of the hot Big Bang that started the universe.

## cosmic rays
Atomic particles, mostly protons, of very high energy moving through space. When they impinge on the Earth's atmosphere they break up air molecules and atoms, and cause showers of other atomic particles.

## cosmic string
A thin string of trapped energy left over from the earliest moments of the Big Bang, with immense mass per unit length. Cosmic strings may have acted as seeds for the formation of galaxies, clusters and superclusters.

## coudé (elbow) focus
The focus of a telescope in which light is brought out in such a way that focus always remains stationary.

## crater
Shallow circular basin found on many bodies of the solar system, thought to have been caused by impact.

## dark matter
Matter not visible either optically or at other wavelengths. If present it will add to the total mass of the universe, and may be sufficient to change its expansion into a contraction.

## declination
The angle between the celestial equator and a celestial body, measured northward or southward. (*See* celestial sphere.)

## decoupling era
A period some 300,000 years after the Big Bang when radiation ceased to be scattered by matter, and became independent of it. This occurred because the temperature had dropped to about 3,000 K, so allowing protons and electrons to form hydrogen atoms, which are transparent to radiation.

## degeneracy
An abnormal state of matter in which, owing to great pressure and temperature, electrons are stripped from atoms to form a mass of nuclei surrounded by an electron gas. Its density can be many tonnes per cubic centimetre. The central regions of white dwarfs and neutron stars are composed of such material.

## diamond-ring effect
A visual effect seen at the close of a total solar eclipse when the Sun's disc begins to reappear behind the Moon, giving the appearance of a ring surmounted by a bright diamond.

## diffraction
The apparent bending of light and radiation of other wavelengths around the edges of an object. This causes bright and dark bands to appear because of the wave nature of the radiation.

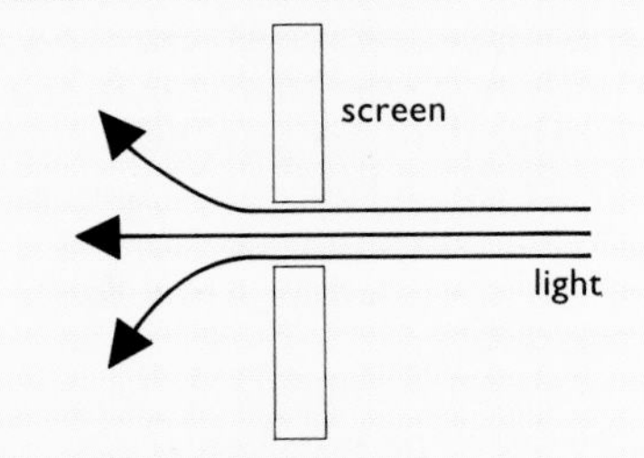


## diffuse nebula
A bright nebula in our own galaxy which spreads irregularly across the sky, unlike a planetary nebula (q.v.).

## Doppler effect
In astronomy the shift of the lines in a spectrum toward either the red or the blue, caused by a motion of the source emitting the spectrum. If the source is moving toward us, the frequency with which the crests of its waves will reach us will increase; this will give an apparent reduction in wavelength and a shift of the spectral lines toward the blue. If the object is moving away the effect is that we receive a longer wavelength and the shift is toward the red.

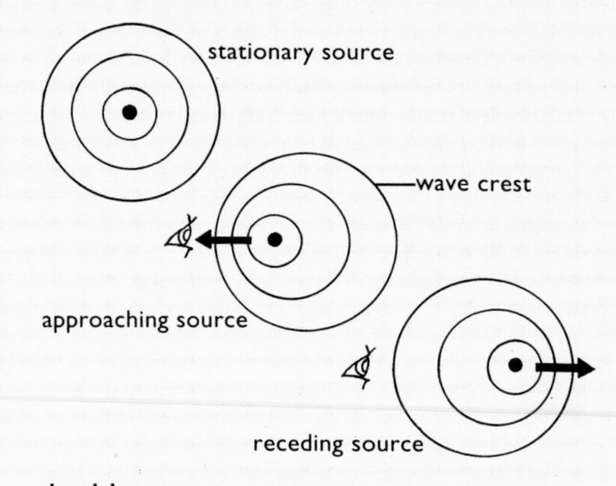


## double star
Stars appearing as a pair because they are close to each other in the line of sight.

# E

## eccentricity
A measure of the amount by which an elliptical orbit differs from a circle. For a circular orbit the eccentricity, *e*, is zero; its maximum value is always less than 1.0.

## eclipses and occultations
Eclipses occur when one celestial body passes into the shadow of another, as in the case of the Moon passing into the Earth's shadow. Eclipses of Jupiter's satellites are also observed from Earth.

An eclipse of the Sun is in fact an occultation (one body passing in front of another), although in astronomy the latter term is usually reserved for the obscuration of a star or planet by an asteroid or satellite such as the Moon.

## electromagnetic radiation
The total radiation band from radio waves to X-rays and gamma-rays, and including ultraviolet, visible and infrared wavelengths.

## electron
A fundamental and stable atomic particle of the class known as leptons. It possesses one negative electric charge, a spin of ½, and a mass of $9.1 \times 10^{-28}$ gm. Its antiparticle is the positron, which possesses one positive electric charge.

## ellipse
A closed elongated curve as traced by planets, asteroids, comets and satellites.

### elliptical galaxy
A galaxy, ellipsoidal in shape, composed primarily of stars with little gas and dust.

### elongation
The angular distance, east or west, between the Sun and Mercury or Venus.

### emission nebula
A nebula that emits visible radiation. Such nebulae may be diffuse or compact.

### emission spectrum
The spectrum given by a glowing gas such as an emission nebula. It is characterized by bright lines against a dark background because the glowing gas radiates only at specific wavelengths, which depend on the chemical elements of which it is composed.

### Encke division
A narrow division in Saturn's ring system. It is a "ripple" rather than a large gap.

### entropy
A quantity which, together with total energy, describes the thermodynamic state of a physical system. It is a measure of the number of ways in which the positions and velocities of the molecules can be rearranged to give the same overall properties. Left to itself, a system evolves so as to maximize the number of equivalent arrangements — that is, the entropy increases as far as possible. (When no further increase can occur, equilibrium has been reached.)

Since, loosely speaking, a disorderly arrangement of objects is one that can be shuffled without making a significant change, entropy is sometimes said to be a measure of disorder.

There are more ways of arranging the molecules of a gas in a large space than in a small one; consequently they maximize their entropy by spreading the energy evenly throughout the container, so that the initial temperature variations are smoothed out.

Variations of density or temperature can always be used to obtain useful work. Thus for a system of given total energy a greater amount can usefully be extracted when the entropy is lower.

### equatorial mounting

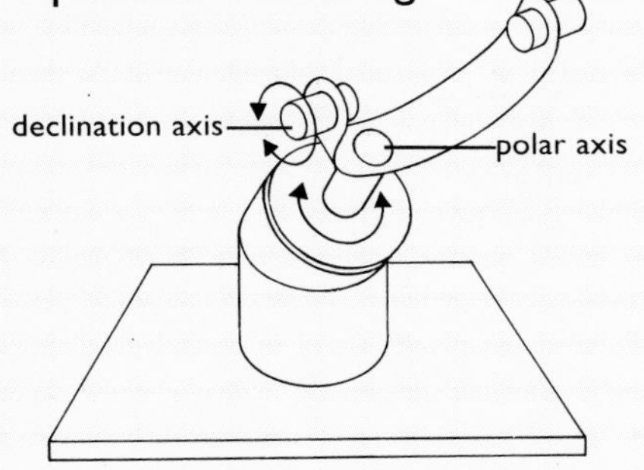


A telescope mounting in which one axis is parallel with the polar axis of the Earth. The other axis about which the telescope rotates is at right angles to the polar axis. This arrangement makes it possible for the observer to track the curved path of a celestial body across the sky by rotating the telescope about only one axis.

### equivalence, principle of
The principle in Einstein's theory of relativity that forces due to gravity and to the inertia of a body (its resistance to change of velocity) are equal. As a result, objects in a box falling freely in a gravitational field will display the same phenomena as if they were in empty space far from any gravitating body. Conversely, objects inside a box accelerating through space will experience the same phenomena as if they were stationary in a gravitational field.

### escape velocity
The velocity which a body must reach if it is to escape into space from a celestial body. The escape velocity depends on the size and mass of the celestial body concerned. For the Earth it is 11.18 km/s, but for the Sun no less than 617.3 km/s.

### event horizon
The boundary of a black hole, within which no event can be observed from the outside universe.

### exclusion principle
The principle in quantum mechanics which states that no two fermions (particles with a spin that is not a whole number, such as electrons, protons and neutrons) may occupy states which have the same quantum conditions (spin, etc.). It is because of the principle that the electrons in an atom do not all gather in the lowest orbit.

## F

### faculae
Bright active areas in the upper layers of the Sun's photosphere, often near sunspots.

### false vacuum
A quantum vacuum state characterized by a vast repulsive force and active during the inflationary period of the Big Bang.

### fermion
A class of atomic particles having a spin that is not a whole number. It includes protons, neutrons and electrons.

### field
A region throughout which a force operates.

### flare
A sudden release of energy visible as a bright light on the Sun. Such a flare, which lasts only a few minutes, actually occurs above active regions of the photosphere, either in the chromosphere (q.v.) or the lower region of the corona. Flares are characterized not only by a visible image but also by the emission of X-rays and sometimes gamma and radio waves.

### Fraunhofer lines
The dark absorption lines which cross the bright continuous spectrum of the Sun. So called because their positions were first carefully plotted by the German optician and astronomer Joseph Fraunhofer in 1814.

### frequency
The number of times per second the crests or troughs of waves of electromagnetic radiation reach an observer. Frequency is obtained by dividing the velocity of light by the wavelength of the radiation.

## G

### galaxy
A celestial island of stars, dust and gas.

### gamma radiation
Extremely energetic radiation, shorter than $10^{-8}$ (one 100-millionth) of a millimetre.

### giant star
A star which is very luminous and usually of large diameter compared with other stars of the same spectral class. The atmospheres of these giants are, by comparison, far more tenuous.

### globular cluster
A comparatively closely packed cluster of stars, spherical in shape. Such clusters can contain anything from a few tens of thousands of stars to over one million. Hundreds of globular clusters form a major component of the halo (q.v.) of the Galaxy.

### gluon
A messenger particle which holds together quarks, which themselves are the basic particles composing the atomic nucleus.

### grand unified theory (GUT)
A theory that aims to unify the basic forces of nature, bringing into one scheme the strong and weak nuclear forces, electromagnetism and gravity. At the extraordinarily high temperatures prevailing in the earliest moments of the Big Bang these forces were in fact indistinguishable from each other.

### gravitational lens
Because bodies cause space-time to curve, a massive body can deflect light and other radiation from a more distant object so that it becomes visible to an observer on Earth, though ordinarily it would remain hidden. It is as if the nearer massive body is acting as a lens. The effect can also lead to the appearance of distorted or multiple images of distant bodies.

### gravitational red shift
A consequence of general relativity resulting in the mass of a body causing a small red shift of spectral lines when light and other radiation is emitted from it.

### graviton
The messenger particle of gravity in theories of quantum gravitation.

### Great Red Spot
The large oval red area observed in the upper cloud layer of Jupiter's atmosphere, which rotates counterclockwise.

### greenhouse effect
The absorption of outgoing infrared radiation by a planetary atmosphere and its reradiation back to the planet's surface, thus helping to raise the average surface temperature. It is responsible for the very high temperature (737 K) on Venus.

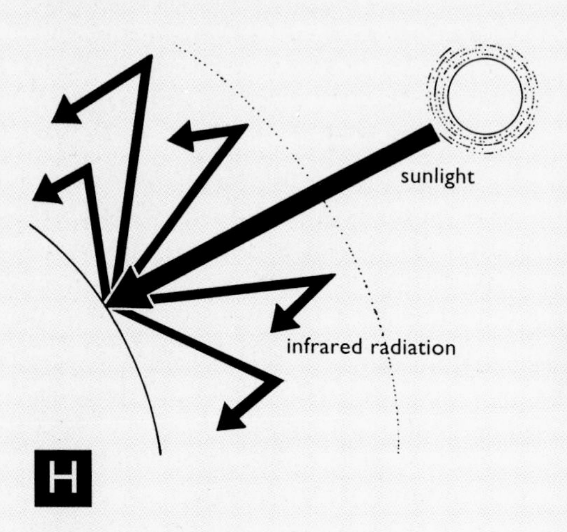


## H

### H I and H II regions
H I regions are those in which neutral hydrogen is present in interstellar space. Such regions are detected because they radiate at a radio wavelength of 21 cm. H II regions are those where singly ionized hydrogen is predominant, such as in the glowing bright nebulae in Orion and other constellations.

### hadron
A fundamental subatomic particle that experiences the strong force (q.v.). Hadrons include protons, neutrons and mesons, and are made of pairs or triplets of quarks.

### halo
A glowing ring observed around a celestial body. Haloes are observed around the Sun and Moon owing to refraction and reflection of their light by the Earth's atmosphere.The term is also used to describe material that is spread spherically around our galaxy.

### Hertzsprung-Russell diagram
A graph in which the absolute magnitudes of stars near the Sun are plotted against spectral class. It shows that there is a relationship between true brightness and spectral class.

### Higgs field
A quantum mechanics field. At its lowest energy state a Higgs field induces spontaneous symmetry-breaking (*see* symmetry) and it is important in theories that attempt to unify the fundamental forces of nature. The field is associated with Higgs particles, which are analogous to photons in electromagnetic fields.

### horizon distance
The maximum distance across which light could have travelled since the origin of the universe.

### hour angle
The angle on the celestial sphere between a celestial object and the observer's meridian (q.v.). It is measured westward along the celestial equator in terms of hours, minutes and seconds.

### Hubble constant
The rate at which the velocity of recession of galaxies increases with distance. At present there is some doubt about its value, which is thought to lie between 17 and 30 km/s per million light-years.

### inertia
The tendency of a body to resist a change in velocity, whether that change be acceleration or deceleration. It is implicit in Newton's laws of motion, which state that a body continues in a state of uniform motion in a straight line unless acted upon by outside forces. The acceleration that a given force can produce in a given body depends on its mass, which is often referred to as the inertial mass.

### inferior planet
A planet whose orbit lies within the Earth's. The two inferior planets are Mercury and Venus.

### infrared radiation
Radiation beyond the red end of the spectrum. Its wavelengths range from 1 mm to .001 mm. It is sometimes called "radiant heat" and is readily absorbed by water vapour in the Earth's atmosphere, which consequently limits the observations that can be made from ground-based observatories.

### interference
(1) In radio technology the degradation of radio signals by other unwanted signals or radio noise, which is why signals from spacecraft are transmitted digitally. (2) In physics, the superposition of one wave on another. The resulting wave caused by the previous waves reinforcing and weakening each other is known as the interference pattern. Optically such a pattern is seen as alternate light and dark bands.

### interferometry
Observations made using the interference of light or radio waves. In astronomy the technique was first applied optically by the American physicist Albert Michelson in the 1920s for measuring the diameters of large nearby stars.

### ion
An ion is an atom or molecule that has either lost one or more of its electrons and thus has a positive electric charge, or gained one or more electrons. In the latter case the ion will have a negative electric charge.

### ionosphere
A region of the Earth's atmosphere that extends from some 60 km to more than 500 km above the ground. Here most of the atoms and molecules are ionized by solar radiation. Being therefore electrified, the ionosphere acts like a mirror in the sky to long, medium and some short radio waves and so permits long-range broadcasting without the aid of a satellite. However, solar magnetic storms upset the ionosphere and so cause radio fade-outs that disrupt such communication.

### irregular galaxy
A galaxy that is too irregular in shape to be classified as a spiral, an elliptical, or a lenticular galaxy (qq.v.).

### isotope

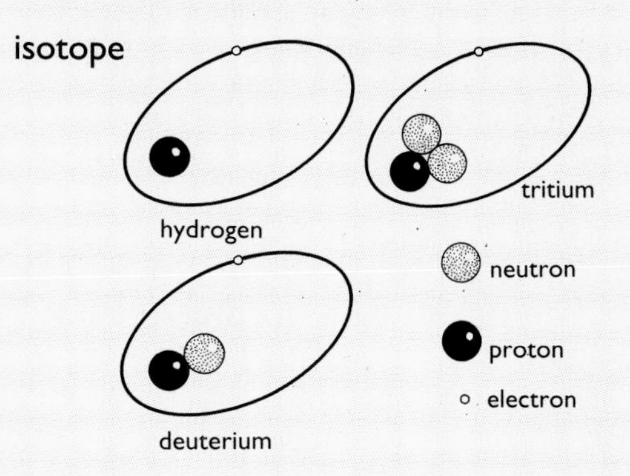

A form of an element in which the atomic nucleus has its characteristic number of protons, but contains a different number of neutrons than the element in its usual form. Chemically isotopes do not differ from the ordinary form of the element, but their behaviour may be different in other ways. For instance, tritium – a form of hydrogen with two neutrons in the nucleus as well as the single proton found in ordinary hydrogen – is radioactive.

# K

## Kelvin scale of temperature
A scale of temperature whose units (called kelvins, symbol K) are equal in size to those of the Celsius scale, and whose zero is fixed at –273.16°C, often known as absolute zero.

## Kepler's laws
Three laws governing the orbital motions of the planets. The laws are: (1) that planets move in elliptical orbits, with the Sun at one of the two "foci" of the ellipse; (2) that the radius vector (line from Sun to planet) sweeps out equal areas of space (areas 1, 2, 3 in diagram) in equal times, thus giving planets greater velocity when closer to the Sun; (3) that the orbital period squared is equal to the cube of that planet's average distance from the Sun.

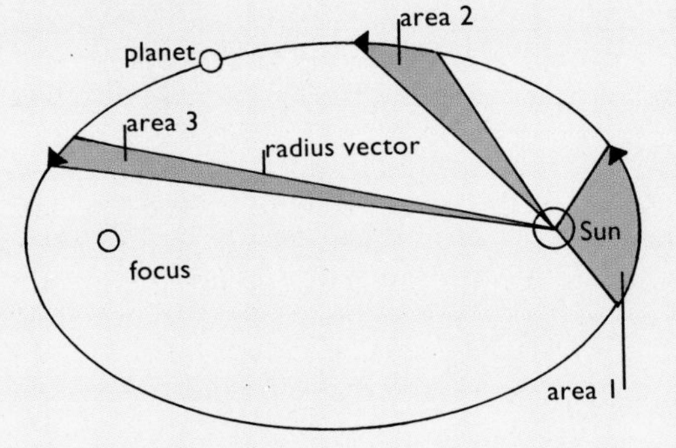


## Kirkwood gaps
Regions within the asteroid belt, between Mars and Jupiter, in which few asteroid orbits occur, owing to the gravitational effects of Jupiter. They were discovered in 1857 by the American mathematician Daniel Kirkwood.

# L

## Lagrangian point
Position in which a small body can remain in a stable orbit in the company of two massive bodies that are in mutual orbit about each other. There are five such points, which were discovered in 1772 by the French mathematician Joseph Louis Lagrange. The so-called Trojan asteroids occupy the Lagrangian points of Jupiter's orbit.

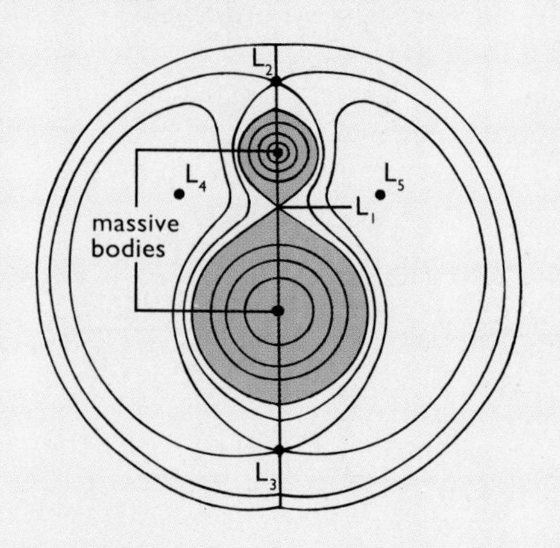


## lenticular galaxy
A galaxy intermediate in shape between an elliptical galaxy and a spiral galaxy (qq.v.).

## lepton
Particles not affected by the strong nuclear force. Those known to date are the electron, muon, neutrino and tau-particle.

## libration
Oscillations in the Moon's motions that reveal a little of its far side. For example, variations of the orbital speed, combined with the constancy of rotational speed, lead to libration "in longitude", or east-west. As a result about 59 percent of the Moon's surface is visible from Earth.

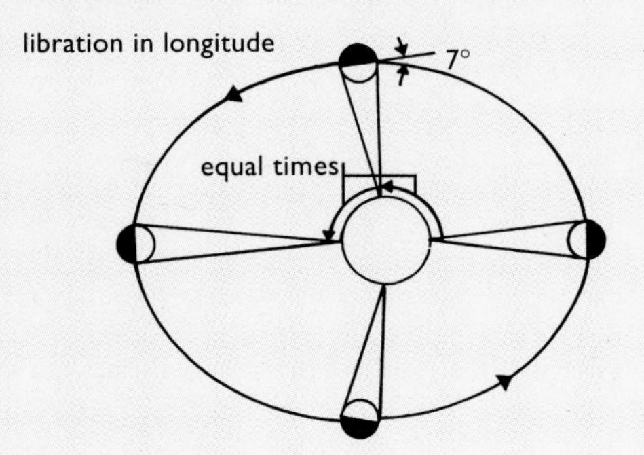


## light-cone
A means of depicting events in space-time by drawing a cone to represent the space over which light from an event will spread with time.

## light-year
A measure of distance, not of time. It is the distance travelled by light or other electromagnetic radiation in one year, and amounts to $9.4607 \times 10^{12}$ (million million) km.

## Local Group
The cluster of galaxies of which our own galaxy is a member. It also includes the Magellanic Clouds and covers a radius of space of some 2.5 million light-years.

## magnetic monopole
A single magnetic pole on its own. Such monopoles do not occur in the everyday world because a magnet always possesses two poles, but their existence was predicted by the English physicist Paul Dirac in 1931; none has yet been observed. However, magnetic monopoles may have existed in the very early stages of the Big Bang.

## magnetosphere
The region around a planet in which ionized particles are under the control of the planet's magnetic field. Its edge is known as the magnetopause.

## magnitude
In astronomy, a term used to specify the brightness of a celestial body. Based on the way the human eye and brain estimate brightness, a difference of one magnitude represents a difference in brightness of 2.512 times. Magnitudes are measured on an ascending scale of faintness, so that a star of magnitude 2 is 2.512 times dimmer than one of magnitude 1. The scale is extended to negative values. (Magnitude zero is 2.512 times fainter than magnitude –1.)

**Absolute magnitude** is the magnitude a celestial body would display if set at a distance of 32.6 light-years (10 parsecs). It permits direct comparison of the brightness of the different bodies.

**Apparent magnitude** is the magnitude a body appears to have to an observer on Earth. It depends upon not only the true brightness of the star but also its distance.

**Bolometric magnitude** is the magnitude taking into account radiation at all wavelengths, not merely visible ones.

**Photographic magnitude** is the magnitude as measured on traditional photographic plates specially prepared for astronomical use, in which the highest sensitivity lies in the blue region of the spectrum.

**Photovisual magnitude** is magnitude measured from a photographic plate with which filters have been used so that the response is similar to that of the human eye.

## main sequence
The area of the Hertzsprung-Russell diagram in which most stars lie. It extends from the lower right, where cool dim red stars are to be found, upward to the top left corner, where very hot bright stars are shown.

## Markarian galaxy
A galaxy that is bright and radiates most strongly at the blue end of the spectrum. They were catalogued by the Soviet astronomer B.E. Markarian in the 1970s.

### mascon
Short for "mass concentration" and used in reference to denser concentrations of mass just below the surface of some lunar plains or "maria" ("seas"); they were located by orbiting Apollo spacecraft.

### meridian
The imaginary circle on the celestial sphere (q.v.) passing through the celestial poles and the zenith (q.v.). It intersects the horizon at points due north and south of the observer.

### meson
A particle composed of a quark (q.v.) and an antiquark.

### meteor
A lump of rocky, metallic or carbonaceous material, possibly no larger than a dust particle, which falls toward the Earth. A meteor burns up owing to friction with the terrestrial atmosphere at heights of between 70 and 115 km. Its velocity is usually between 11 and 74 km/s; it therefore appears as a bright streak in the sky lasting for a fraction of a second.

### meteorite
A lump of rock or metal, or a mixture of the two, that lands on the Earth because its original size was too great for it to be consumed completely in the terrestrial atmosphere. If large enough, it will leave a crater where it falls.

### Milky Way
A hazy band of light crossing the entire sky in both northern and southern hemispheres. So named after its appearance, it is now known to be caused by myriads of stars as well as dust and gas lying in the central plane of our galaxy.

### molecule
The smallest unit of a pure substance that will retain its composition and so its chemical properties. It may be a single atom, or a collection of atoms.

### multimirror telescope
A design of optical telescope with a number of main mirrors, which bring their light to a common focus. The purpose is to construct a telescope with light-gathering power equivalent to that of a single much larger mirror. The multimirror telescope on Mount Hopkins, south of Tucson, has six mirrors, each 183 cm in diameter, and is equivalent to a single mirror 4.5 m in diameter. Such relatively small mirrors are less prone to distortion under their own weight.

### multiple star
A star of three or more components in mutual orbit about their centre of mass.

### muon
A lepton (q.v.) having an electric charge equal to that of the electron but 207 times heavier.

### nadir
The point on the celestial sphere directly underneath the observer, opposite the zenith (q.v.).

### nebula
A cloud of dust or gas in space. Nebulae can be either dark or bright, diffuse or compact.

### neutrino
A lepton (q.v.) with no mass and no electric charge; it enters only into reactions involving the weak nuclear force.

### neutron
A fermion (q.v.) that has no electric charge but a mass just a little greater than a proton. It is a constituent of many atomic nuclei.

### neutron star
A massive star toward the end of its life, whose degenerate material is composed of tightly packed neutrons. Their diameters are about 20 km and their masses vast, because their density is $10^{15}$ (one thousand million million) times that of water. When magnetized and rotating they emit pulses of radiation and are known as pulsars. In X-ray binary stars one component is a neutron star and the X-rays are emitted as material is attracted to this massive body.

### nova
Meaning "new", this is an ageing star which suddenly flares up in brightness — perhaps by 10,000 times — and so suddenly appears noticeable in the sky. Novae seem to be associated with binary systems in which one member is a white dwarf. The flare-up occurs on the white dwarf as material from the companion builds up.

Novae flare up within days or weeks at the most, and then sink back to about their original brightness over months or even years.

### nucleosynthesis
The building up of heavier atomic nuclei from the nuclei of lighter atoms. Conditions for this are found in the central regions of stars, where hydrogen is changed to helium and, in the most massive stars, helium is further built up through successive elements into carbon. In explosions of supernovae (q.v.), conditions occur that make it possible for nuclei heavier than iron to be synthesized. Nucleosynthesis was also a feature of the early Big Bang universe.

### nutation
A nodding motion of the Earth's axis of rotation due to the varying distances of the Sun and Moon, and their changing gravitational effects on the Earth.

### occultation
(*See* eclipses and occultations.)

### Olbers' paradox
A so-called paradox discussed in 1826 by the German amateur astronomer Heinrich Olbers, but first recognized long before. It concerns the question why, if the universe is static and infinite, and the stars uniformly distributed in it, the sky appears dark at night; under such conditions it should appear bright. It is now known that the universe is not static, nor are the stars uniformly distributed, but grouped into galaxies, all of which are moving away from each other. The red shift of their radiation, coupled with the fact that the universe has a definite age and so does not extend infinitely, explains why the night sky is dark.

### Oort cloud
In 1950 the Dutch astronomer Jan Oort suggested that comets originate from a cloud of cometary material spread out at the edge of the solar system at a distance between about 0.47 and 1.6 light-years. When comets in this cloud were disturbed by a passing star, some of them could be deflected into orbits that pass close to the Sun. If such a comet also makes a close encounter with a massive planet such as Jupiter, it will be thrown into a short-period orbit.

### open cluster
A cluster of stars within our own galaxy in which the members number less than 100 and are comparatively widely spaced. The Hyades and Pleiades clusters are notable examples.

### open universe
A universe that expands for ever.

### opposition
(*See* conjunction and opposition.)

### orbit
The path of one body around another. Depending on the velocity of the orbiting body and the mass of the body about which it orbits, its path will either be an ellipse — and therefore closed — or follow the open curves of a parabola or hyperbola. In the latter two cases an orbiting body will not return to make future circuits.

## orbital elements

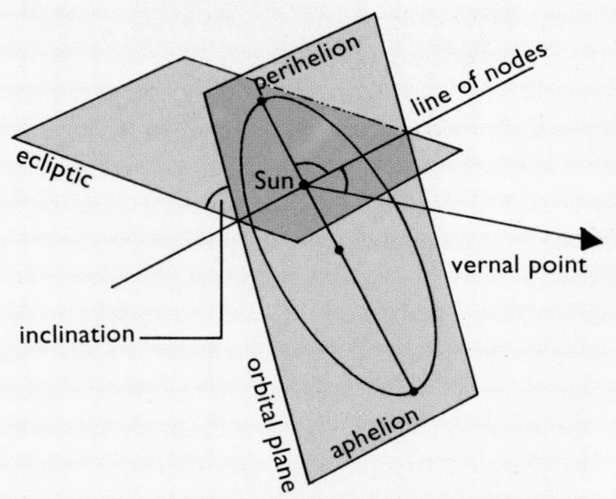


The various parameters that describe the details of an orbit. Its inclination is its tilt in relation to the ecliptic (the plane of the Earth's orbit); its orientation is shown by the angle between the line of nodes (where the plane of the orbit crosses the ecliptic) and the direction of a position called the vernal point; the distances from the Sun of its perihelion (q.v.) and aphelion (q.v.) define its shape.

## ozone layer

A layer of ozone that lies between 12 and 50 km above the Earth's surface. Formed by the action of solar ultraviolet radiation on atmospheric oxygen, it is of vital importance to life on Earth because it absorbs damaging ultraviolet radiation from the Sun.

# P

## parallax

The apparent shift of a nearby object against a more distant background. In astronomy parallax is used for obtaining distances in space.

## parhelion

An atmospheric effect, due to the refraction and reflection of sunlight, in which luminous spots appear on each side of the Sun, 22° away from it.

## parity, non-conservation of

Parity is the conversion of a view into its mirror image. Thus the mirror image of the left hand is that of a right hand. In atomic physics a mathematical transformation is used to achieve parity.

It was once believed that overall parity is always conserved in a system of fundamental atomic particles, but it has been shown that such parity is not conserved in interactions involving the weak nuclear force. In this respect there seemed to be a left- and right-handedness built into the universe. However, it later became evident that if antiparticles were taken into account (charge conjugation) and events were considered also as reversing in time (time reversal), another quantity, involving parity, is conserved.

## parsec

The distance at which a body would have an annual parallax of one second of arc. It equals 3.26 light-years. No star is as near to the Sun as this.

## perigee

The nearest point to the Earth in the orbit of the Moon or of an Earth-orbiting artificial satellite.

## perihelion

The nearest point to the Sun in the orbit of a body revolving around it.

## period-luminosity relation

The relationship between period of variation and true brightness of Cepheid variable stars (q.v.). Discovered in 1912 by the American astronomer Henrietta Leavitt.

## photoelectric effect

The emission of electrons by certain metals, such as selenium, when exposed to electromagnetic radiation. The number of electrons emitted depends upon the intensity of the radiation received, and their velocity on the frequency of that radiation. In 1905 this led Albert Einstein to propose that light and other electromagnetic radiation consists of particles that have a wavelike behaviour, and with this he laid some of the foundations for quantum theory.

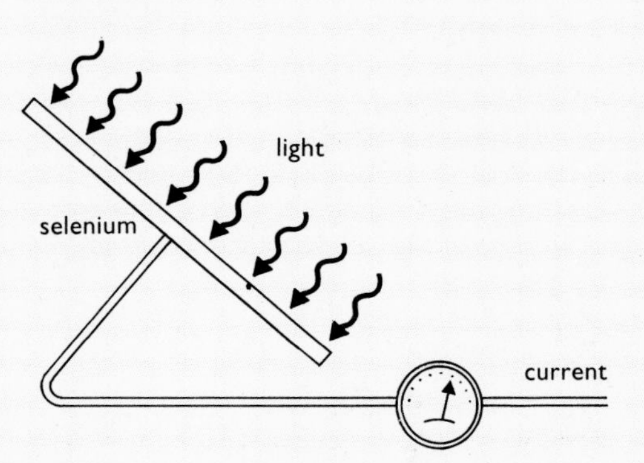


## photon

The quantum particle of light and the messenger particle of electromagnetic radiation.

## photosphere

The "surface" or apparent disc of the Sun and the source of the Sun's absorption spectrum. Its temperature is about 6,000 K.

## planet

A body that orbits the Sun and which shines only by reflected sunlight. Also applies to any such bodies orbiting other stars.

## planetary nebula

A nebula that appears visually in a telescope as a small greenish disc, rather like that of a distant planet. It is now known to be an expanding symmetrical shell of gas around a star that is close to the end of its life.

## planisphere

A flat (two-dimensional) map of part of the celestial sphere centred on either the north or south celestial pole. With an overlay or rotating mask, it can be used to indicate which constellations are visible at any hour of the night throughout the year.

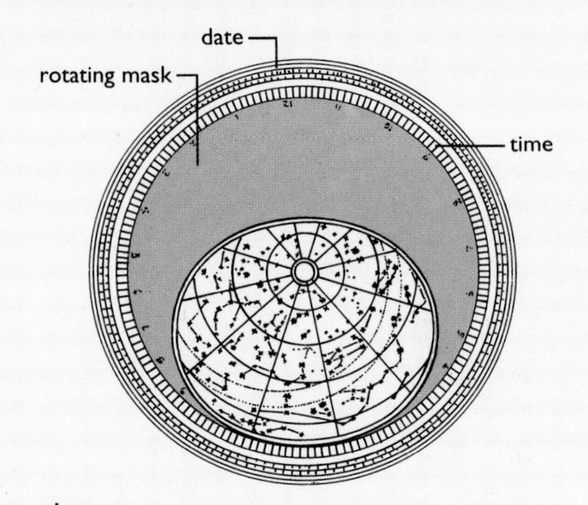


## plasma

An ionized gas consisting of ions and electrons moving freely. Plasmas are affected by electric and magnetic fields, and are to be found in stars and interstellar gas.

## polarization

Waves of radiation lie at right angles to the direction in which they are moving. They are at all angles perpendicular to this direction unless the radiation is linearly polarized, when the waves lie in one plane. However, these planes can rotate; the wave is then said to be circularly polarized.

unpolarized light

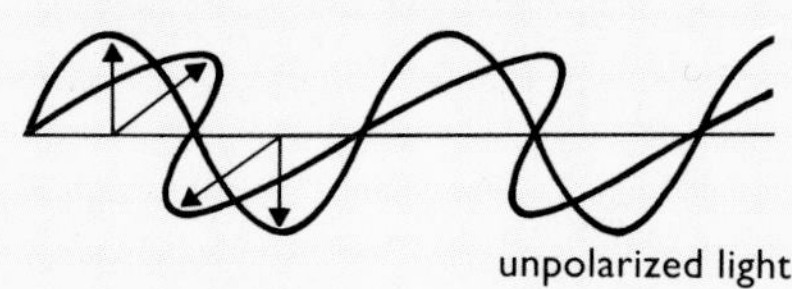

linearly polarized light

## Populations I and II

Population I stars are young and are to be found in the spiral arms in the plane of the Galaxy. They tend to be hot and bright, with a comparatively high metal content. Population II objects are older and found throughout the Galaxy, those in the halo being readily observable.

### precession
The continual movement of the axis of a rotating body due to external gravitational forces. In the case of the Earth, precession causes the celestial poles to describe a small circle on the celestial sphere. Motion around this circle is "backward" (i.e. in an east to west direction and so opposite to the direction of the Earth's daily rotation). It causes the equinoxes to move westward along the ecliptic. Thus precession causes a continual change in right ascension and declination, as well as in the celestial latitude and longitude of celestial objects. One complete circuit of the poles and thus of the equinoxes takes 25,800 years.

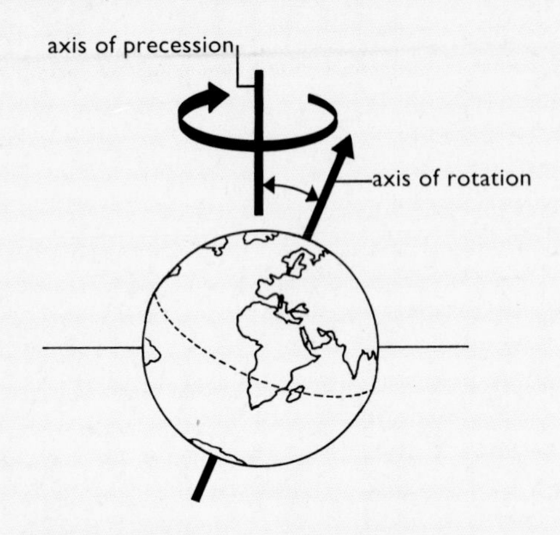


### prime focus
The "first" focus of a telescope, before the radiation has passed through secondary reflectors or lenses. In an optical reflector it is situated at the top of the telescope tube; in a refractor, at the eye-end of the tube.

### prominence
A cloud of gas lying in the upper chromosphere or lower part of the corona of the Sun. Against a dark background, as during a total solar eclipse, they appear as bright clouds or loops; those of hydrogen are pink in colour. Against the solar disc they appear as dark filaments. Prominences are classified as active – in which case they display comparatively rapid changes – or quiescent, when they may last for months. They are most frequent a few years after the onset of a solar cycle minimum.

### proper motion
The apparent motion of a star across the celestial sphere due to its real motion in space. This motion alters its coordinates (q.v.).

### proton
A hadron (q.v.) residing in the nucleus of an atom and possessing one positive electric charge. It is composed of three quarks and its mass is about 1,836 times that of an electron.

### protostar
A primitive stage in the formation of a star, following its condensation from gaseous material and before it is dense enough for nuclear reactions to cause it to shine.

### pulsar
A rotating neutron star (q.v.) that emits regular pulses of radiation.

### quark
The fundamental particle that forms all hadrons, or particles subject to the strong force (q.v.). Six kinds (or "flavours") are known. These are the "up", "top" and "charm" type quarks, each with an electric charge of $+\frac{2}{3}$, and the "down", "bottom" and "strange" kinds, each having a charge of $-\frac{1}{3}$. All have a spin of $\frac{1}{2}$. A neutron, for instance, is composed of one up quark and two downs (charges $+\frac{2}{3}, -\frac{1}{3}, -\frac{1}{3}$) giving it its characteristic total charge of zero.

### quasar
Acronym for a quasi-stellar radio object. Quasars are starlike in appearance and display prodigious red shifts. They are now believed to be very distant objects, probably the cores of active galaxies.

### radioactivity
The spontaneous breakdown of the nuclei of isotopes (q.v.) into the nuclei of other atoms, accompanied by the emission of energy in the form of atomic particles and sometimes very short-wave radiation as well. The time such atoms take to reduce their numbers by half is known as their half-life.

### radio telescope

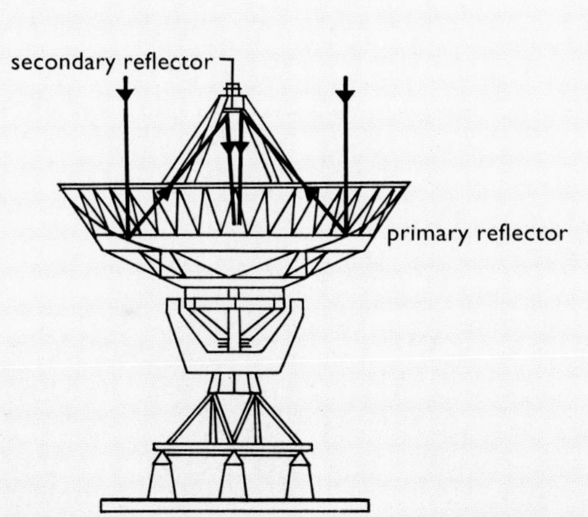


A telescope for receiving and imaging celestial radio sources. Its form can range from arrays of antennae (aerials) arranged to observe interferometrically, as in aperture synthesis telescopes and very large arrays (VLAs), to a dish type of telescope, analogous to the optical reflector. Such dishes can be linked to make a large interferometer.

### red dwarf
A dim red star at the lower end of the main sequence (q.v.) with surface temperatures of 2,500–5,000 K.

### red giants and supergiants
Bright red stars of large size – 10 to 100 times the diameter of the Sun – which are to be found in the upper right-hand part of the Hertzsprung-Russell diagram (q.v.).

### red shift
The shift of spectral lines toward the red, or long-wavelength, end of the spectrum, generally indicating that the source is moving away from the observer, but sometimes due to the gravitational field of the source.

### reflecting telescope
An optical telescope in which light is focused by means of a curved primary mirror.

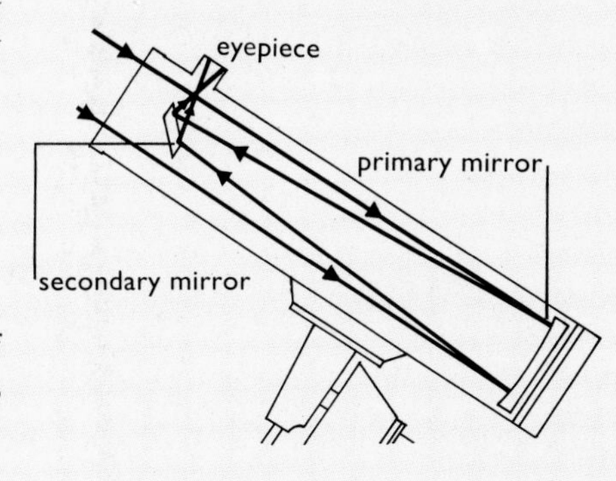


### reflection nebula
A glowing cloud of gas that shines because it scatters light from one or more stars embedded in it.

### refracting telescope
An optical telescope in which light is gathered and brought to a focus by a lens.

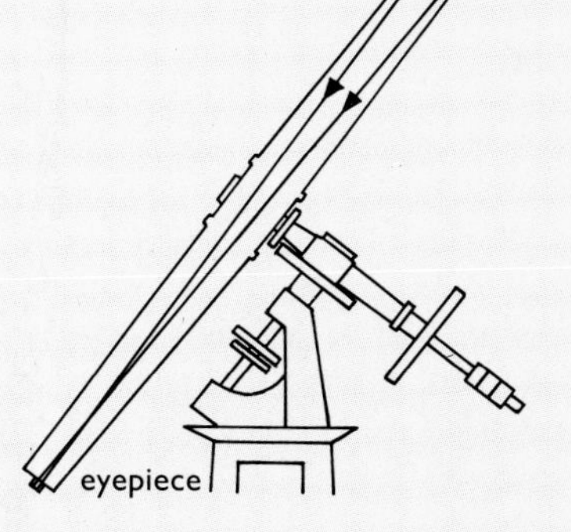

### refraction
The bending or deflection of a beam of radiation when it passes from one medium to another, which is due to a change in velocity of the waves that compose the radiation as they enter a medium of different density.

### regolith
A layer of rock and dust fragments covering the surface of the Moon, Mars and other similar bodies with little or no atmosphere.

### resolving power
The power of a telescope to resolve detail. Technically it is the angle between the closest objects that can be distinguished. It depends on the ratio of the aperture of the main mirror, lens or other light-gathering device to the wavelength of the radiation observed.

### resonance
This occurs when the orbital period of one body is an exact fraction of that of another nearby orbiting body of larger mass. It results in a series of pulls by the larger on the smaller and gives rise, for instance, to Kirkwood gaps (q.v.) in the asteroid belt.

### retrograde motion
Motion of a planet in an east to west direction, relative to the background stars, as observed from the Earth. It also refers to motion in a clockwise direction, as viewed from the north celestial pole – whether of comets, asteroids and satellites in their orbits, or of a body rotating on its axis.

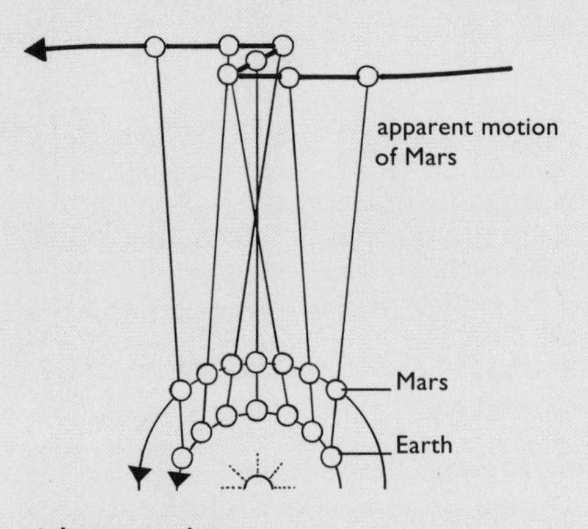


### right ascension
The angle between the meridians (q.v.) of the vernal equinox and of a celestial body (*see* celestial sphere).

### Roche limit
The least distance at which a satellite can orbit a planet. If closer than this, gravitational forces will tear the satellite apart.

### RR Lyrae stars
Very old giant stars that have turned into variables whose light variation is caused by pulsation, as occurs with Cepheid variables (q.v.). Almost all vary within 9 to 17 hours though a few take up to around 29 hours, and most have periods of 13 hours. Their general absolute magnitude is about +0.5. They are a large group; more than 2,000 are known, and half are to be found in globular clusters (q.v.). They are named after the star RR Lyrae, although it was not the first to be discovered. Like the Cepheids, these stars are important for determining distances in space.

## S

### satellite
A body which orbits another. In astronomy, usually applied to planetary moons and artificial satellites. The latter are placed in various orbits: 24-hour equatorial for communciations, polar for global coverage, and tilted low-altitude for mapping and surveillance.

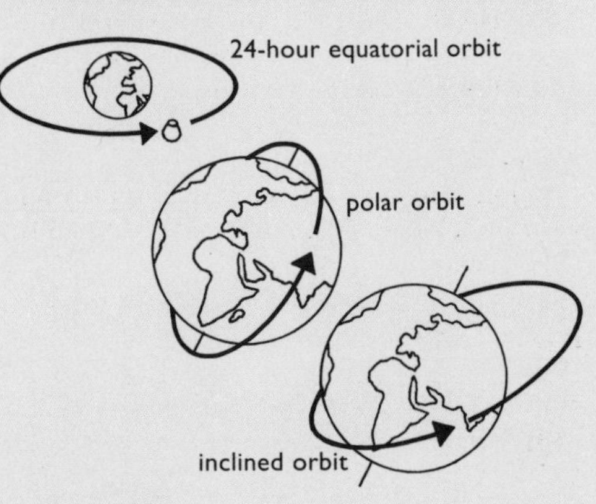


### scintillation
The twinkling or rapid variation in intensity of radiation from a distant source. Scintillation of stars is due to refraction caused by movements within the Earth's atmosphere. Scintillation of radio sources is due to the motion of interstellar matter between the source and the Earth.

### Seyfert galaxy
A spiral or barred galaxy with an intensely bright central region. Their radiation covers all wavelengths and is not thought to be a permanent feature, but rather a stage in their development. They show some similarities to quasars (q.v.).

### sidereal day
The period of rotation of the Earth with respect to the vernal equinox (*see* celestial sphere). Its value is 23h 56m 4.09s.

### sidereal period
A period of revolution measured with respect to the stars. Used to express the time taken by a planet to orbit the Sun, or a satellite to orbit a planet.

### simple mounting

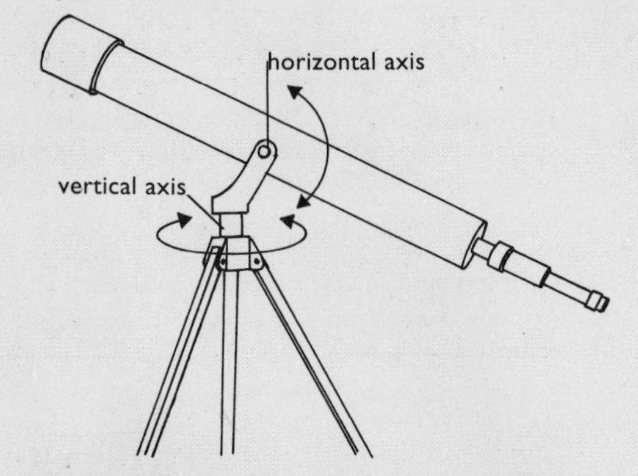


A telescope mounting in which one axis is vertical and the other horizontal. In consequence the instrument pivots up and down (in altitude) and around (in azimuth). Its disadvantage is that it requires two simultaneous motions, one in altitude and the other in azimuth, to follow the curved paths of celestial bodies

### solar apex
A point in the heavens toward which the Sun appears to be moving with respect to the stars in our local region of the Galaxy. It lies in the constellation Hercules. The Sun's velocity in that direction is 19.5 km/s.

### solar constant
The total radiation energy per unit area received on Earth from the Sun. It is measured as radiation received at the top of the Earth's atmosphere and is equivalent to 1.367 kilowatts per square metre.

### solar system
The system of planets and their natural satellites, together with asteroids, comets and cometary debris which are all in orbit about the Sun. The outermost limit of the system is that of the Oort cloud (q.v.).

### solar wind
A stream of protons, electrons and some helium nuclei sent in all directions from the Sun's corona into interplanetary space.

### space-time
A combination of three dimensions of space and one of time, allowing events to be described mathematically using four coordinates.

### speckle interferometry
A technique to improve the resolving power of a telescope by making many millisecond exposures of a celestial object in order to "freeze" the turbulence of the Earth's atmosphere and then combining the images, which show an interference pattern. Subsequent analysis allows a composite picture to be formed.

### spectral class
A classification of stars by their spectra. Originally intended to be an alphabetical sequence of types, the classification now has the order O, B, A, F, G, K, M. It is a sequence of colours and temperatures from O – very hot, bright, blue-white stars – to M, which are much dimmer, cool red stars.

### spectral line
A bright or dark line in the spectrum of a body emitting radiation.

### spectroscope
An instrument for producing an optical image of a spectrum and its spectral lines (q.v.).

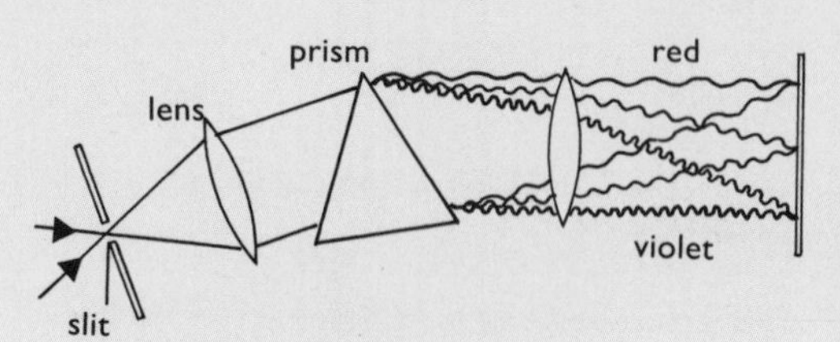


### spectrum
A band of radiation comprising different wavelengths. The optical spectrum runs from violet to deep red, and its wavelength range extends from 380 to 750 millionths of a millimetre.

### spinar
A hypothetical very massive object lying at the centre of an active galaxy and giving rise to the emission of energy in a similar way to a pulsar (q.v.).

### spiral galaxy
A galaxy that has a central bulge and spiral arms in its equatorial plane. It is composed of dust, gas and stars.

### star
A self-radiating celestial body in which energy is generated in the central regions by thermonuclear reactions.

### steady-state theory
A theory of the universe originally proposed by Hermann Bondi, Fred Hoyle and Thomas Gold in 1948, in which the universe never had a beginning nor will ever have an end but always remains in a steady state. After the discovery in 1965 of microwave background radiation in the universe, the hot Big Bang theory became dominant.

### strong force
The force that holds together quarks (q.v.). It is thought to operate by means of messenger particles known as gluons.

### sunspot
A dark patch visible on the Sun's photosphere. Each spot consists of a dark central area – the umbra – and a slightly brighter surrounding region – the penumbra. They are caused by magnetic fields on the Sun, and indicate areas that are cooler than their surroundings. They vary in size from minute patches to areas that may cover several billion square km. Their apparent darkness is a contrast effect due to the brightness of the photosphere. Their numbers vary in an 11-year cycle.

### supercluster
A cluster of clusters of galaxies, some of which are as much as 360 million light-years in diameter.

### superforce
A term used to express the idea that the universe is ruled by a single force. Such a force involves the achievement of a grand unified theory (GUT) of the four fundamental forces – the strong and weak nuclear forces, electromagnetism and gravity.

### supergiant star
A red giant star that is larger and brighter than other red giants. It involves stars with absolute magnitudes of –5 to –8, that is, stars at least 8,000 times brighter than the Sun.

### superior planet
A planet in the solar system whose orbit lies beyond that of the Earth.

### supernova
A star close to the end of its life that undergoes an explosion which ejects most of its material into space.

### superstring theory
A theory that suggests that atomic particles are vibrating strings within a 10- or 11-dimensional universe.

### symmetry and supersymmetry
A theory or atomic process is said to possess symmetry when it undergoes no change when certain operations are performed on it.

**Geometrical symmetry** occurs when, for instance, the operations of reflection and of rotation are performed on a circle, because the circle remains unchanged.

**Gauge symmetry** concerns mathematical operations concerned with rescaling or regauging certain quantities without affecting their essential relationships.

**Symmetry breaking** refers to a lower state of symmetry than previously occurred.

**Supersymmetry** is an extension of symmetry that unites particles and messenger particles (i.e. fermions and bosons, qq.v.), and thus matter and force.

### synchrotron radiation

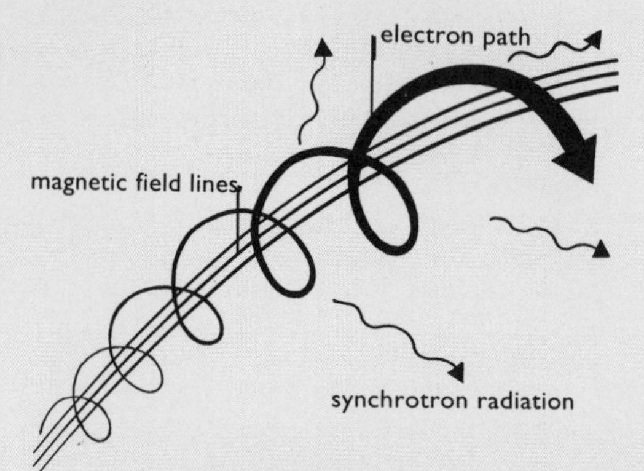


A synchrotron is a machine built to accelerate atomic particles to very high velocities by driving them around a circular magnetic field. Electrons and other charged particles accelerated in this way emit electromagnetic radiation at right angles to the spiral paths that they pursue. The wavelength of such synchrotron radiation depends upon the mass, electric charge and velocity of the particles involved.

### synodic period
The time taken by a celestial body to return to a given position in relation to the Earth. For planets this is the period between successive conjunctions or oppositions (*see* conjunction and opposition).

## T

### tektite
A small round glassy body, usually aero--dynamically shaped and about the size of a button. There are four major locations in which thay are found. It is not clear whether they are of extraterrestrial origin.

### terminator
The boundary between the sunlit and dark hemispheres of a moon or a planet.

### thermonuclear reaction
A high-temperature reaction between atomic nuclei that results in their fusion to form heavier nuclei. One such reaction, the conversion of hydrogen nuclei into helium nuclei, is the predominant source of the stars' energy.

### time dilation
The slowing of the passage of time on a body moving with respect to an observer.

### Titius-Bode Law
An arithmetic relationship first discovered in 1766 by the German physicist Johann Titius

or Tietz, but only made widely known by the German astronomer Johann Bode in 1772.

If 4 is added to each number in the sequence 0, 3, 6, 12 , 24, 48, 96, the sequence 4, 7, 10, 16, 28, 52, 100 is obtained. If the distance from the Sun to the Earth is taken as 10, then the other numbers closely correspond to the distances between the Sun and the other major planets out to Saturn. The continued sequence was found to be valid when Uranus was discovered, but it breaks down for Neptune and Pluto. It led to the recognition of a gap between Mars and Jupiter (i.e. at the number 28 in the sequence) and so to the discovery of the asteroids.

### transit
The passage of a celestial body across the observer's meridian. Also used for the passage of the inferior planets across the Sun's disc. Such planetary transits are infrequent: those of Mercury occur at intervals of 3, 13 and 46 years; those of Venus fall in pairs 8 years apart, with successive pairs separated by a long interval of well over a century. The two transits in the 18th century were used for determining the distance from the Earth to the Sun.

### T Tauri stars
Very young rapidly rotating stars, probably the last stages of a protostar before it joins the main sequence. T Tauri is the prototype. Less massive than the Sun, they have extended and active gaseous atmospheres.

## U

### ultraviolet radiation
Radiation lying beyond the violet end of the visual spectrum. Its wavelength ranges from 380 down to 25 millionths of a millimetre. Only the longer wavelengths can penetrate the Earth's atmosphere.

### umbra and penumbra
The umbra is the darker interior part of a shadow, and the penumbra the lighter outer region. The terms are also used in describing eclipses and sunspots (q.v.).

### uncertainty principle
In 1927 the German physicist Werner Heisenberg published his principle of indeterminacy or uncertainty. It arose out of his mathematical work on quantum theory and states that the position and momentum (the mass multiplied by the velocity) of a particle cannot both be specified precisely at the same time. There is an indeterminacy or uncertainty about every particle or, to put it another way, the numbers specifying its position and motion have a statistical spread.

### Van Allen belts
Two regions in the Earth's magnetic field or magnetosphere in which electrically charged atomic particles become trapped.

### variable star
A star whose apparent radiation varies in intensity. This may be due to variation in output from the star, or because it is a binary system in which each component eclipses the other. The latter is known as an eclipsing binary.

### wavelength
The distance between successive crests or troughs of a wave, particularly one of electromagnetic radiation. It is related to frequency (q.v.).

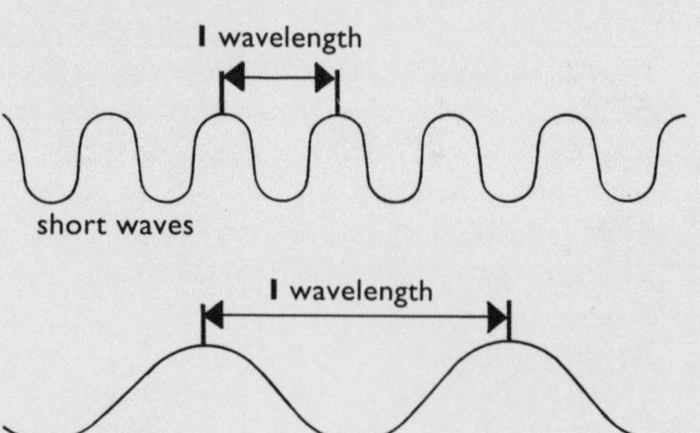


### weak force
The nuclear force involved in radioactivity (q.v.) and some neutrino reactions. It is 100,000 times weaker than the strong nuclear force and operates over a distance of less than $10^{-14}$ mm.

### white dwarf
A superdense state to which a star not more than 1.4 times the mass of the Sun eventually shrinks, most of its matter entering a state of degeneracy (q.v.).

### window
In astronomy, a range of wavelengths that can penetrate a planet's atmosphere. In spaceflight, the period in which a spacecraft can be launched on a particular mission.

### Wolf-Rayet star
Stars whose spectral lines are all bright and therefore emission lines instead of the customary dark absorption lines. These are very hot bright stars with surface temperatures between 25,000 and 50,000 K; they are between 100,000 and one million times as bright as the Sun and also between 10 and 50 times more massive.

### X-rays
Very short-wavelength highly penetrating electromagnetic radiation. The range of wavelengths runs from $10^{-5}$ to $10^{-8}$ mm (one hundred-thousandth to 10 billionths of a millimetre). X-radiation from space is evidence of highly energetic reactions on celestial bodies.

### zenith
The point directly above an observer.

### zodiacal light
A faint glow seen under good conditions in the west just after sunset and in the east just before sunrise. It is due to sunlight reflected from a belt of dust lying in the plane of the solar system.

# Biographies

**Bohr, Niels Henrik David 1885-1962**
Danish physicist, who studied atomic structure in Copenhagen and with Ernest Rutherford at Manchester University, England. He is noted for his model of the atom (1913) with a central nucleus and orbiting electrons. This he tied in with the then novel quantum theory, which he helped to develop further. He was awarded the Nobel Prize for Physics in 1922.

**Brahe, Tycho 1546-1601**
Danish astronomer, who was given the island of Hven, where he built a large observatory for measuring celestial positions, though later he was forced to move to Prague. Designing the instruments himself and taking account of any inherent error they might have – a totally new approach – Brahe obtained a degree of precision unprecedented for the pre-telescope era. His observations led Kepler to his laws of planetary motion. He also devised his own planetary system , with the Earth fixed at the centre but with the planets orbiting the moving Sun.

**deBroglie, Louis 1892-1987**
French physicist, born into the French aristocracy, who studied history and entered physics only after World War I. In 1924, following a suggestion by Einstein 20 years before, he proposed that not only light but also the electron and other atomic particles all possessed wavelike properties, and supported this by mathematical reasoning. Experimental confirmation came in 1927, and the Nobel Prize for Physics in 1929.

**Copernicus, Nicholas 1473-1543**
Polish astronomer. Educated at Cracow, Bologna and Padua Universities, Copernicus was an expert in Greek, canon law, medicine, mathematics and astronomy. Appointed to a non-ecclesiastical canonry in Frauenburg cathedral, he received a lifelong stipend and high administrative position. Copernicus is now remembered for his book *De Revolutionibus,* published in 1543 as *De Revolutionibus Orbium Coelestium* (On the Revolutions of the Celestial Spheres). In this he proposed that the Sun, not the Earth, was the centre of the universe, thus completely breaking with tradition by relegating the Earth to the position of a mere planet.

**Eddington, Arthur Stanley 1882-1944**
English astronomer and mathematician. A staunch supporter of the theory of relativity, he was also an expert on the constitution, mass and luminosity of the stars, and was the first to suggest that spiral galaxies were similar to our Milky Way system.

**Einstein, Albert 1879-1955**

German physicist. Between 1905 and 1916 he developed the theory of relativity, which provided a radically new outlook on space, time and gravitation, and for the first time proposed the equivalence of mass and energy. The theory affected the whole of astronomy and physics. Einstein also proposed the dual particle and wavelike nature of light and all electromagnetic radiation, and laid some of the foundations of quantum theory, which he later opposed. In 1933 he fled from Europe to the United States where he remained for the rest of his life.

**Faraday, Michael 1791-1867**
English physicist and chemist. A largely self-taught scientist, whose most significant research was on the relationships between electricity and magnetism. His investigations led him to the invention of the dynamo, electric motor, and transformer, while he also investigated the effects of electromagnetism on light. He expressed his results using the concept of electric and magnetic fields, which exerted a profound effect on later physics and astronomy.

**Feynman, Richard Phillips 1918-1988**

American physicist who developed quantum theory, ironing out many inconsistencies. With Murray Gell-Mann he formulated a theory to explain most phenomena concerned with the weak nuclear force. Feynman also developed simple diagrams to help understand the complex reactions of particles. With Julian Schwinger and Tomonaga Shin'ichiro he shared the Nobel Prize for Physics in 1965 for correcting previous work on reactions between photons and electrically charged atomic particles such as electrons and positrons (quantum electrodynamics).

**Galilei, Galileo 1564-1642**

Italian physicist and astronomer best known for his pioneering use of the telescope in astronomy, with which he discovered mountains on the Moon, Jupiter's four largest satellites, sunspots and the phases of Venus, and that the Milky Way is composed of myriads of stars. His famous advocacy of the Copernican theory, which put the Sun at the centre of the universe, brought him into conflict with the Roman Catholic Church.

**Gamow, George 1904-1968**
Russian-American physicist. He helped develop the theory of one type of radioactive disintegration, then turned his attention to nuclear processes in stellar evolution. In 1948, with Ralph Alpher and Hans Bethe, he produced a paper on the origin of chemical elements in stars; in it the hot Big Bang beginning of the universe is suggested. Gamow also suggested that there was a genetic code based on DNA.

**Halley, Edmond 1656-1742**
Britain's second Astronomer Royal. He made a chart of the southern hemisphere stars, and followed this with many other astronomical observations, including a complete 18-year cycle of the Moon's motion. He suggested that nebulae were gaseous material in space, discovered that the stars have their own motions, and advocated using transits of Venus for determining the Sun's distance.

Above all he was instrumental in persuading Newton to write his famous *Principia* and, having done so, edited it and paid for its publication. It was Halley who first applied Newton's laws of planetary motion to specific comets and predicted the return of the comet of 1682; it reappeared in 1758 as he had calculated, and is now known as Halley's Comet.

**Heisenberg, Werner Karl 1901-1976**
German physicist, best known for his principle of indeterminacy whereby there is an inherent uncertainty in the position and momentum of atomic particles. He received the Nobel Prize for Physics in 1932.

**Herschel, Frederick William 1738-1822**
Hanoverian-English musician, astronomer, and builder of astronomical telescopes of excellent quality, who constructed the world's largest instrument in 1785. He was also the greatest of all observational astronomers. Herschel discovered Uranus and catalogued celestial objects, particularly nebulae and galaxies. He was the first to suggest that the Milky Way marked the boundaries of an island of stars.

**Hipparchos** active **146-127 BC**
A Greek astronomer who discovered precession, calculated the length of the year correct to within 6½ minutes, and was the first to complete a catalogue of the stars. He also improved knowledge of the Moon's motion, determined the inclination of the ecliptic to within 5 arc minutes and measured the distance and size of the Moon and the Sun. However, his values for the latter were far too small, as were all such measures in antiquity. Hipparchos was certainly the greatest observational astronomer of the ancient world.

**Hubble, Edwin Powell 1889-1953**
American lawyer turned astronomer, Hubble is noted for his epoch-making work on galaxies, which in 1924 he showed were external to our own Milky Way. He also discovered that galaxies recede from us at a velocity which depends upon their distance, which implies that the universe is expanding.

**Jansky, Karl Guthe 1905-1950**
American radio engineer who in 1928 joined the Bell Telephone Laboratories where he was commissioned to study radio interference. Serendipitously he discovered the existence of radio waves from space in 1931 but although he published his results the next year, did not follow the matter any further. Nevertheless, his discovery lies at the foundations of modern radio astronomy.

**Kepler, Johannes 1571-1630**

German mathematician and astronomer, who first made his name when he published a theory about the distances of the planets and their relationship to the regular solids of Euclid's geometry. He is noted above all for his discovery of three laws of planetary motion with which he displaced the old belief that planets orbit in circles at an unvarying pace.

**Leavitt, Henrietta Swan 1868-1921**
American astronomer who in 1908 discovered the relationship between the period of variation and intrinsic brightness of Cepheid variable stars. She also carried out fundamental work on determining stellar magnitudes.

**Maxwell, James Clerk 1841-1879**
Scottish mathematical physicist who discovered the nature of Saturn's rings. Most significantly, he discovered electromagnetic radiation.

**Messier, Charles 1730-1817**
French astronomer. In order to avoid confusing other objects with the hazy appearance in a telescope of a comet before it came close to the Sun, Messier catalogued all star clusters and nebulae he could detect visually. In 1771 he published a list of 45 such objects, and by 1784 had extended this to 103. They are the basis of the Messier numbers still used today.

**Newton, Isaac 1643-1727**

English mathematician and physicist educated at Cambridge University, where he became Lucasian Professor of Mathematics in 1669. Newton discovered that white light was composed of light of all colours, devised a theory that light is composed of "corpuscles" or particles, constructed the first successful reflecting telescope, and independently laid the foundations of the calculus. He is best known for his laws of motion and his theory of universal gravitation, both published in 1687. These were two of the most important contributions to the development of modern science.

**Planck, Max 1858-1947**
A German physicist. In order to explain the way heat is radiated by a totally absorbing body (a "black body"), Planck suggested in 1900 that its oscillating atoms receive and emit radiation only in discrete quantities or "quanta". This is the basis of quantum theory, which, with the theory of relativity, has revolutionized 20th-century physics.

**Ptolemy** active **161-180 AD**
Greek astronomer, geographer and mathematician, who seems to have worked at Alexandria. Ptolemy's most important work was his *Almagest*, which explained Greek mathematical astronomy and was the standard reference for the next 1,500 years. He was also a geographer of note and one of the world's first true mapmakers.

**Rutherford, Ernest 1871-1937**
New Zealand physicist. He studied ionization and radioactivity, reaching the correct conclusion that atoms of one element spontaneously disintegrated to form others. This work led to his receiving the Nobel prize for chemistry in 1908. In 1911 he made his greatest contribution with his theory that an atom is composed of a tiny nucleus accompanied by electrons.

**Schrödinger, Erwin 1887-1961**
German physicist, who formulated a special equation in quantum mechanics which does for atoms what Newton's equations of motion do for planetary behaviour.

**Wegener, Alfred Lothar 1880-1930**
German geologist, meteorologist and explorer. Wegener originated the doctrine of continental drift in its modern form, which he made public between 1912 and 1915. It caused great controversy at the time but was subsequently shelved by most geophysicists. Studies of rock magnetism in the 1950s have since led to a general acceptance and further refinement of Wegener's theory.

# Index

## N



## O



## P

# Acknowledgments

The author is most grateful to Dr Peter Cattermole, Dr Merton Davies, Professor Michael Green, Dr David Malin, Dr Patrick Moore and Sir Brian Pippard FRS for their advice and assistance, but must exonerate them from any errors and omissions which the book may contain.

The lower illustration on p. 133 is based on the work of Marc Buie (Space Telescope Science Institute, Baltimore) and David Tholen (University of Hawaii). The illustration on pp. 162–63 is based on work of Walter M. Fish and Emanuel Margoliash (Northwestern University).

### Illustration credits

Gary Thompson
pp. 30/31, 38/39, 165, 166/167, 170/171, 174/175.

David Fathers
pp. 10 (nucleon), 26/27 (nucleus), 28/29, 34/35, 82/83.

Sue Sharples
pp. 22/23, 24, 46/47, 53, 59, 64/65, 71 (graphs), 77, 78/79, 84/85, 86/87, 88, 90/91, 98, 100/101, 102/103, 110, 122 (Io), 152/153, 162/163.

Mainline Design
pp. 17, 18/19, 20/21, 32/33, 36/37, 51, 56/57, 94/95, 106/107, 108–133, 143, 144, 172/173, 178, 186/187.

David Wood
pp. 67, 68, 70/71, 92/93, 184/185.

Ed Stuart
pp. 150, 154, 160/161.

Dave Ashby
pp. 152/153, 158/159, 192–205.

Mark Iley
pp. 108–135 (planet friezes).

Technical Art Services
pp. 15, 43, 45, 105, 108, 135, 188–191.

### Picture credits

*t*=top; *c*=centre; *b*=bottom; *l*=left; *r*=right
1–3 David Malin/Anglo-Australian Telescope Board; 4 NASA/Science Photo Library; 5 Dr Bradford A. Smith/National Space Science Data Center; 6–7 G. Deichmann/Planet Earth Pictures; 9 Patrice Loiez, CERN/Science Photo Library; 11*t* Cath Ellis, Dept. of Zoology, University of Hull/Science Photo Library; 11*b* Earth Satellite Corporation/Science Photo Library; 12*t* David Malin/Anglo-Australian Telescope Board; 12*c* Science Photo Library; 12*b* John Sandford/Science Photo Library; 13*t* NASA/Science Photo Library; 13*c–b* David Malin/Royal Observatory Edinburgh & Anglo-Australian Telescope Board; 14–15 AEA Technology; 16–17*t* Gerolf Kalt/Zefa Picture Library; 16–17*b* Jean Pottier/Rapho; 25 Doris Haselhurst/The Dance Library; 26–27 David Parker/Science Photo Library; 36–37 Douglas Kirkland/Colorific!; 39 Adrian L. Melott, University of Pittsburgh; 40–41 Margaret J. Geller & John P. Huchra/Harvard-Smithsonian Center for Astrophysics; 42–43*t* Smithsonian Institution/Science Photo Library; 42–43*b* Lund Observatory; 43 Max-Planck-Institut for Radio Astronomy/Science Photo Library; 44 David Parker/Science Photo Library; 45 Peter Menzel/Science Photo Library; 46–47 National Optical Astronomy Observatories; 50*l* Royal Greenwich Observatory/Science Photo Library; 50*r* David Malin/Anglo-Australian Telescope Board; 52 David Malin/Anglo-Australian Telescope Board; 53 Dr Rudolph Schild/Smithsonian Astrophysical Observatory/Science Photo Library; 54 David Malin/Anglo-Australian Telescope Board; 54–55 Dr Jean Lorre/Science Photo Library; 56–57 David Malin/Royal Observatory Edinburgh; 58 Royal Greenwich Observatory/Science Photo Library; 59 Dr Marshall Joy (Marshall Space Flight Center), Victor Blanco (Cerro Tololo Interamerican Observatory) & Jim Higdon (University of Texas at Austin); 60–61 David Malin/Anglo-Australian Telescope Board; 67*l* Royal Greenwich Observatory/Science Photo Library; 67*r* The Observatories of the Carnegie Institution of Washington/Science Photo Library; 68 NRAO/AUI/Science Photo Library; 69 Dr Jean Lorre/Science Photo Library; 71 X-ray Astronomy Group Leicester University/Science Photo Library; 72 Jean Arnaud/Observatoire de Midi-Pyrénées; 73 NRAO/AUI/Science Photo Library; 74–77 David Malin/Royal Observatory Edinburgh & Anglo-Australian Telescope Board; 79–80 David Malin/Anglo-Australian Telescope Board; 81 David Malin/Royal Observatory Edinburgh; 87 Yerkes Observatory/University of Chicago; 89–91 David Malin/Anglo-Australian Telescope Board; 92 Harvard-Smithsonian Center for Astrophysics; 93*t* Palomar Observatory, © C.I.T.; 93*b* Harvard-Smithsonian Center for Astrophysics; 96–97 NASA; 97 National Optical Astronomy Observatories; 98*t* NASA; 98*b* Royal Greenwich Observatory; 99 NASA; 100 S. Koutchmy *et al.*/Mission de l'Institut d'Astrophysique (CNRS); 101 Jack Finch/Science Photo Library; 104 NASA/Science Photo Library; 109 NASA; 111 NASA/Science Photo Library; 112 NASA; 112–13 Novosti/Science Photo Library; 113–15 NASA/Science Photo Library; 116*t* U.S. Geological Survey/Science Photo Library; 116*b* NASA/Science Photo Library; 117 Dr Jean Lorre/Science Photo Library; 118–19 NASA/Science Photo Library; 120–23 NASA; 123 NRAO/AUI/Science Photo Library; 124–27 NASA; 127–29 NASA/Science Photo Library; 131–32 NASA; 134 Dr Bradford A. Smith/National Space Science Data Center; 136 Martin Dohrn/Science Photo Library; 137*t* NASA; 137*b* Don Davis/D. W. Wilhelms/Academic Press; 138*t* NASA; 138*b* NASA/Science Photo Library; 139 NASA; 140 NASA/Science Photo Library; 141 NASA; 142 NASA/Science Photo Library; 145*t* Fred Espenar/Science Photo Library; 145*b* NASA/Science Photo Library; 146–47 John Sanford/Science Photo Library; 149 NASA/Hansen Planetarium; 151 George I. Bernard/Oxford Scientific Films; 155*t–c* Peter Gould/Oxford Scientific Films; 155*b* CNRI/Science Photo Library; 156–57 Jan Hinsch/Science Photo Library; 157 Robert Hessler/Planet Earth Pictures; 159*t* Biocompatibles Ltd; 159*b* Royal Greenwich Observatory/Science Photo Library; 161 Carolina Biological Supply Co./Oxford Scientific Films; 163 Sinclair Stammers/Science Photo Library; 164 Dr Richard K. La Val/Animals Animals/Oxford Scientific Films; 168–69 CERN/Science Photo Library; 176–77 Dr David Blair/University of Western Australia; 178–79 Mauna Kea Observatory, University of Hawaii/S. Wykoff & P. Wehinger/Science Photo Library; 180 Science Museum/Michael Holford; 181 Science Graphics Inc, Bend, Oregon, USA; 182–83 Zefa Picture Library